Ambient Assisted Living

Sauro Longhi · Pietro Siciliano
Michele Germani · Andrea Monteriù
Editors

Ambient Assisted Living

Italian Forum 2013

Springer

Editors
Sauro Longhi
Department of Information Engineering
Università Politecnica delle Marche
Ancona
Italy

Michele Germani
Industrial Engineering
Università Politecnica delle Marche
Ancona
Italy

Pietro Siciliano
Department of Information Engineering
Institute for Microelectronics
and Microsystems IMM-CNR
Lecce
Italy

Andrea Monteriù
Information Engineering
Università Politecnica delle Marche
Ancona
Italy

ISBN 978-3-319-01118-9 ISBN 978-3-319-01119-6 (eBook)
DOI 10.1007/978-3-319-01119-6
Springer Cham Heidelberg New York Dordrecht London

Library of Congress Control Number: 2013954025

Printed on acid-free paper

Springer is part of Springer Science+Business Media (www.springer.com)

Preface

The concept of "Ambient Assisted Living" (AAL) has become deeply relevant for the present and future challenges. This is strongly motivated by the fact that the annual growth of the older population is significantly higher than that of the total population. For this reason, good aging and AAL activities are the focus of many national and international R&D projects. In this regard, the European Commission has proposed extended funding for doing research on Ambient Assisted Living, which will run alongside the new Horizon 2020 program. The new proposal suggests that Ambient Assisted Living Joint Partnership (AAL JP) will receive over €700 million between 2014 and 2020.

AAL needs to involve and combine multidisciplinary research fields, such as cognitive sciences, computer science, industrial design, user interfaces, electrical engineering, etc., in order to extend the time older people can live in their preferred home environment. Their use of ICT products and remote services will allow them to be more autonomous and assist them in carrying out activities of daily living, thereby enhancing their quality of life.

The Fourth Italian Forum on Ambient Assisted Living (ForItAAL2013), held in Ancona, Italy, in October 2013, was the annual showcase event, which brought together developers, producers, service providers, carriers, and end user organizations working in the different fields of technology and applications of AAL. This book summarizes the main results of the Italian Forum and addresses the issues and new technological developments, which support the autonomy and independence of individuals with special needs through an innovative and integrated approach, designed to respond to the socioeconomic challenges of an aging population. AAL is seen here from different perspectives and within different topical areas. The knowledge and insights provided in this book can not only help researchers but also all people involved in the AAL to understand the new societal trends, the new technological developments and pressing and future challenges concerning Ambient Assisted Living.

Sauro Longhi
Pietro Siciliano
Michele Germani
Andrea Monteriù

Committee

Scientific Committee

Valerio Aisa (Indesit Company, Fabriano, AN)
Bruno Andò (Università di Catania)
Laura Burzagli (IFAC-CNR, Firenze)
Niccolò Casiddu (Università di Genova)
Filippo Cavallo (Scuola Superiore di Sant'Anna, Pisa)
Maria Gabriella Ceravolo (Università Politecnica delle Marche)
Paolo Ciampolini (Università di Parma)
Fiorenzo Conti (Università Politecnica delle Marche)
Assunta D'Innocenzo (Abitare e Anziani, Roma)
Paolo Dario (Scuola Superiore di Sant'Anna, Pisa)
Monica De Angelis (Università Politecnica delle Marche)
Paolo Lorenzo Dessì Fulgheri (Università Politecnica delle Marche)
Ilaria De Munari (Università di Parma)
Natale Giuseppe Frega (Università Politecnica delle Marche)
Ennio Gambi (Università Politecnica delle Marche)
Fabrizia Lattanzio (INRCA, Ancona)
Alessandro Leone (IMM-CNR, Lecce)
Giuliana Trisorio Liuzzi (Agenzia regionale Puglia per la Tecnologia e l'innovazione)
Massimiliano Malavasi (Centro Reg. Ausili, Bologna)
Piero Malcovati (Università di Pavia)
Luciano Malfer (PAT, Trento)
Fabio Pianesi (FBK-irst, Trento)
Alberto Pilotto (IRCCS Casa Sollievo della Sofferenza, San Giovanni Rotondo)
Massimo Pistoia (eResults, Cesena)
Antonio Domenico Procopio (Università Politecnica delle Marche)
Lorena Rossi (INRCA, Ancona)
Ludovico Scortichini (Confindustria, Ancona)
Fiorenza Scotti (Università di Trieste)
Pietro Siciliano (IMM-CNR, Lecce)
Maria Giovanna Vicarelli (Università Politecnica delle Marche)

Organizing Committee

Sauro Longhi (Università Politecnica delle Marche)
Michele Germani (Università Politecnica delle Marche)
Andrea Monteriù (Università Politecnica delle Marche)
Silvia Ceccacci (Università Politecnica delle Marche)
Alessandro Freddi (Università Politecnica delle Marche)
Susanna Spinsante (Università Politecnica delle Marche)

Contents

Part VI ICT Tools in AAL

Part VII Smart Housing

Part I
Sensor Technologies in AAL

A Low-Cost Sensor for Real-Time Monitoring of Indoor Thermal Comfort for Ambient Assisted Living

Gian Marco Revel, Marco Arnesano and Filippo Pietroni

Abstract The present paper illustrates an innovative low cost solution for the monitoring of indoor thermal comfort by means of Predictive Mean Vote (*PMV*) index for multiple positions. This is particularly interesting in an Ambient Assisted Living environment as replacement of typical thermostat used for the climate control. In fact, the system proposed considers also personal parameters, as metabolic rate (M) and clothing level (I_{cl}), instead of the merely environmental parameters. If this is important for normal living conditions, it becomes crucial in case of elderly people and long-term care patients where a reduction of M or I_{cl} causes a high sensitivity to thermal conditions (especially for cold sensation), or where the disability does not allow the subject re-action (e.g. shading opening/closing when solar radiation occurs). The device proposed uses a set of low-cost non-contact sensors to determine, based on algorithms provided by ISO 7726 and 7730, Mean Radiant Temperature (*MRT*) and *PMV*, which are provided as output of the device through wireless or wired connection. The capability of predicting thermal comfort conditions for multiple positions of the occupant in the room has been tested and validated in a real case study: it resulted in a discrepancy of $\pm 0.5\,^{\circ}\mathrm{C}$ in the *MRT* measurement and ± 0.1 for the *PMV* with respect to a reference measurement system (microclimate station). The sensitivity to the metabolic rate and clothing level for AAL applications is also discussed together with a procedure for an estimation of these parameters. The accuracy achieved allows a better measurement of the real thermal sensation for a more comfortable environment with lower energy consumption.

G. M. Revel · M. Arnesano (✉) · F. Pietroni
DIISM, Università Politecnica delle Marche, 60131 Ancona, Italy
e-mail: m.arnesano@univpm.it

G. M. Revel
e-mail: gm.revel@univpm.it

F. Pietroni
e-mail: f.pietroni@univpm.it

S. Longhi et al. (eds.), *Ambient Assisted Living*, DOI: 10.1007/978-3-319-01119-6_1,

1 Introduction

Assisted living environments concern the capability of assuring the well-being of occupants including the provision of the right thermal comfort. Traditionally the control of HVAC (Heating, ventilation and air conditioning) systems is based on a thermostat which is able to measure the air temperature at one point in the space and sends a feedback to the control system in order to switch on or off the heating/cooling system in relation to the selected set-point. In recent years many examples have demonstrated how this method is not fully accurate because the merely air temperature cannot represent the thermal sensation of the occupants. If this is important for normal environments, it becomes crucial in case of elderly people and long-term care patients for several reasons as suggested in [21]. As stated in [13] and [8], older adults do not perceive thermal comfort differently from younger adults but there is an effect related to personal parameters such as activity and clothing level. Older adults have a lower activity level, and thus metabolic rate, than younger. While different considerations have to be done for long-term care patients where disease can cause changes in the thermoregulation or thermal sensitivity, as shown in [22] for older people with dementia. Some studies [6, 19, 20] found discrepancies in the heat balance or preferences for higher or lower temperatures between the old and the young. Besides, the *PMV-PPD* (*PMV*: Predicted Mean Vote, *PPD*: Predicted Percentage of Dissatisfied) model supposes the same thermal preference for all ages. This means that a dedicated calibration of the *PMV-PPD* model for this particular application is needed, following the example of [5].

The present paper describes a low cost real-time tool for monitoring indoor thermal comfort that represents an innovative solution in this field. The heart of the device is a simple array of thermopiles to be assembled and installed on the ceiling of the occupied room. The embedded microcontroller together with the software implemented allows the automatic scanning of each indoor surface to evaluate the temperature distribution. Algorithms provided by [9] and [10] are also embedded so that thermal and comfort parameters (as *PMV*) can be estimated for several positions in the environment. The system is accompanied by a software tool based on Android platform for remote control and data processing and is able to communicate with commercial BMS (Building Management Systems). The results can be given in terms of real-time *PMV* maps, suitable to provide feedbacks for modular control not achievable by standard thermostats. Thus, the solution proposed could substitute the classic thermostat avoiding its limits and providing new eyes to the BMS in order to make it able to actuate an optimal control of the indoor environment. Basics of the approach were presented by the authors in [16] with an initial validation to demonstrate its feasibility, while the final system is shown here with the description of the methodology applied and its validation process. The accuracy of the device is compared with a reference system (a microclimate station) and the sensitivity to the metabolic rate and clothing level for AAL application is discussed. The present work is developed as a part of the FP7 UE project CETIEB (Cost Effective Tools for Better Indoor Environment in Retrofitted Energy Efficient Buildings). The system is currently patent pending.

2 Comfort Model

The control and the monitoring of the indoor thermal conditions represent an important task with the aim of ensuring suitable working and living spaces for people. Several rules and standards have been released to provide technicians with suitable design tools, effective indexes and parameters for checking of the indoor microclimate. Among them, *PMV* index is often adopted to assess the thermal comfort conditions of moderate environments [2, 10]. From the sensitivity analysis of all the variables affecting *PMV* estimation, as presented in [1], *MRT* resulted as one of the most influential among the measuring parameters. This shows that the conventional indoor temperature and humidity control could not guarantee an adequate indoor comfort condition [24]. The model here considered is based on the *PMV-PPD* model [7]. *PMV* is an index that predicts the mean value of votes of a large group of people on a seven-point thermal sensation scale, which goes from −3.0 (Cold) to +3.0 (Hot). *PPD* index provides information on thermal discomfort or thermal dissatisfaction by predicting the percentage of people likely to feel too warm or too cool in a given environment. The mathematical expressions of *PMV* model proposed by Fanger is function of:

$$PMV = f(T_a, RH, V_a, MRT, M, I_{cl}) \tag{1}$$

where T_a (°C) is the indoor mean air temperature, *RH* (%) is the relative humidity, V_a (*m/s*) is the air velocity, *MRT* (°C) is the mean radiant temperature, *M* (*met*) is the metabolic rate, I_{cl} (*clo*) is the clothing level. ISO 7730 provides an algorithm for the calculation of all the variables needed for a correct estimation of *PMV* and *PPD* parameters. A sensitivity analysis of the *PMV* has been here performed according to the methodology described in [17] in order to investigate the *PMV* variation related to the metabolic rate and clothing level. *PMV* was computed at a given thermo-hygrometric condition for different values of *M* and I_{cl} together with its variations $dPMV = \partial PMV / \partial M \cdot dM$ and $dPMV = \partial PMV / \partial I_{cl} \cdot dI_{cl}$, where $dM = \pm 0.1$ *met* and $dI_{cl} = \pm 0.1$ *clo* according to the accuracies required in ISO 7730. As shown in Fig. 1a, for low metabolic rates (<1 *met*) the sensitivity of *PMV* becomes higher and the subject tends to feel a colder sensation (i.e. *PMV* decreases). These results confirm that monitoring the thermal sensation is crucial for subjects at low metabolic rate as older adults, long-term care patients or generally for people with low mobility.

3 Method Functionalities and Test

The solution proposed is a low-cost sensing system (about 200€) for real-time monitoring of indoor thermal comfort. It is based on a microcontroller and a series of embedded algorithms to derive thermal comfort indices for multiple positions in the room. The configuration of the overall system can be performed through a simpli-

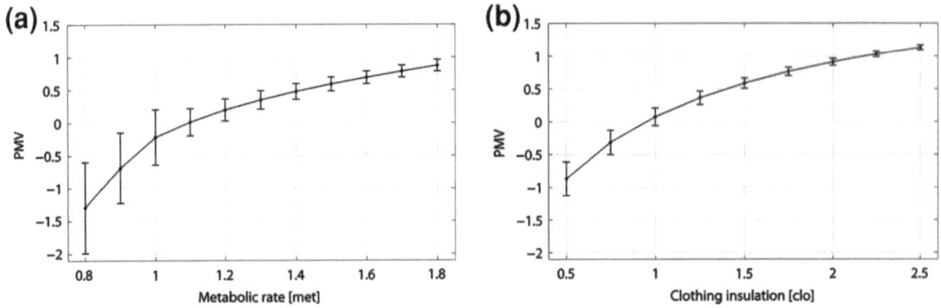

Fig. 1 *PMV* as a function of the metabolic rate M (**a**) and clothing level (**b**) with variation intervals associated to $dM = \pm 0.1$ *met* and $dI_{cl} = \pm 0.1$ *clo*

fied graphical user interface developed for Android devices. Strengths of the device are non-invasive measurement performances, interoperability and ease of integration, as demonstrated within the CETIEB project, due to open-source communication protocols. The overall measuring procedure will be described in the following section. Then, the results of validation will be presented.

3.1 System Functionalities

A description of the system functionalities is here reported. The initial configuration can be performed during the installation phase with any Android device (tablet, smartphone). In practice input parameters required for the embedded algorithms are room geometry (length, width, height) and personal parameters (metabolic rate and clothing level, expressed in *met* and *clo* respectively). In case of users unable to interact with the user interface this configuration is performed by the technician and stored directly into the microcontroller via EEPROM so that the system is able to work continuously without further actions to be performed by the end user. In addition, further development will be aimed to allow the remote configuration through the Android device. Once that all the parameters are updated, the system performs an automatic scanning of all the surfaces inside the room using two servos assembled together with the sensor. Surface temperatures are measured by a thermal array of thermopiles: the algorithm implemented on the microcontroller manipulates the temperatures $T_{1(h,v)}$, $T_{2(h,v)}$,...$T_{8(h,v)}$, where h and v represents the horizontal and vertical orientation of the sensor during the acquisition. Data are processed to generate thermal images with low resolution of 10×17 pixels for the horizontal surfaces (about 1 min is required for the scanning and reconstruction of each surface image) and of 10×8 pixels for the floor (30 s). The air temperature and relative humidity are measured in a single point with an integrated *T-RH* sensor as traditionally done by thermostats. After the room interior surface temperatures measurement, the mean radiant temperature is determined at time t for different subjects *s1, s2, si* in the room.

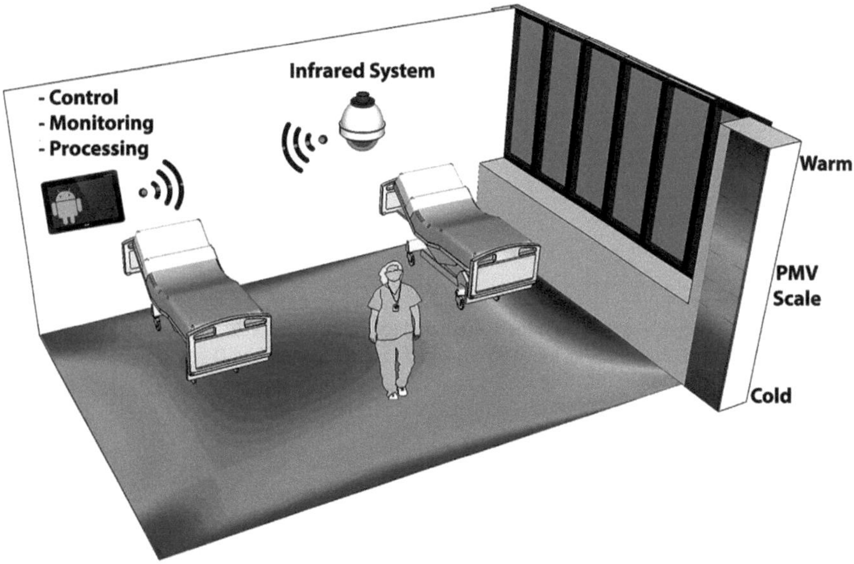

Fig. 2 Concept of the proposed solution for the monitoring of the *PMV* on multiple positions in the space (e.g. bedding patient or nurse with different personal parameters)

To perform this task, an embedded algorithm has been implemented to automatically apply the calculation of *MRT* using the angle factors methodology according to the absolute coordinates (*Xi, Yi*) of the *i-th* person [3, 4]. The algorithms proposed have been implemented in a C++ programming language and turned into a library that is called by the main program uploaded into the microcontroller. Details about the methodology applied to evaluate and use the personal parameters metabolic rate (M) and clothing level (I_{cl}) are discussed in Sect. 3.3. Based on the spatial information of the *MRT*, embedded libraries implemented on microcontroller allow the computation of thermal comfort index *PMV* for multiple positions of the subject in the environment.

The methodology applied to compute the *MRT* and the consequent *PMV* is developed according to [9] and [10]. Finally, the measured values (air temperature and humidity, surface temperatures, *MRT* and *PMV* for different positions) are collected into a single string (45 bytes plus 10 bytes for each additional position monitored) and sent back to the Android device via Bluetooth communication protocol. Timing required for the entire acquisition process was carefully programmed and it resulted in a minimum measuring interval of 1 sample every 5 min. The prototype makes use of a Bluetooth communication, but other low-rate, low-power wireless transmission techniques could be easily applied in order to reduce the energy consumption and avoid the issues related to co-existence and compatibility among different devices within the same indoor environment.

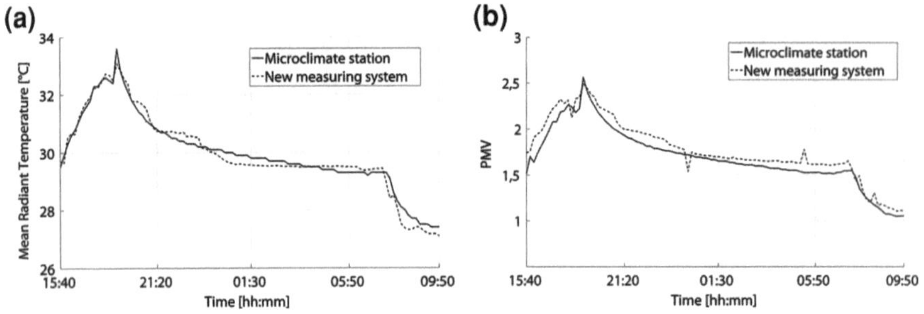

Fig. 3 **a** *MRT* comparison between the system and the microclimate station resulted in discrepancies of 0 ±0.5 °C; **b** comparison for *PMV* resulted in discrepancies of 0.1 ± 0.1 (coverage factor $k = 2$)

3.2 Validation

The system has been tested into an indoor environment at Università Politecnica delle Marche, which consists of a 6.6 × 4.4 × 3 m room. The validation of the system was performed through the comparison with a microclimate station, whose measurement accuracies are respectively: ±0.1 °C for the air temperature and mean radiant temperature, ±2.5 % for the relative humidity and ±0.02 m/s for the air velocity. According to data-sheet information, the accuracy of the integrated *T-RH* sensor resulted in ±0.5 °C and ± 2 % respectively. As for the mean radiant temperature, during a laboratory calibration procedure of the device developed the accuracy obtained was ±1.2 °C, with a coverage factor $k = 2$. An example of comparison between the measuring systems in terms of *MRT* and *PMV* is reported in Fig. 3a, b, which is related to a measurement campaign conducted during August 2012. The two parameters were calculated for the same position where the micro-climate station was placed (at the center of the room). Results show the possibility to provide thermal comfort data in a non-intrusive way and with a potential cost reduction of 1/10 of the actual price of a microclimate station.

3.3 Considerations on Personal Parameters in AAL Applications

As presented in [1], the *PMV* sensitivity to personal parameters is high, especially for the metabolic rate *M*. Despite the importance of providing an accurate measurement of this parameter, generally this value is provided by referring to table values, which are provided in literature [10] as resulting from statistical classifications according to occupation or activity level of the subject. ISO 8996 [11] provides some approaches with the aim of reducing the risk of error caused by different aspects, as individual variability, gender, anthropometric characteristics, etc., by applying procedures based on direct measurements to be performed in field. One method allows an estimation of metabolic rate (M) using continuous heart rate (HR) monitoring. The relationship between these two parameters is provided in the following equation:

$$M = \frac{1}{RM} \cdot (HR - HR_0) + M_0 \tag{2}$$

where RM ($\frac{bpm}{W \cdot m^{-2}}$) is the increase in heart rate per unit of metabolic rate, HR_0 (bpm) is the heart rate at rest, under neutral thermal conditions, and $M_0 = 55$ (Wm^{-2}) is the metabolic rate at rest. The increase in HR due to metabolic rate M is estimated by the following equation:

$$RM = \frac{205 - 0.62 \cdot A - HR_0}{(K - 0.22 \cdot A) \cdot p^{0.666} - M_0} \tag{3}$$

where A is the age of subject (years), P is the weight (kg) and K is a constant value, which is 41.7 for men and 35.0 for women. Heart rate measurements need to be performed during a representative period, which has to be defined. The method proposed to estimate the metabolic rate through HR measurements requires an accurate sensor for the continuous heart rate monitoring: as from the previous equations, an increase of 1 bpm in the HR induces a $+0.08$ *met* in M. As a consequence the heart rate measurement accuracy should not exceed ± 2 bpm for a good estimation of M (± 0.16 *met*) and to apply this estimation procedure. This method proves to be more accurate with respect to assuming M as a constant taken from the ISO tables, because the obtained value is not based on a statistical approach but takes into account the subject variability by means of its HR. Besides, other cases or situations that are not assessed in the tables can be evaluated with direct approaches. For example, long-term care patients or elderly persons can be constantly monitored and these information can be integrated into the main computing algorithm implemented on the microcontroller to improve the thermal comfort assessment in terms of PMV. For these subjects the heart rate does not differ from HR_0 which is measured at rest, so M is generally about $0.9 \div 1$ *met*.

As a demonstration of what assessed in the Sect. 2 in relation to the sensitivity to M, in another measurement session the PMV was computed with a metabolic rate of 0.9 *met* (Fig. 4) and compared to that of a subject with $M = 1.2$ *met* (the table value for office worker). PMV and so thermal perception shows negligible discrepancies for high values of PMV due to a warmer environment, while this deviation increases for a lower level of PMV assessing the increased thermal sensitivity of low metabolic rate subjects. These results demonstrate the potential of using the system proposed to monitor comfort for AAL applications.

Concerning the clothing level, ISO 7730 provides several tables to refer to, with a different level of details (which are based on a standard set of clothes or by computing the insulation of each single cloth). However, this solution may not be applicable in every situation and some specific considerations have to be done for this specific application. For example, if the subject is unable to move and is placed in bed all the

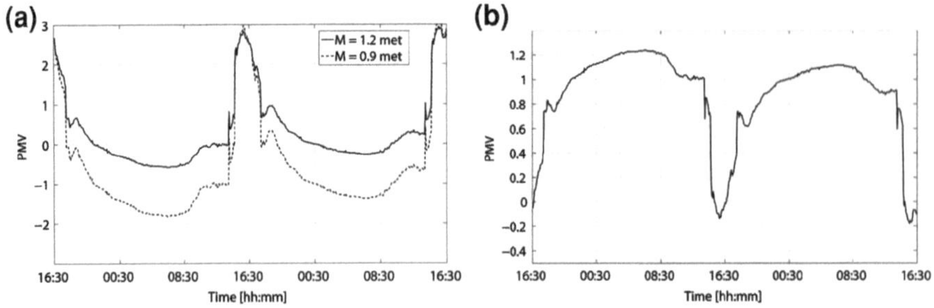

Fig. 4 **a** *PMV* profiles calculated with two different metabolic rate; **b** Deviation between the two *PMV* profiles

time, the total insulation has to take into consideration the effects of bed, blankets, quilts. In [14] and [15] a mathematical model has been proposed and validated in [23] for people in a bedding system. The total insulation is computed as a function of the thickness of the item (i.e. blanket), H_{fab}, and the percentage of body area covered by the blanket, according to the following equation:

$$\frac{1}{r_T} = \frac{\frac{3}{5} * (\alpha - 0.233)}{\left(\left(0.03984 * H_{fab}\right) + \frac{1}{8.335}\right)} + \left(\left(\frac{2}{5} * (\alpha - 0.233)\right) * 5.1\right) \left[\frac{\left(\frac{2}{\left(0.03984 * H_{fab}\right) + \frac{1}{8.335}}\right)}{\frac{2}{\left(0.03984 * H_{fab}\right) + \frac{1}{8.335}} + 5.1}\right] + \frac{(1 - \alpha)}{\frac{1}{8.335}} \tag{4}$$

This method allows the estimation of clothing level for bedding subjects, which could be the case for example of people in hospitals or for AAL. The input configuration is able to provide different values of clothing level for different positions of subjects in the room, allowing a more accurate measurement of the comfort level for the particular subject in exam. Seasonal values will be used also to adapt to the different climatic conditions. However, as stated by Fanger in [7], the PMV-PPD model is a static model, so it is not possible to consider transient variations of the considered parameters, i.e. changing of the clothing level during the day. So, if more people are present in the same environment, potentially with different personal parameters (e.g. an older adult and a younger), supposed that they are most of the time located in different areas, the system can use different values for different zones. Moreover, recent research trends in AAL focus on the use of computer vision systems [18] that could facilitate the real-time identification of position, activity and clothing level for several occupants in the environment by applying image processing algorithms. Thus, the solution proposed could benefit of this kind of information to provide dynamic personal parameters.

4 Conclusions

Traditional HVAC control systems based on single-point measurements of air temperature and relative humidity sometimes may not be enough accurate because these parameters are not fully representative of the thermal sensation perceived by subjects. Moreover, for older people, long-term care patients and generally speaking in AAL applications, personal parameters (metabolic rate and clothing level) are significantly different with respect to standard levels and so the perceived thermal comfort cannot be predicted without taking them into consideration. An innovative solution for measuring indoor thermal comfort has been proposed by authors which can potentially improve the indoor thermal comfort for AAL users by means of accurate estimation of comfort level. The system is capable of providing continuous monitoring of environmental parameters with enough accuracy (discrepancies in the order of ± 0.1 for *PMV*) for multiple positions and signicant reduction of costs. The system has been assembled with open-source hardware for potential integration with traditional BMS and other devices. The same choice has been done for the software (both the microcontroller main program and the Android user interface) which can be easily updated and modified: the overall *PMV-PPD* model, which is adopted to assess indoor thermal comfort, can be adapted to the requirements of specific AAL applications (e.g. long-term care patients where the thermal sensitivity is affected by the specific disease), for example by providing a dedicated calibration, as observed in [5], or by coupling the system with heart rate measurement or vision systems.

Acknowledgments The work is part of the FP7 number 285623 European Project CETIEB (Cost Effective Tools for Better Indoor Environment in Retrofitted Energy Efficient Buildings) co-founded by the European Commission.

References

1. Alfano, F., Palella, B., Riccio, G.: The role of measurement accuracy on the thermal environment assessment by means of pmv index. Build. Environ. **46**, 1361–1369 (2011)
2. ASHRAE: ASHRAE Handbook-Fundamentals. American Society of Heating Refrigeration and Air Conditioning Engineer Inc, Atlanta (2005)
3. Butera, F.M.: Principles of thermal comfort. Renew. Sustain. Energy Rev. **2**, 39–66 (1998)
4. Cannistraro, G., Franzitta, G., Giaconia, C., et al.: Algorithms for the calculation of the view factors between human body and rectangular surfaces in parallelepiped environments. Energy Build. **19**(1), 51–60 (1992)
5. Corgnati, S., Filippi, M., Viazzo, S.: Perception of the thermal environment in high school and university classrooms: subjective preferences and thermal comfort. Build. Environ. **42**, 951–959 (2007)
6. Enomoto-Koshimizu, H., Kubo, H., Isoda, N., et al.: Effect of the radiant heating on the elderly. In: Proceedings 13th Triennial Congress of the International Ergonomics Association, Tampere, vol. 5, pp. 433–435 (1997)
7. Fanger, P.O.: Calculations of thermal comfort: introduction of a basic comfort equation. ASHRAE Trans. **73** (1973)
8. Havenith, G.: Temperature regulation and technology. Gerontechnology **1**(1), 41–49 (2001)

9. ISO7726: ergonomics of the thermal environment—instruments for measuring physical quantities. International Standardization Organization, Geneva (2002)
10. ISO7730: ergonomics of the thermal environment—analytical determination and interpretation of thermal comfort using calculation of the PMV and PPD indices and local thermal comfort criteria. International Standardization Organization, Geneva (2005)
11. ISO8996: ergonomics of the thermal environment—determination of metabolic rate. International Standardization Organization, Geneva (2004)
12. ISO9920: ergonomics of the thermal environment—Estimation of the thermal insulation and evaporative resistance of a clothing ensemble. International Standardization Organization, Geneva (1995)
13. Korhonen, I., Pärkkä, J., Gils, M.: Health monitoring in the home of the future. IEEE Eng. Med. Biol. Mag. **22**(3), 66–73 (2003)
14. Lin, Z., Deng, S.: A study on the thermal comfort in sleeping environments in the subtropics-measuring the total insulation values for the bedding systems commonly used in the subtropics. Build. Environ. **43**, 905–916 (2008)
15. Pan, D., Lin, Z., Deng, S.: A mathematical model for predicting the total insulation value of a bedding system. Build. Environ. **45**, 1866–1872 (2010)
16. Revel, G., Sabbatini, E., Arnesano, M.: Development and experimental evaluation of a thermography measurement system for real-time monitoring of comfort and heat rate exchange in the built environment. Meas. Sci. Technol. **23**, 035005 (2012)
17. Saltelli, A.: Sensitivity Analysis in Practice: A Guide to Assessing Scientific Model. Wiley, Chichester (2004)
18. Spasova, V., Iliev, I.: Computer vision and wireless sensor networks in ambient assisted living: state of the art and challenges. J. Emerg. Trends Comput. Inf. Sci. **3**(4), 1204–1995 (2012)
19. Tsuzuki, K., Ohfuku, T., Iwata, T. et al.: Thermal comfort for the elderly people. In: Proceedings of International Workshop on Gerontechnology: Age-Related Change in Human Factors and Application Technology. National Institute of Bioscience and Human Technology, Tsukuba (2001)
20. Tsuzuki, K., Ohfuku, T.: Thermal sensation and thermoregulation in elderly compared to young people in Japanese winter season. In: Proceedings of, Indoor Air 02, Monterey (2002)
21. Van Hoof, J., Hensen, J.: Thermal comfort and older adults. Gerontchnology **4**(4), 223–228 (2006)
22. Van Hoof, J., Kort, H.S.M., Hensen, J.L.M., et al.: Thermal comfort and the integrated design of homes for older people with dementia. Build. Environ. **45**, 358–370 (2010)
23. Verheyen, J., Theys, N., Allonsius, L., Descamps, F.: Thermal comfort of patients: objective and subjective measurements in patient rooms of a belgian healthcare facility. Build. Environ. **46**, 1204–1995 (2011)
24. Yang, K.H., Su, C.H.: An approach to building energy savings using the pmv index. Build. Environ. **32**(1), 25–30 (1997)

A Novel Device for Contactless Detection of Small Body Movements Dedicated to People with Severe Mobility Impairments

L. Fanucci, M. Malavasi, F. Iacopetti, M. Turturici, F. Sciarra and F. Lupi

Abstract In the present work a novel system aiming at the contactless detection and recognition of small gestures of the body is proposed. Gestures can be subsequently associated to actions performed by a generic device, via a PC or an external actuator. The present work is related to the advanced stages of neuromuscular diseases, when the capabilities to control voluntary muscles are very poor and limited to a few muscles such as the facial ones. The proposed system uses a video acquisition unit, processing algorithms able to detect and recognize movements, a hardware-software platform to translate gestures into PC application controls or interaction with other devices, like cellular phones or ringing bells. The application of the system to a real scenario consisting in the detection of small facial movements of a user affected by Duchenne's syndrome is described. The target application is raising a voluntary remote call for assistance by means of a predefined set of user gestures.

1 Introduction and Background

The topic of this paper is related to neuromuscular diseases causing the progressive disruption of muscle integrity and strength. Among them, in particular, the considered pathologies are muscular dystrophies, including Duchenne's and Becker's,

L. Fanucci (✉) · F. Iacopetti · M. Turturici
Dipartimento di Ingegneria dell'Informazione, Università di Pisa, Via Caruso, 16, I-56122 Pisa, Italy
e-mail: luca.fanucci@iet.unipi.it

M. Malavasi
Centro Regionale Ausili Regione Emilia Romagna, via S. Isaia, 90, I-40123 Bologna, Italy
e-mail: mmalavasi@ausilioteca.org

F. Sciarra · F. Lupi
BK s.r.l. , via XXIV Maggio 331, I-19125 La Spezia (SP), Italy
e-mail: fabrizio.sciarra@b-k.it

S. Longhi et al. (eds.), *Ambient Assisted Living*, DOI: 10.1007/978-3-319-01119-6_2,

which may lead to the progressive lack of movement of the voluntary muscles, with the possibility of ending into the locked-in syndrome, consisting in tetraplegia and complete lacking of speaking capability in persons maintaining cognitive integrity. The present work addresses the advanced stages of the considered pathologies, in which users have lost most of their capability to control voluntary muscles and the residual movements are substantially limited to the facial ones, in particular to nose, eyes and mouth. The proposed system is capable to detect and recognize small movements, in the order of millimeters, of any body district. If these movements are voluntarily repeatable in a quite accurate way by the user, so that they are almost unambiguous, we call them gestures. Hence the system is capable to recognize gestures (or sequences of gestures) and use them to control any electronic device, by means of proper output interfaces connected to the processing PC. Examples of system applications are the activation of a ringing bell and an emergency phone call.

In the present paper, the specialization of the system to the detection of small facial movements of a user affected by Duchenne's syndrome is described, where the target application is to raise a voluntary remote call for assistance by means of a predefined sequence of user gestures. The particular pathology and its stage are considered significant for the application of the system to a wider set of less severe motor skill disabilities. Furthermore, due to the scarce availability of commercial solutions for persons with very reduced capabilities, the proposed work aims at providing a novel and important contribution in the field of Assistive Technology for people with severe disabilities.

2 State of the Art

People with severe mobility impairments have very few possibilities to interact with the surrounding environment due to the lack of muscular capability. Several diseases, like Amyotrophic Lateral Sclerosis (ALS) and Duchenne's syndrome, lead to a dramatic reduction of strength and functionality of muscles.

In the last years, thanks to recent innovations in the field of Assistive Technology, several devices have been developed and are commercially available allowing to exploit nearly every single residual movement and allowing to interact with the surrounding environment or to communicate with other people.

For the reference users, the available solutions [1–3] for the control of one or more functions of an electronic device or a PC are essentially limited to eye gaze systems, devices able to detect small face movements, switch devices able to detect small movements or muscular electrical activity, voice recognition systems, brain-computer interface systems [4].

For instance persons affected by amyotrophic lateral sclerosis, even in advanced stages, can proficiently communicate and interact with the surrounding environment by using a gaze system. These systems have become very useful and reliable but have some important issues that prevent a simple usage in any given daily scenario.

In fact they must be calibrated very accurately with the help of a skilled caregiver before use, and if the user has been moved from the current position, typically for a cleaning or medical action, they must be calibrated again.

For different types of disease and mobility impairment some other solutions exist, such as switch sensors designed to exploit any residual capability of the user, even very small and imperceptible ones. For instance if the user can breathe independently, a sip and puff sensor can be used to produce two independent controls; if the user can move one or both eyelids, eye-blink sensors can be exploited; if the person has a residual nervous activity but very reduced capabilities in muscle fibers, then neuromuscular activities can be captured by specific sensors. Voice recognition systems are divided into Speaker Independent (SI) and Speaker Dependent (SD) systems; SI systems allow the user the control of the PC, smartphones and other devices without any training, but they generally need a good speaking capability, which is not present in many situations in the reference scenario, where the user may have difficulties in speaking clearly, or with a loud voice, etc.. Moreover, SI systems suffer from environmental noise and generally need a good quality audio chain, consisting in a microphone placed at a closed distance from the mouth, which is not always possible or comfortable for the user. SD systems allow the recognition of the user's specific voice after a training phase, allowing word recognition even of badly pronounced words, but the word set is generally limited to a few or some words; moreover, SD systems suffer from user's voice alteration due for example to temporary illness, fatigue, etc..

In recent years, brain-computer interfaces (BCI) mostly based on electroencephalography (EEG) have been developed [3], but they generally require the use of an electrode cap that must be worn by the user and specific user's training, which not always allows for a comfortable and simple use.

The illustrated devices are very useful and reliable but have in common an important issue: they must either be worn by the person or they somehow need to be put in contact with the user. The "contact" characteristic makes switch sensors generally very simple to use, but on the other hand it makes them somehow uncomfortable in many daily scenarios, like during the night sleep and when the person moves or need to be moved by a caregiver. Furthermore they must be put on and off by an external person, and if their position is wrong these sensors cannot work properly.

The system proposed in this paper has instead a different approach, based on infrared cameras, that is completely contactless and does not even need to be put close to the user but has instead to be positioned at a distance of about half a meter from the body part having residual mobility. Furthermore, thanks to being sensitive to IR light, it can be used even during night. The only part of the system in contact with the body consists in a couple of small adhesive markers that have to be applied on the body district whose gesture must be detected.

3 System Architecture

The proposed system architecture is composed of three fundamental blocks:

- a video unit for the real-time image acquisition and an IR light source to light up the body part having residual mobility
- a processing unit with adequate computing power, and a specific software able to detect and recognize the residual movements and extract appropriate gestures
- a software to translate the recognized gestures into application controls or commands to other external electronic devices

The proposed system is shown in Fig. 1. The first block consists of a pair of cameras for the acquisition of images and a lighting system composed of several tens of IR LEDs. The lighting system is completely adjustable both in the average intensity of the light and the peak value of the emitted power. For this purpose a PWM approach and commercial LED drivers are used.

The cameras are sensible both to IR and visible light, in fact using different setups of the video acquisition and of the lighting systems, and proper different processing algorithms, it is possible working either in the visible or in the near infrared part of the spectrum, allowing several different applications. For example IR light allows a safe and reliable usage in night scenarios, while visible light allows an usage during the day and in presence of strong daylight. To be noted that in presence of intense sunlight a great quantity of IR radiation causes a reduced SNR on the processing unit, so the preferable scenario for the IR setup is the night or a dark environment. However the software is capable to process in an acceptable way the IR signals even with a normal daylight. The light emission system may be controlled in terms of light intensity and duty cycle, also opening to the possibility to differentiate its contribution from the natural ambient light.

The video acquisition system is embedded in a dedicated case and lenses are surrounded by the IR lighting system. For IR operation the two lenses are properly modified in order to reject the visible part of the spectrum by means of two absorption filters.

The second block consists of a medium-high performance personal computer with a Linux operating system and a custom designed software able to capture video stream and recognize user's gestures. Depending on the specific gestures to be recognized, the software is capable to process one or two video streams by combining data from the two cameras; this allows both two and three dimensional reconstruction of the movement of the involved body district. This software is the core of the entire system, and further details about the processing algorithms will be provided in dedicated Sects. 3.1 and 3.2.

The last part of the system provides interaction with the application context. This part may vary depending on the target application; however it consists in general of a software, running on the same PC, for translating the recognized gestures into commands, such as data via the serial port or via a Bluetooth interface to a dedicated hardware able to perform the desired action. For instance if the target application is

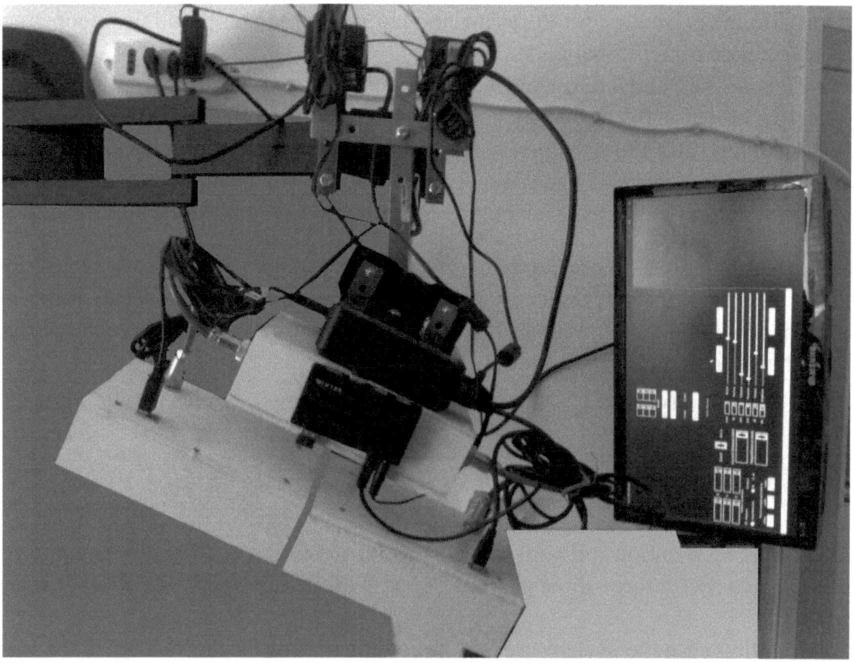

Fig. 1 A picture of the video acquisition hardware and the graphical user interface (GUI) of the detection software

to ring an alarm, an external board with electronic switches connected to a ring bell may be used, communicating with the PC by means of a serial protocol. Another important example, described in Sect. 4, is a phone call: in this case a Bluetooth dongle and a common mobile phone are used.

From the user's side, as previously mentioned, the system does not need any specific device to be worn nor any particular operation, except for the correct system positioning, helped by the program graphical interface and, in case of night usage, the application of two IR reflective markers with the approximate dimension of a quarter of square centimeter. These markers are not needed if visible light is used.

3.1 Processing Software

The first important block of the processing software is a multi-programmable filter chain, which is needed in order to clean the acquired signal as much as possible, so to accurately reconstruct the user's gesture. The most important filters used by the processing chain are listed below; further details are omitted for readability reasons.

- space-time correlators, for the association of measured points and the reconstruction of three dimensional positions
- high-pass filter, and a frequency variable gain
- low-pass filter, and a frequency variable gain
- Kalman filter, for the reduction of noise and the characterization of motion
- dynamic filters in speed and space for resistance to small shifts in the dynamics
- projection matrices, for the adaptation of the control volume to the control surface

The combination of these components can reduce some typical noises (such as the flicker at about 4–6 Hz typical of Parkinson's disease), can reject spasms resulting from the involuntary contraction of a muscle and can follow small involuntary drifts of the body districts.

3.2 User Gesture Recognition Software

The gesture recognition software is made up of different processing blocks currently used alternately for three-dimensional (IR) or two-dimensional (visible light) applications.

In the first case the system uses the following approaches to recognize the user gestures:

- a classifier based on neural networks trained on gestures previously loaded
- an evaluator of the distance between two tracks in 3D
- inertia of movement regenerator with x/y

In the two-dimensional applications different processing policies are used instead, such as:

- analyzer of the state of the eye (either open or closed)
- analyzer of the state of the fingers of one hand, and particularly a software that evaluates whether fingers are close or not to each other

4 Application in a Real Context

The system proposed in this paper has been developed thanks to the collaboration of Assistive Technology (AT) experts and end-users. By this proficient collaboration two important application scenarios have been identified since early stages of the project.

The first application aims at the recognition of user gestures or movements through the detection of the pattern described by two different points of the user's body. The second application is based on image processing aiming at the recognition of user gestures, like the eye-blink and the particular configuration assumed by the fingers of one hand. In the first application scenario, IR reflective markers are placed on two body districts of the user, assuming that these body districts can be observed at the same time by the cameras and that have the possibility of stretch or shorten their relative distance by voluntary muscle movements. These districts can for example be constituted by a point on the forehead and one on the chin, or one on another point of the face and one on the chin, or even by a point above the upper lip and one on the lower one or on another point on the face. All the listed cases require the possibility of movements of the facial musculature, and have been chosen according to user's condition and the fatigue that this entails. The two markers are observed by the two cameras and their relative position in the three dimensional space is evaluated. When the distance between the two points either increases or decreases according to a pre-defined pattern, the software recognizes a gesture; that gesture or a particular set of gestures, e.g. three repetitions within a given time, can be interpreted as a command to perform a specific action. It is important to note that the distance between the two points is evaluated in the three dimensional space, allowing for a reliable detection when the two points lie on a plane that is not parallel to the cameras one. The processing software features several configuration parameters such as distance thresholds, overcome thresholds (a Schmitt's trigger-like approach), number of repetitions and more. All these parameters can be set by means of a simple and intuitive graphical user interface provided by the PC application itself.

For the test of the application, a user suffering from Duchenne's syndrome has been involved. His residual movements are essentially limited to the facial muscles and in particular to eyes, nose and mouth. The user has a quite good control of the tongue.

Following a number of experiments carried out with a previous similar system, based on a different technology, the movement of the lower lip by means of the tongue push has been identified as usable gesture without excessive fatigue for the user.

The magnitude of the variation of the distance between the markers associated with voluntary movements executable by the user is substantially in the order of 10 mm, in comparison with a rest distance between the markers from 20 to 30 mm.

In the first test campaign, according to the user will, the system was configured to recognize and count the shortenings of the distance between the two markers, and if that gesture was repeated by the user for three times within few seconds, the system raised an alarm. When the time span between two gestures overcame a pre-defined threshold the count was reset by the system. This policy has been adopted in order to better recognize voluntary actions and minimize the raise of unwanted alarms. In order to allow the user to be aware of the system condition, a visual pre-alarm on the display and an audible warning are raised when a first gesture is recognized, so that the user is informed of the status of early warning and may decide either not to make further movements in the next few seconds to get out of this condition or to make two more gestures to raise the alarm. The final application is to make a phone call to one or more pre-defined numbers, in order to warn a relative or a caregiver. To this aim a Bluetooth interface is used to communicate with an appropriate mobile phone.

The detection system is suspended above the user by means of a special articulated arm attached to the wall; as already mentioned cameras are mounted at a distance of about $40 \div 50$ cm from the face of the user, allowing a caregiver to operate on him for cleaning or medical activities.

Once the system is installed near the user, the service personnel must only apply the markers on the face of the user, place the detection system in the working position, launch the application on the PC and set thresholds through a semi-automatic procedure. At that point, the application is ready to activate the call to the mobile phone in the case of recognition of the gesture pattern set. Obviously, the parameters can be saved between sessions, thus minimizing the number of operations to be performed by the assistant.

5 Results

The test campaign conducted in a real operating environment highlighted the possibility to reconstruct with sufficient accuracy the position of the two markers and the consequent possibility of detecting the variation of the distance between them with a few millimeter accuracy, even at distances between cameras and the face of some tens of centimeters. It has been tested that the user can easily raise an alarm or make an emergency phone call by using a given gesture. Some issues of the IR application were highlighted in presence of strong ambient sun lighting or temporary transit in the scene of detection of objects that cause strong reflections, such as some particular clothing, or as a result of the intervention of an assistant to perform some actions in the field of detection.

The system was tested for the detection of movements of facial muscles, however the developed technology make plausible the use of the system for the detection of movements of other body parts, thus expanding the application scenario to other disabilities and potentially also to a more sophisticate pattern recognition for the control of applications requiring more than one control input.

6 Conclusion

In this paper a novel system allowing people with severe mobility impairments to interact with the surrounding environment is presented. The proposed system is able to recognize several kinds of user's gesture by processing an image stream of the body districts having residual mobility. The developed technology can be used both during the day and the night, by means of a near infra-red lighting system and infrared and visible spectrum cameras. Furthermore the processing software can be set up in order to adapt to a relevant number of user gestures, even small ones, such as mouth opening-closing, eye blink, chin movement and many more. Unlike most of other Assistive Technology devices the proposed system is completely contactless, allowing a comfortable user experience.

The proposed device is specifically designed for a few cases of persons with severe mobility reduction, but it represents however an important result since those persons have very few possibilities to interact with the surrounding environment. In the near future some other tests with users will be carried out, both by developers' team and independently by the caregivers.

References

1. EASTIN, European Assistive Technology Information Network, http://www.eastin.eu/en-GB/searches/products/index
2. SIVA portal, the Italian Web Portal providing information and guidance on Assistive Technologies for independence, quality of life and participation of persons with disabilities, http://www.portale.siva.it/en-GB/home/default
3. Cook, A.M.: Cook and Hussey's Assistive Technologies—Principles and Practice. Elsevier, St. Louis (2008)
4. Pasqualotto, E., Federici, S., Olivetti Belradinelli, M.: Toward functioning and usable brain-computer interfaces (bcis): a literature review. Disabil. Rehabil. Assistive Technol. **7**(2), 89–103 (2012)

Improved UWB Radar Signal Processing for the Extraction of Vital Parameters

Giovanni Pelliccioni, Susanna Spinsante and Ennio Gambi

Abstract The availability of contactless sensors capable of detecting vital parameters is of particular interest, especially when it is requested to ensure the patient's freedom of movement, or when the patient's physical conditions do not allow the application of sensing devices on the skin. Thanks to its characteristics of high spatial resolution and tissues penetration, an Ultra Wide Band radar system can be used for the measurement of respiratory and heart rates of hospitalized patients. As typical of radar systems, however, the useful echo is superimposed to a multiplicity of unwanted echoes, due to reflection by obstacles, normally fixed, that may be present in the environment considered. Among the various techniques proposed in the literature, this work presents an enhanced processing of the signal received by an Ultra Wide Band radar, in the presence of static echoes, the magnitude of which may be also considerably higher than the level of the signals reflected by the target. The simulation results show the effectiveness of the proposed processing method, and its sensitivity to the radar system design parameters.

1 Introduction

The detection and estimation of vital parameters through the use of Ultra Wide Band (UWB) radar devices are receiving increasing attention, as they allow monitoring a patient without contact [5, 9]. On one hand, this condition avoids the need to connect

G. Pelliccioni · S. Spinsante (✉) · E. Gambi
Dipartimento di Ingegneria dell'Informazione (DII), Università Politecnica delle Marche, via Brecce Bianche 12, 60131 Ancona, Italy
e-mail: s.spinsante@univpm.it

G. Pelliccioni
e-mail: g.pelliccioni@univpm.it

E. Gambi
e-mail: e.gambi@univpm.it

S. Longhi et al. (eds.), *Ambient Assisted Living*, DOI: 10.1007/978-3-319-01119-6_3,

cables to the patient, thus also eliminating the risk of unintentional disconnection of the monitoring device [7]. On the other hand, there are some situations in which a contactless monitoring is the only possible solution, such as in the case of patients who have sustained severe burns, for whom it is essential to avoid any contact with the burnt skin [3].

Thanks to its penetration capabilities, the signal generated by an UWB radar equipment can be also used to search for humans, under certain conditions: as an example, it may be adopted to reveal the presence of kidnapped people, to locate people hidden behind walls, or buried under rubble, during search and rescue operations [6]. The problem of detecting cardiac activity through a radar-based approach is addressed in [2]: static reflections, summing up to the useful signal, make it difficult to extract the heart rate, and should be eliminated by means of a processing step, based on the use of a motion filter. An alternative to the use of UWB radar for non-invasive detection of vital signs is given by the microwave Doppler radar, proposed in [1], which, however, suffers from low penetration capability into solid materials. On the contrary, the penetration capacity peculiar of UWB signals, allows to perform measurements through obstacles, even if they generate a static clutter that must be eliminated, in order to properly detect the useful signal [4]. As an example, in the case of a subject buried under ruins, the echoes generated by the fixed rubbles may be several dB's over the amplitude of the useful signal, that must be detected to locate the person. The same conditions may apply also to people moving around in an indoor home environment, where possible stationary obstacles may be present in the way of the UWB radar signals.

A thorough analysis of the heart and breathing rate detection problem has been developed by Venkatesh et al. in [8], in which an accurate procedure to derive the breathing rate is also described. The results provided highlight the difficulty of obtaining the exact heart rate value; moreover, the proposed method does not allow to get the breathing profile in the time domain, that would enable a more accurate analysis of the signal, to study breathing-related pathologies in the monitored subjects. At the same time, however, this work is of interest, especially from an engineering perspective, thanks to the simplicity of the proposed approach. For this reason, the limitations evidenced in [8] are addressed in this paper, through the introduction of a new signal processing methodology, that enables the detection of vital parameters even in the presence of a multiplicity of echoes generating a significant clutter. The system capabilities and limitations in detecting thoracic movements are discussed; through an appropriate choice of the system parameters, it is shown how to extract information about the motion profile, which allows the study of breathing-related diseases, as well as the identification of respiratory and heart rates.

The paper is organized as follows: Section 2 describes the main components of the radar system, and the signals used: it presents the mathematical model of the detection process, and the motion filter algorithm. Section 3 discusses the simulation results obtained, in different operational conditions. Finally, Sect. 4 concludes the paper.

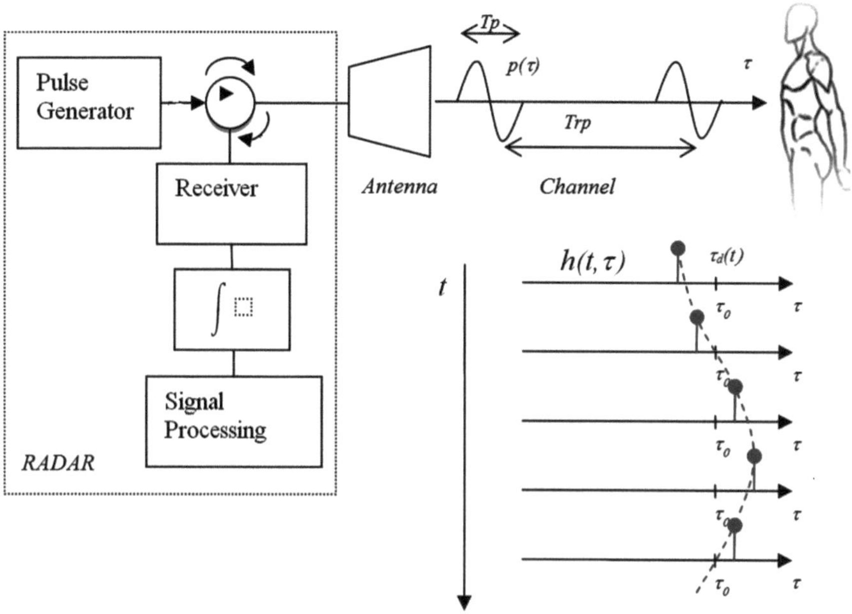

Fig. 1 Reference UWB radar system architecture

2 Radar System

The reference model for the radar system considered in this paper is shown in Fig. 1. The computer program developed to simulate the system involves the generation, every T_{rp} seconds, of UWB pulses which are reflected at a distance d from the transmitter (corresponding to the position of the patient's thoracic surface with respect to the radar sensor), and a receiver block that processes returning echoes. The movement of the patient's thoracic surface, due to joint breathing and heartbeat, produces a deviation around the distance value d_0, that may be described as:

$$d(t) = d_0 + m_b sin(2\pi f_b t) + m_h sin(2\pi f_h t) \quad (1)$$

As shown in Eq. (1), the distance $d(t)$ is defined as the sum of an average distance term, d_0, and two sinusoidal components, the half-amplitudes of which are m_b and m_h, respectively. They represent half of the change in position of the subject's thoracic surface, due to breathing and heartbeat movements, at frequencies f_b and f_h, respectively. The distance $d(t)$, as a function of the vital signals we want to detect, depends on the time variable t: this dependence is called *slow time*, that is different from the *fast time*, denoted by τ, which, on the contrary, describes the time variation of the UWB signal. The propagation delay along the path *antenna—thoracic surface—antenna* equals $\tau_d(t) = 2d(t)/c$, where c is the speed of light, and $\tau_d(t)$ varies between $\tau_0 + \tau_b$ and $\tau_0 - \tau_b$. In ideal conditions, the channel may be described by the following expression:

$$h(t,\tau) = A\delta\left(\tau - \tau_d(t)\right) + \sum_i A_i \delta(t - \tau_i) \tag{2}$$

where A represents the attenuation of the useful signal along the propagation path, whereas coefficients A_i represent the attenuation affecting signal replicas, due to reflections from static objects located in the propagation environment.

About the radar resolution capability, i.e. the radar performance in detecting small movements, of very limited amplitude, it is basically affected by the sampling time in the *fast time* domain. In order to detect a very small change in position of the thoracic surface, it is necessary to apply a quite high-frequency sampling process. As an example, if the time duration in the *fast time* domain is $T_{ft} = 2.5\ ps$, according to the relationship among space, velocity, and time, it is not possible to detect variations in distance smaller than $0.75\ mm$. On the other hand, by applying a sampling process with $T_{ft} = 0.4167\ ps$, variations in distance smaller than $0.125\ mm$ will not be revealed. It is evident that the very small movements of the thoracic surface due to heartbeat are very difficult to reveal, unless a very heavy sampling processing in the *fast time* domain is used. On the other hand, the radar capability of distinguishing among different objects depends on the time duration of the pulses used. The waveform selected for the transmitted signal is a Gaussian monocycle:

$$p(\tau) = -C \cdot \tau \cdot e^{\frac{-\tau^2}{2\sigma^2}}, \sigma = \frac{T_p}{2\pi} \tag{3}$$

When the constant value C equals $\sqrt{e}/\sigma$, the maximum value of the Gaussian monocycle waveform is one; T_p is the pulse duration and it affects, together with the propagation speed ν_p, the spatial resolution of the radar, according to the following equation:

$$D_{sr} = \nu_p T_p / 2 \tag{4}$$

Very short pulses may allow to detect and discriminate reflections originating from objects spatially close to each other.

2.1 Analytical Model of Detection

As introduced above, assuming the propagation channel described by Eq. (2), the received signal may be represented as: $r_s(t,\tau) = p(\tau) \otimes h(t,\tau) = Ap\left(\tau - \tau_d(t)\right)$, where operator $\otimes$ represents the correlation integral. By using a matched filter at the receiver, the expression of the received signal at the output of the correlator is given by:

$$r(t,\tau) = p(\tau)\otimes r_s(t,\tau) = p(\tau)\otimes p(\tau)\otimes h(t,\tau) = c(\tau)\otimes h(t,\tau) = Ac\left(\tau - \tau_d(t)\right) \quad (5)$$

where $c(\tau)$ denotes the autocorrelation function of pulse $p(\tau)$. The propagation delay is obtained by estimating the amount of delay which shall be introduced at the receiver in order to get the correlation peak value A, normalized with respect to the maximum value of $c(\tau)$, as discussed also in [3, 6]. The noise contribution added to the useful signal, or the presence of false echoes generated by undesired reflections, degrades the precision attainable in the estimation of the distance information. When a static clutter is present, i.e. generated by fixed, stationary objects, it is possible to apply processing techniques aimed at its removal [4, 8]. However, it is necessary to differentiate the condition of small replicas, the amplitude of which is lower than the level of the useful signal, from the case of replicas featuring amplitudes comparable to the useful signal level, or even greater ones. In the latter condition, solutions proposed in the literature do not perform effectively when dealing with the extraction of observed vital parameters, and it is necessary to introduce a suitable filtering operation to obtain the desired information.

2.2 Motion Filter

The most commonly used Motion Filter (MF) is based on the Mean Subtraction (MS) method, that features an easy implementation and limited computational complexity [8]. This type of filtering, named MSMF, is able to eliminate the static signal component in the *slow time* domain, but at the expenses of a distortion of the useful signal components. The filter acts by subtracting the mean value of the useful signal, computed over all the *slow time* instants, from the received signal, for each time instant in the *fast time* domain:

$$r_{MS}(t,\tau) = r(t,\tau) - \lim_{T\to\infty}\frac{1}{2T}\int_{-T}^{T} r(t,\tau)dt \quad (6)$$

This obviously implies the knowledge of the signal for all the observation duration, namely the signal samples must be stored for a reasonable time, or for the duration of the measurement. As evident, the MS filtering also removes a DC component of the useful signal, that is consequently altered. This term influences the detection of movement since the estimate is based on the correlation peak of the received signal.

With the aim of eliminating all the clutter components, but avoiding the distortion of the useful signal due to the DC component cancellation, an algorithm that only deletes the fixed replicas of the signal is proposed, named Distortionless Motion Filter (DMF). The proposed approach is based on pinpointing, in the *fast time* domain, the time interval $\tau_0 \pm (\tau_b + T_p)$ at which the thoracic surface movement takes place. The received signal, that includes both fixed and variable echoes due to the thoracic surface movements, is correlated to the locally generated signal at the receiver. The correlation signals are then represented in matrix form, thus providing the range as a function of the *slow time* domain. Such a way of representing signals allows to locate and distinguish the echoes at a fixed position from the echoes due to thoracic movements, by computing the variance of the correlation matrix in the *slow time* domain, for each time instant in the *fast time* domain, and by taking into account the only *fast time* instants at which the variance overcomes the noise power. The variance term is always zero for the echoes generated by fixed obstacles, irrespective of their amplitude with respect to the level of the useful signal. The proposed solution basically consists in a *range gating* able to extract from the received signal the only components due to movements, that are consequently not affected by the subtraction of the mean value.

According to the discussion presented above, the noise impact may degrade the quality of the useful signal, when the noise variance is bigger than the variance of the signal related to the observed movement. However, it is important to notice that the adoption of classical radar processing techniques, as pulse integration, that increase the signal-to-noise ratio, may reduce the possibility such a condition takes place.

3 Simulations and Results

As previously discussed in Sect. 2.1, the UWB signal, to which termal noise is added, is processed by a correlator receiver, followed by a decision stage that allows to estimate the delay associated to the maximum correlation peak. It is well known from the theoretical development that the frequency resolution of the FFT is inversely proportional to the time duration of the measurement T_{meas}; for this reason, the signal is processed over a time duration of $800\,s$ (which provides a frequency resolution of $1/T_{meas} = 0.0013\ Hz$), and the computation of the rates associated to vital parameters is performed every $40\,s$. By this way, after an initial transitory of $800\,s$, a high spectral resolution data may be obtained (i.e. high precision), which is frequenctly updated (i.e. high reliability). The simulations performed deal with a signal representing an ideal thoracic movement, generated by the superimposition of two sinusoidal signals accounting for the respiration effect and the heartbeat, according to Eq. (1).

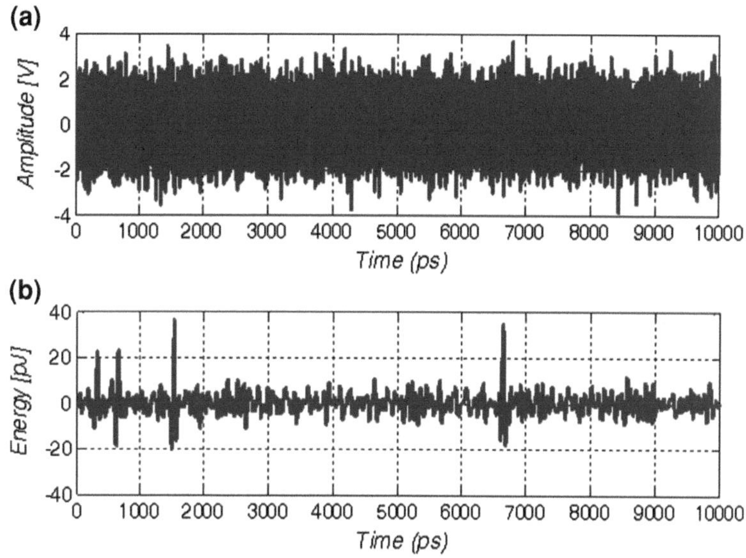

Fig. 2 **a** Received signal $r_s(t, \tau)$ for $t = 120\,s$, **b** Signal output from the correlator at the receiver

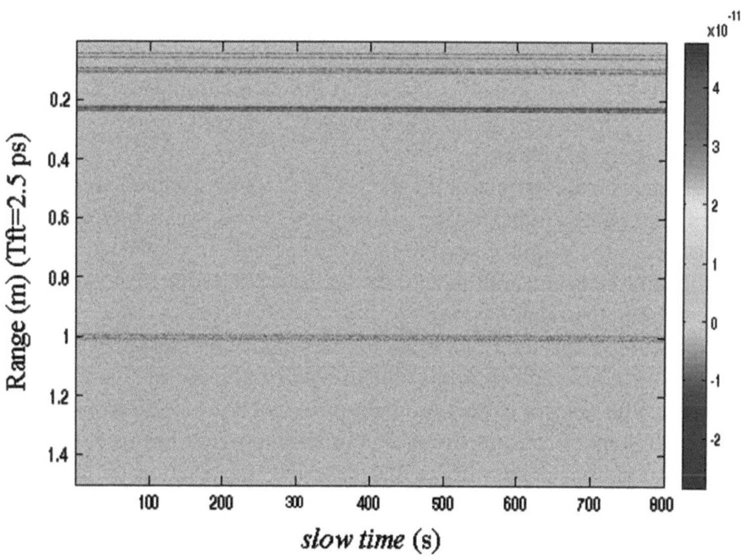

Fig. 3 Received signal matrix, correlated to the locally generated signal

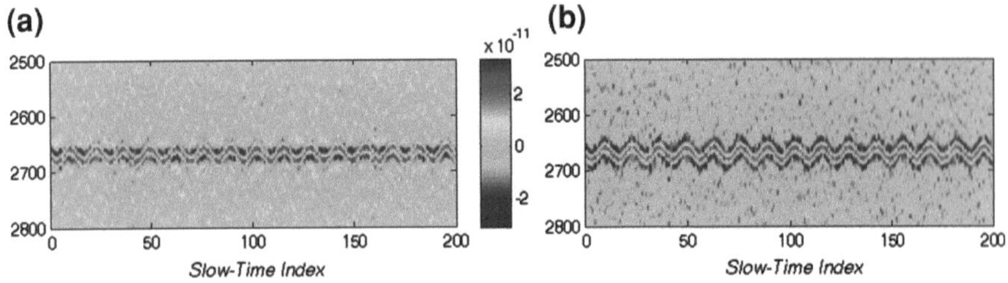

Fig. 4 Effect of filtering on the correlated signal $r(t, \tau)$: **a** MSMF, **b** DMF

Two simulation campaigns have been carried out: in the former, a sample time $T_{ft} = 2.5\,ps$ is configured in the *fast time* domain, in the latter the value is $T_{ft} = 0.4167\,ps$. The following parameters have been adopted in simulatios:

- signal related to thoracic movement: $f_b = 0.189\,Hz$, $f_h = 0.889\,Hz$, $m_b = 4\,mm$, $m_h = 0.09\,mm$, $A = 0.73$, $d_0 = 1\,m$
- static clutter: $A_1 = 0.61$, $A_2 = 0.7$, $A_3 = 0.95$, $d_1 = 0.05\,m$, $d_2 = 0.1\,m$, $d_3 = 0.23\,m$

where d_i represents the distance corresponding to the *i-th* delay term τ_i.

In Fig. 2a, the received signal $r_s(t = 120\,s, \tau)$ is represented, affected by noise $(SNR = -20\,dB)$. In Fig. 2b), the same scan is shown, correlated by the matched filter $r(t = 120\,s, \tau)$. It is possible to see that the correlation peaks are emphasized (due to the strong reduction of the noise impact), thus making it possible to reveal the useful UWB signal echoes.

Finally, Fig. 3 shows the matrix of the received and correlated signal $r(t, \tau)$, in which the signal related to the thoracic movement (at a distance of around $1\,m$), and the static echoes (at a distance smaller than $0.3\,m$) may be evidenced. It is possible to notice the quite relevant amplitude of the signal replicas due to clutter, to the point that they can blind the radar device.

The effects of the MS and DMF motion filters are shown, respectively, in Figs. 4a and 4b. From the color scale, it is possible to see how the useful signal level reaches higher values, with respect to background noise, when the DMF technique is used. This technique is able to ensure stronger immunity to noise. In Fig. 5, the frequency spectra corresponding to the curves that describe the position of the correlation peaks are shown. The reduction of the sampling time T_{ft} in the *fast time* domain, from $2.5\,ps$ (Fig. 5a) to $0.4167\,ps$ (Fig. 5b), allows to obtain also the frequency component corresponding to the heartbeat. It is important to evidence that this result is attained even in the presence of a breathing signal (m_b) of amplitude $17\,dB$ greater than the amplitude of the heartbeat signal (m_h).

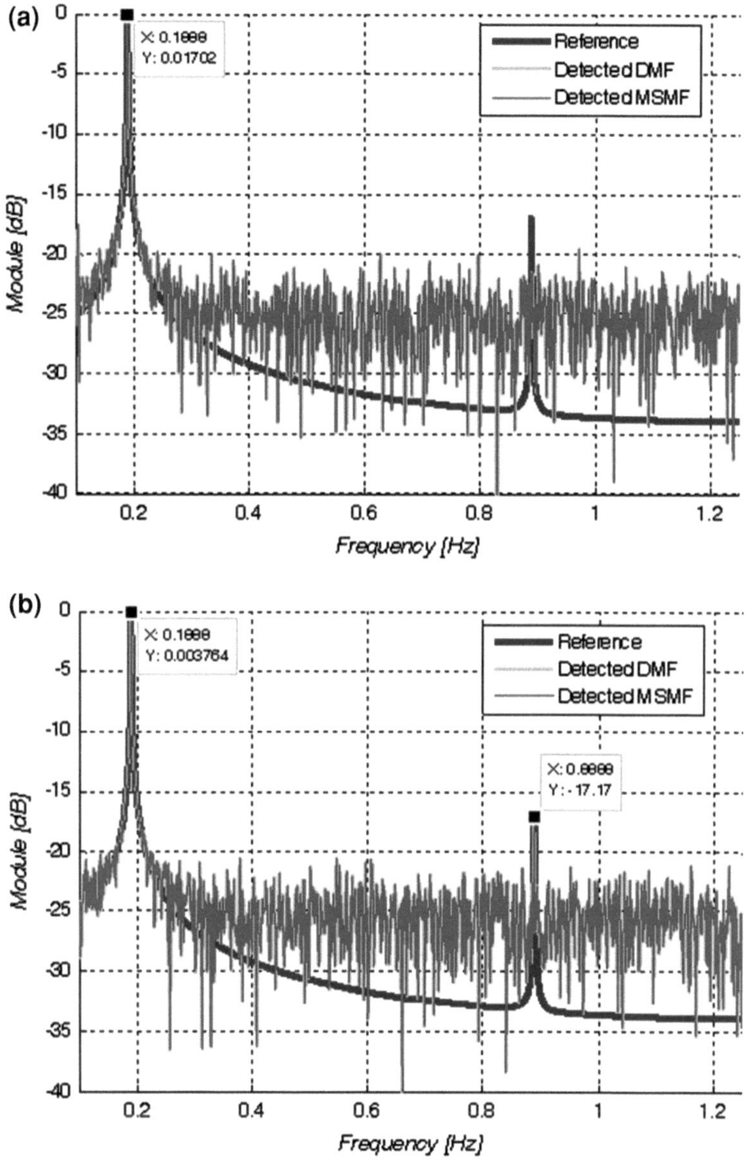

Fig. 5 Comparison of signal spectra: **a** $T_{ft} = 2.5\ ps$, **b** $T_{ft} = 0.4167\ ps$

4 Conclusion

This paper presented a new methodology to filter UWB radar signals, with the aim of extracting information about the vital parameters of a subject monitored by means of a contactless UWB sensor. By comparing two simulated conditions featuring the same configuration, but different sampling time in the *fast time* domain, the limits of the system, due to the sampling process itself, have been shown. The adoption of a higher sampling frequency allows to reveal weak frequency components of very low amplitude, corresponding to the thoracic movements, thus enabling also the detection of the subject's heartbeat rate.

References

1. Host-Madsen, A., Petrochilos, N., Lubecke, V., Boric-Lubecke, O., Park, B.-K., Zhou, Q.: Signal Processing Methods for Doppler Radar Heart Rate Monitoring. In Signal Processing Techniques for Knowledge Extraction and Information Fusion. Springer, Berlin (2008)
2. Lazaro, A., Girbau, D., Villarino, R.: Analysis of vital signs monitoring using an ir-uwb radar. Prog. Electromagnet. Res. **100**, 265–84 (2010)
3. Leib, M., Menzel, W., Schleicher, B., Schumacher, H.: Vital signs monitoring with a uwb radar based on a correlation receiver. Proc. EuCAP **2010**, 3425–9 (2010)
4. Nezirovic, A.: Stationary clutter and linear-trend suppression in impulse-radar-based respiratory motion detection. In: Proceedings of IEEE International Conference on Ultra-Wideband (2011)
5. Nijsure, Y., Tay, W.P., Gunawan, E., Wen, F., Yang, Z., Guan, Y.L., Chua, A.P.: An impulse radio ultrawideband system for contactless noninvasive respiratory monitoring. IEEE Trans Biomed. Eng. **60**(6), 1509–1517 (2013)
6. Ossberger, G., Buchegger, T., Schimback, B., Stelzer, A., Weigel, R.: Non-invasive respiratory movement detection and monitoring of hidden humans using ultra wideband pulse radar. In: Proceedings of 2004 Joint UWBST & IWUWBS, pp. 395–399 (2004)
7. Pelliccioni, G., Borioni, F., Gambi, E., Spinsante, S.: A PPAM technique for energy efficient UWB sensor Body Area Networks. In; Proceedings of 20th International Conference on Software, Telecomunication and Computer Networks (2012)
8. Venkatesh, S., Anderson, C., Rivera, N.V., Buehrer, R.M.: Implementation and analysis of respiration-rate estimation using impulsebased uwb. Proc. IEEE Mil. Comm. Conf. **5**, 3314–3320 (2005)
9. Zito, D., Pepe, D., Mincica, M., Zito, F., De Rossi, D., Lanata, A., Scilingo, E.P., Tognetti, A.: Wearable system-on-a-chip UWB radar for contact-less cardiopulmonary monitoring: Present status. In: Proceedings of 30th Annual International Conference of the IEEE Engineering in Medicine and Biology Society, pp. 5274–5277 (2008)

Near Field Communication Technology for AAL

Valeria di Credico, Simone Orcioni and Massimo Conti

Abstract A common problem for ageing people is that they forget to take drugs at the correct moment. This can be a problem if they live in an independent way without other persons who can help them. In this work we propose the use of smartphones with Near Field Communication technology for medical care. The idea is to apply a tag in the box of the drug to be taken. The application in the smart phone indicates the time when the drug should be taken. When the user takes the drug, he approaches the smartphone to the box of the drug, and the smartphone tracks the event. With a simple interactions, which involves touching tags with the mobile phone, it is possible to manage the information easily. A user friendly application has been developed.

1 Introduction

The increment in Europe of the percentage of ageing people, the increment of the number of people with high disability, the reduction of the components in the family, cause a risk of a great degradation of the quality of the life of a great part of people and their exclusion in the active participation in the economy and in society. For many people the complexity and lack of utility, accessibility and usability of ICT is a major barrier.

This work is dedicated to my father. Massimo

V. di Credico · S. Orcioni · M. Conti (✉)
Dipartimento di Ingegneria dell'Informazione, Università Politecnica delle Marche,
via Brecce Bianche 12, 60131 Ancona, Italy
e-mail: m.conti@univpm.it

S. Orcioni
e-mail: s.orcioni@univpm.it

S. Longhi et al. (eds.), *Ambient Assisted Living*, DOI: 10.1007/978-3-319-01119-6_4,

In EU research frameworks (FP7 and Horizon 2020) some guidelines are given to respond to these trends by mainstreaming and radically improving the accessibility and usability of new ICT solutions, especially for people with disabilities, functional limitations or lacking digital competences.

A great effort is spend in the creation of advanced prototypes of systemic solutions for independent living and active ageing, including reorganization of integrated care and rehabilitation processes, leading to a significant prolongation of personal autonomy.

Many EU projects have been developed and are under development in the field of ambient assisted living, between them:

- the initiative "Ambient Assited Living" (AAL) [1] aims to develop of a European Research Area in the field of "AAL for the Ageing";
- "eCAALYX" [2], is devoted to develop a wearable light device able to measure specific vital signs of the elderly;
- "SENSACTION-AAL" [3] worked on the development of ambulatory assisting devices for enhancing safety and security in balance and movement;
- "NETCARITY" [4] proposed a new integrated paradigm for supporting independence in elderly people living alone at their own home;
- "SMILING" FP7 and "Happy Ageing" [5], coordinated by INRCA of Ancona (IT), devoted to the definition of new instruments for rehabilitation.

An ambient for assisted living can be reached with a seamless integration and plug-and play operation of sensors, devices, sub-systems and integrated care services into cost-effective, self-maintaining, reliable, self-learning and adapting, privacy-respecting and trusted systems. A key point should be the simplicity to use in order to ensure the acceptability by the end user.

Near Field Communication (NFC) is a contactless or proximity communication medium, which is based on magnetic induction. The availability of smartphones and tablet at low cost with NFC facility allows the diffusion of this technology and the development of applications in many different fields, in particular in AAL and healthcare.

Vergara et al. [6] presented a proposal to enable patients getting prescriptions from home by using mobile phones and NFC, avoiding to go to the health-care center.

In [7] a model for the development of context awareness applications using NFC is presented. As an example, the authors presented an application in which the user in emergency situation could ask for help in an intuitive way, just by touching with the NFC smartphone an image with a tag associated, for example a picture of an ambulance.

In [8] the authors proposed an NFC-enabled medical devices to automatically collect clinical data. The doctor touches his NFC card to launch the application installed on the mobile phone. Then, he touches the medical device to obtain the measured data, and finally he touches the patient's NFC tag to identify the user.

In [9] NFC has been used to support Alzheimer patients. The patient measures his vital signs touching the monitoring device's NFC tag with the mobile phone. The application on the mobile phone is automatically launched, it activates the monitoring device obtaining the measure through a Bluetooth connection. Then, the application analyzes the measure and, if necessary, it sends alert messages to the healthcare center.

In [10] a smartphone integrated with an NFC has been used to help the nurse to detect and update drug allergies and drug interactions during medication administration.

Many apps are available for smartphone related to drug assumptions, but they do not use NFC. Mainly the apps remind to the patient when medications need to be taken (e.g. Med Helper Pill Reminder [11]), or they give comprehensive drug information and they are used by doctors, nurses and other healthcare professionals for clinical information (e.g. Micromedex Drug Information [12] and Medscape [13]).

The application for smartphone with NFC, proposed in this work, helps the patients in taking drugs and records in the database the history of the drug assumption, signaling if the correct drug assumption is not followed.

Section 1 will briefly present the NFC technology and Sect. 3 will present the application developed.

2 Near Field Communication

In recent years, new standards related to Radio-frequency Identification (RFId) technology have been defined, among them the Near Field Communication (NFC) is extremely important. NFC protocol is defined by the NFC Forum [14], created in 2004 by Nokia , Philips and Sony. Today NFC Forum has more than 160 members and partners among which Samsung, Visa, MasterCard , Google , Intel etc.

NFC is a radio technology that allows two devices to communicate at short distance, about 4 cm. The two devices are called respectively "Iniziator" and "Target":

- **Initiator**: it starts the commutation and controls and manages the data exchange, for example a reader.
- **Target**: the device that responds to the requirements of the initiator, for example a card or a tag.

NFC devices operate in two modes, passive or active.

- **Passive mode**: in this mode, the reader generates an electromagnetic field and the passive tag responds by modulating the antenna load for data transmission. In addition, NFC technology allows a smart phone to emulate a passive tag, or it can act as a reader of tags.
- **Active mode**: both devices generate a magnetic field and modulate the opposite magnetic field. A communication between a reader and a smartphone, or between two smartphones can be established.

The NFC protocol and message formats are based on standards that are already in use for RFId to maintain compatibility with them (allowing applications such as reading information from food labels), but they seek to extend the operational capabilities thanks to peer-to-peer among “intelligent” devices, for example two smartphones.

The standard of communication between NFC tags and reader works at a single operating frequency of 13.56 MHz, with a bandwidth of 2 MHz. The reasons that led to the choice of such frequency are related to the fact that the size of the tag antenna decreases as the frequency increases. Furthermore, the transmission at higher frequencies than 13.56 MHz requires more energy. The speed of data exchange can reach an intermediate value between 106 Kbit/s and the maximum peak of 424 Kbit/s.

The standards [15, 16] are defined in ISO, ECMA and ETSI, are divided into:

- The NFC IP-1 describes the radio interface, initialization, anti-collision, a texture format and a protocol for exchanging data block with error handling. Both active communication and passive communication are defined. The standards ETSI TS 102.190 , ISO/IEC 18092 and ECMA 340 define (in the same way) NFC IP-1.
- The NFC IP-2 specifies a mechanism for selection between different modes of communication. This protocol refers to the fact that the same devices can communicate as NFC IP-1, or as ISO/IEC 14443 or as ISO/IEC 15693, in all cases operating at 13.56 MHz.
- The NFC-WI (wired interface) controls the exchange of data between NFC device and the “front-end” of the apparatus in which the NFC device is located.

Using NFC, the communication between smartphone and tag automatically starts when the mobile phone gets close to the tag. The interaction between user and the object is therefore fast and intuitive. This is a fundamental requirement in many different applications [17]. The NFC technology allows cover a fundamental requirement imposed by AAL: the ease of use especially for elderly people.

3 Farm Alert Application

The idea to design an application for the management of drugs is born from the observation that the majority of people, even young people, do not easily remember to take the prescribed drugs. Many people, especially elderly persons or person with disability, need to take often many drugs. In these cases, it usually happens that the person forget to take the drug in the correct moment or he is not sure that he has already taken the drug.

The application, that we named “FarmAlert”, proposed in this work, helps the patients in taking drugs in extremely easy way. The idea is the use of smartphone with NFC to assist at home people who must take drugs, as shown in Fig. 1. The application of the smartphone will remind the person that he must take the drug and it will trace the drugs that the user has already taken.

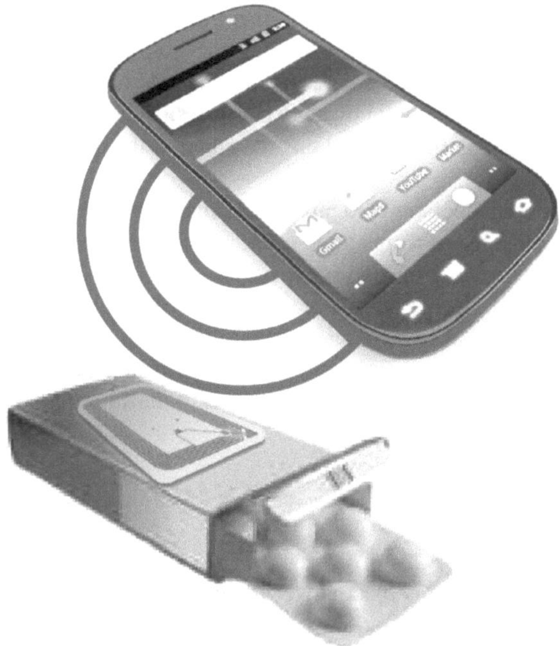

Fig. 1 FarmAlert: application on smartphone with NFC and drug with RFId NFC

The application is partitioned in two sections: everyday use and drug list update. This allows the use to persons not skilled at all in mobile phones.

- **Everyday use**: the procedures, that are used everyday, are performed in an extremely easy way: the user just touches the RFId tag placed in the box of the drug with the smartphone with NFC in different moments of the day to know which of them must be taken and when, and the application registers in the smartphone that the drug has been taken.
- **Update drug database**: The critical procedure are writing the tag and writing or changing the drug prescription database on the smartphone. These procedures are performed only when a change in the prescription is needed and they can be performed by the user itself, if skilled enough, or by a person with a normal knowledge on smartphone helping the person in the normal life, or by the doctor who makes the prescription. These procedures are protected by password so that a not skilled users cannot modify the prescription.

The application has been developed on a low cost Samsung Nexus S smartphone with NFC and Android OS using the NetBeans development environment, used by Oracle. Other smartphones with NFC or development environment, such as Eclipse, can be used. The Android platform gives free tools and library, allowing and easy and fast development of the desired application.

As a first step, in each box containing the drugs an RFId tag is inserted. In some cases this tag is already inserted in the box for traceability reason during the production and storage of the drug. If the tag is not already present, it can be inserted in the box by the user, it is programmed by an expert user using the application TagWriter in the smartphone so that the tag contains the name of the drug, expiration date and other information.

As a second step the person who must take drugs, or somebody who take care of him/her, will create a list of drugs to be taken and the time schedule during the day. The creation of the list of drugs is very easy.

Finally, the user places the mobile the smartphone over the drug tag to verify if it is the correct time and the smartphone register that the drug has been taken.

The functionalities of the application can be activated through the NFC (NFC activation) or manually activating the application in the smartphone (Menu).

3.1 NFC Activation

As soon as the smartphone is placed over the drug box, the menu shown in Fig. 2 will appear with the name of the drug written in the tag. Then the user can press:

- **Take drug**: This functionality is activated only if the drug is already present in the database, a password is not required. This procedure is performed in the everyday use. If the actual time is close (with a tolerance that can be fixed, for example: 10 min before and 30 min after) to the one registered in the database for that drug, a new menu will appear, and the actual date and time will be written in the database, registering that the drug has been taken and when. If the actual time is not in the range allowed or the drug has already been taken in that time period another menu will appear, indicating an error.
- **Add/remove drug**: a password is required, to avoid that a drug could be inserted by mistake. Provided the password, the menu of Fig. 3 is shown. The name of the drug written in the tag is shown, the user can insert the name of the drug that will be shown in the list and at what time the drug must be taken. Then the data inserted can be saved pressing "save" or discarded pressing "delete".
- **Quit**: the registration procedure of the drug is discarded.

The flow diagram of the NFC activation procedure is summarized in Fig. 4.

Once the drug list has been filled, a short time (e.g. 5 min) after the prescription time, the smartphone will inform the user with an acoustic signal that he must take the drug. The acoustic signal will be repeated periodically (for example every 5 min for the successive 30 min), if the user does not perform the procedure "take drug".

Fig. 2 NFC activation menu

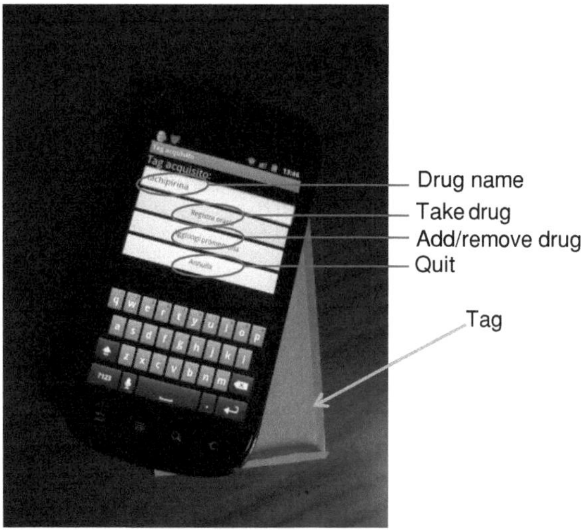

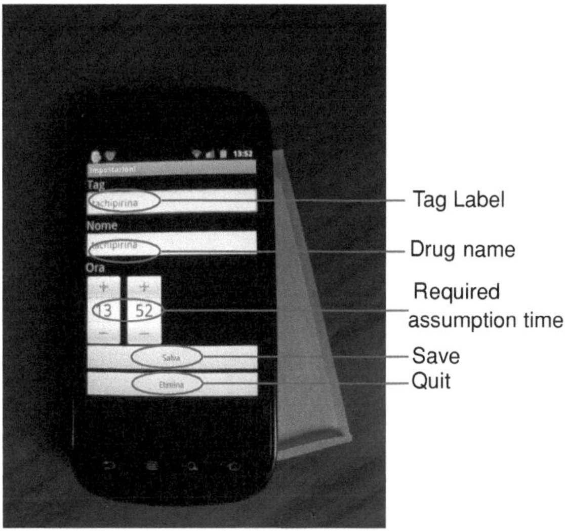

Fig. 3 Menu to add a new drug

3.2 Menu Activation

At any time, with a click on the main application, without the use of the NFC of the drug box, the user can visualize the menu shown in Fig. 5, with the following procedures:

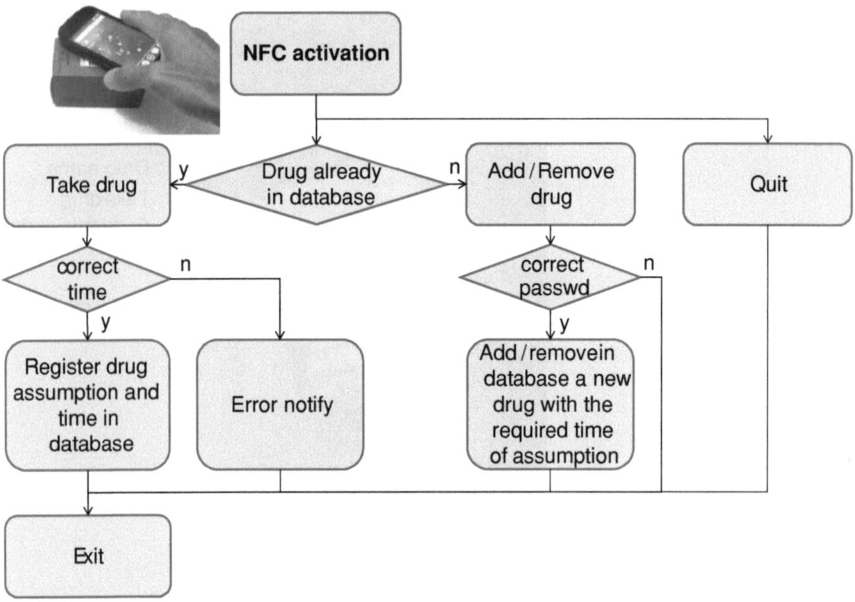

Fig. 4 Flow diagram of the NFC activation procedure

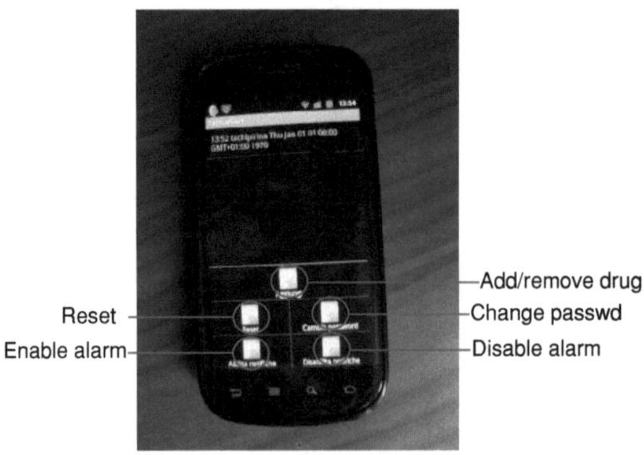

Fig. 5 Main menu

- **Add/remove drug**: open a page for the manual insertion or removal of a drug in the user database, this page is protected by a password.
- **Reset**: delete the entire drug database, this page is protected by a password.
- **Change password**: allows to change the password, provided the old password.
- **Enable alarm**: enable the acoustic alarm at the time the drugs must be taken.
- **Disable alarm**: disable the acoustic alarm.

The flow diagram of the Menu activation procedure is summarized in Fig. 6.

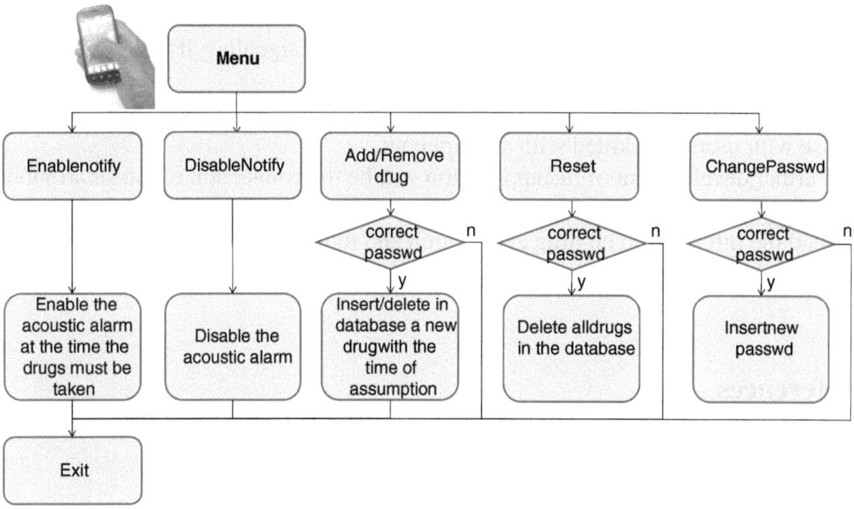

Fig. 6 Flow diagram of the menu procedure

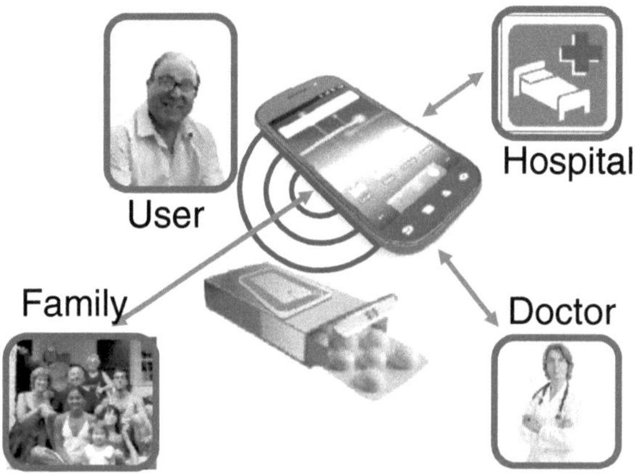

Fig. 7 Interaction of FarmAlert with external environment

4 Conclusions

We proposed an easy to use application on smartphone for drug assumption at home, just touching the drug with the smartphone the user can:

- know when he must take the drug,
- the last time he has taken the drug,
- register that he is taking the drug.

The smartphone signal will remind the drug assumption, if necessary, and register in the database the history of the drug assumption, signaling if the correct drug assumption is not followed.

The application developed is under test with elderly people to improve the easiness of use with users not skilled with smartphones.

Further development of the application will be the connection of the smartphone with a remote superuser (a familiar of the user, a doctor, a pharmacy or the hospital) to send the information on drug assumption and to notify immediately if something wrong happens in the drug assumption, as summarized in Fig. 7.

References

1. http://www.aal-europe.eu/
2. http://ecaalyx.org/ecaalyx.org/index.html 2009–2012
3. http://www.sensaction-aal.eu/ 2007–2009
4. http://www.netcarity.org 2009–2012
5. http://www.smilingproject.eu/ 2008–10, http://happyageing.info/project.html 2009–2012
6. Vergara, M., Díaz Hellín, P., Fontecha, J., Hervás, R., Sánchez-Barba, C., Fuentes, C., Bravo, J.: Mobile Prescription: an NFC-based proposal for AAL. In: Second International Workshop on Near Field Communication, pp. 27–32, IEEE Computer society (2010) doi:10.1109/NFC.2010.13
7. Garrido, P.C., Miraz, G.M., Ruiz, I.L., Gómez-Nieto, M.Á.: A model for the development of NFC context-awareness applications on internet of things. In: Second International Workshop on Near Field Communication, pp. 9–14. IEEE Computer society (2010)
8. Morak, J., Hayn, D., Kastner, P., Drobics, M., Schreier, G.: Near field communication technology as the key for data acquisition in clinical research. In: First International IEE Workshop on Near Field Communication (NFC '09), Hagenberg, Austria, Feb 2009
9. Bravo, J., López-de-Ipaña, D., Fuentes, C., Hervás, R., Peña, R., Vergara, M., Casero, G.: Enabling NFC technology for supporting chronic diseases: a proposal for Alzheimer Caregivers. In: European Conference, Am I 2008, pp. 109–125, LNCS, Springer, Nuremberg, Germany, Nov 2008. ISBN:978-3-540-89616-6
10. Alabdulhafith, M., Sampangi, R.V., Sampalli, S.: NFC-enabled smartphone application for drug interaction and drug allergy detection. In: Near Field Communication (NFC), 2013 5th International Workshop on, pp. 1–6. Zurich, Switzerland, 5 Feb 2013
11. Med Helper Pill Reminder: EarthFlare. http://earthflare.com/
12. Micromedex Drug Information: Truven Health Analytics. http://www.micromedex.com/
13. Medscape: WebMD. http://www.webmd.com/
14. http://www.nfc-forum.org/specs/spec_dashboard/ "NFC Data Exchange Format (NDEF) V-1.0"; "NFC Record Type Definition (RTD) V-1.0"; "NFC Text Record Type Definition V-1.0"; "NFC URI Record Type Definition V-1.0"; "NFC Smart Poster Record Type Definition V-1.0"
15. Di Vedat C., Kerem O., Busra O.: Near Field Communication: From Theory to Practice, J. Wiley, 13 Feb 2012. ISBN-10: 1119971098
16. Ahson, S.A., Ilyas, M.: Near Field Communications Handbook (Internet and Communications), CRC Press Taylor and Francis, 23 Sept 2011. ISBN-10: 1420088149
17. Conti, M., Fedeli, D., Virgulti, M.: B4V2G: Bluetooth for electric vehicle to smart grid connection. In: Proceedings of the Ninth Workshop on Intelligent Solutions in Embedded Systems (WISES 2011), Regensburg, Germany, 7-8 June 2011, pp. 13–18. ISBN: 978-3-00-033401-6

Remote Measurement of the Respiration by Electromagnetic Sensing

Lorenzo Scalise, Valerio Petrini, Paola Russo, Alfredo De Leo, Valter Mariani Primiani, Valentina Di Mattia and Graziano Cerri

Abstract Respiration activity is one of the fundamental functions of the human being (also known as vital signs) which is monitored for health purposes. Often it is required to operate the sensing of the respiration activity out of hospitals or clinical environments and without contact with the patient and possibly from a distance i.e. for the home monitoring of patients or for the ambient assisted living, AAL). Unfortunately, at present the sensing apparatus for respiration monitoring purposes are mainly based on the use of contact methods (chest belt, nasal temperature transducers, etc.) which appear to be not suited for such tasks. In this paper, the authors propose a novel approach based on the use of electromagnetic sensing of the respiration activity of subject. The proposed approach provide a remote, continuous measurement of the inspiration/expiration acts of a subject without requiring a contact; these features of the proposed system are suited for domestic monitoring and AAL.

1 Introduction

Respiration activity is one of the fundamental functions of the human being (also known as vital signs) monitored for heath purposes. The presence of respiration acts, their frequency and eventual suspension/rapid variation of their rate are the quantities typically monitored. It is a quantity of primary importance (as it is the heart rate, the

L. Scalise (✉)
Dipartimento di Ingegneria Industriale e Scienze Matematiche, Università Politecnica delle Marche, via Brecce Bianche, 60131 Ancona, Italy
e-mail: l.scalise@univpm.it

V. Petrini · P. Russo · A. De Leo · V. M. Primiani · V. Di Mattia · G. Cerri
Dipartimento di Ingegneria dell'Informazione, Università Politecnica delle Marche, via Brecce Bianche, 60131 Ancona, Italy
e-mail: g.cerri@univpm.it

S. Longhi et al. (eds.), *Ambient Assisted Living*, DOI: 10.1007/978-3-319-01119-6_5,

arterial pressure, etc.) not only for patients recovered in hospitals, but also for subjects living at home and requiring a remote control of their physiological status. This is for example the case of elders living at their homes and needing to be kept under observation in order to undergo potentially risky conditions (respiration rate is an important predictive parameter for many pathologies of the elder). For example, the Chronic obstructive pulmonary disease is a common condition occurring in 17 % of men and 8 % of women between the ages of 45 and 64 years [1]. The obstructing sleep apnea syndrome (OSAS) is also an example of a clinical disorder (prevalence: 1–6 % in adult population [1]) is a clear example of respiratory disorder which could relate also to cardiovascular diseases. More in general it is of large interest, particularly for ambient assisted living (AAL), to collect information of a subject on the respiration activity during rest and sleep, at home, in order to observe the rising of dangerous respiration patterns (such as respiration suspensions, prolonged apnea, etc.).

For such aims, it is fundamental to operate the sensing of the respiration activity out of specialized clinical environments, without contact with the patient and possibly from a distance. Unfortunately, at present the sensing apparatus for respiration monitoring purposes are mainly based on the use of contact methods (chest belt, nasal temperature transducers, etc.). They require the presence of a specialized care giver to be correctly installed and cannot be used for long period of time due to their degree of invasiveness. Therefore, the attention of researchers has been put on the proposition on novel non-invasive methods mainly based on methods based on image acquisition and processing [2–5]. They are promising, but more suited to the use in specialized, clinical environments instead of use in domotic area. In fact the acceptability of video system in domestic environments is very low for privacy reasons.

In this paper, the authors propose a novel approach based on the use of electromagnetic sensing of the respiration activity of a subject. In adjunction to its non contact nature, the use of electromagnetic waves present the advantage respect to the image-based methods to be easily installable at home without the need to have in direct view with the target (some materials are transparent to the electromagnetic waves) and therefore they result to be particularly suited for use in AAL.

2 Measurement Method and Experimental Set-up

The measurement method proposed is based on the measurement of the phase variations of the reflection coefficient (S_{11}). In our experimental set-up, the EM signal is launched by an horn antenna with half power beam of about 30° in H plane and 38° in the E plane. A vectorial network analyser (VNA) is used to feed the antenna and to measure the phase reflection coefficient of the same antenna (S_{11}), defined as the ratio of the amplitude of the emitted wave to the amplitude of the reflected wave (Fig. 1).

It is possible to observe that the phase of S_{11} is related to the distance D (Fig. 1) between the antenna and the subject and that eventual variation of the value of D

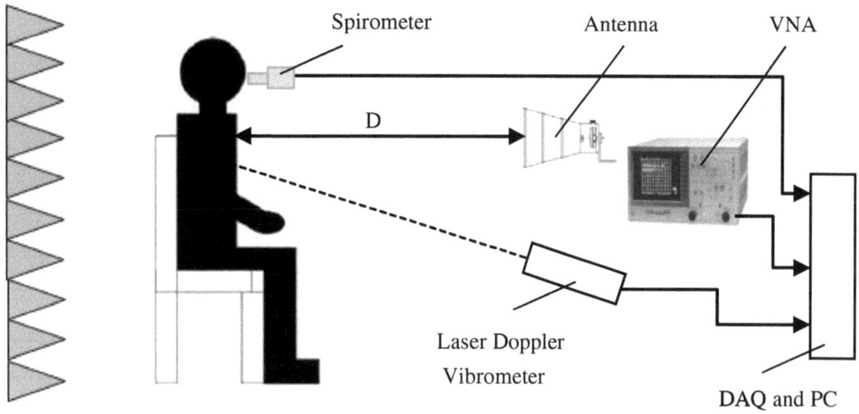

Fig. 1 Sketch of the experimental set-up

produce variation on the phase of S_{11} [6]. The inspiration/expiration acts are therefore assessable from the S_{11} phase signal (examples of the acquired signals are reported in Fig. 3). In our experiments the frequency of the EM signal was 6 GHz allowing to operate with a wavelength of 5 cm: From our tests and numerical simulations, such wavelength has been selected because it has been demonstrated to be the most adequate to detect the thorax excursions caused by the respiration activity. The S_{11} signal is sampled at 26 Hz by a data acquisition board and recorded on a PC.

As a reference for the evaluation of the respiration activity, a laser Doppler vibrometer (LDVi) and a spirometer (only for testing the apneas) have been used in parallel with the EM system [6–9]. Data from the reference system have been sampled at 1 kHz and successfully decimated to 26 Hz.

On both the reference and the EM signals, a band pass filter (Butterworth, 5th order, 0.5–5 Hz) was applied in order to reduce the effect of motion artifacts, cardiac activity and high frequency noise.

The subject is positioned inside a space delimited by pyramidal absorbers to avoid unwanted reflections; the distance between the subject and the absorbers is about 1.5 m, while the distance from the antenna and the chest of the subject (D in Fig. 1) can be varied as well as the position of the subject with respect to the main lobe.

In Fig. 1, a sketch of the experimental set-up is reported. A detailed description of the experimental set-up and the equipment used is reported in [6, 9]. All test have been carried out on the same volunteer (male, age 24) and in Fig. 2 the three postures tested in our experiments are reported: standing in front of the antenna, sitting in front of the antenna and laying on a side. Distance D between subject and antenna were: 0.25, 0.5, 1, 1.5, 2 and 2.5 m (for standing and sitting postures) and 1 m for the subject laying.

For what concern the exposure limit, respect to the International Commission on Non-Ionizing Radiation Protection (ICNIRP) [10], the value of the electrical field (E) calculated by the use of a simulation software [11] are reported in Table 1.

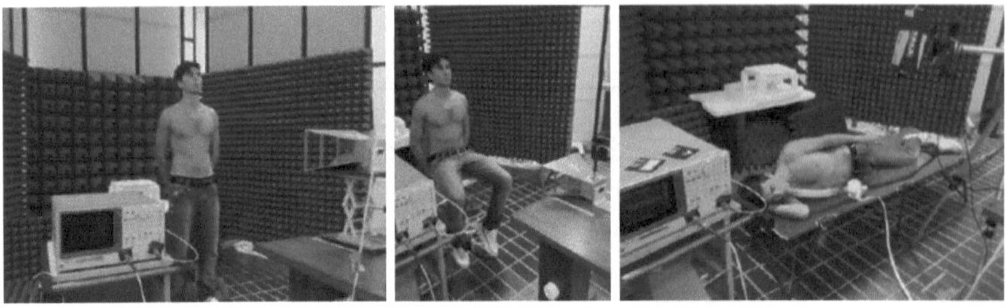

Fig. 2 Subject postures during the test: standing, sitting and laying on a side

Table 1 Exposure values in function of the distance D

Distance, D (cm)	E (simulated) (V/m)
25	1.60
50	1.04
100	0.59
150	0.41
200	0.33
250	0.26

From Table 1, it is possible to observe that the predicted values are well below the recommended limits for occupational exposure (137 V/m) and general purpose exposure (61 V/m) [10]. For what concern, the Italian regulations, [12] also in this case such values are below the exposure limit for occupational and general purpose (40 V/m) and for living places (6 V/m).

3 Results

An example of a set of acquired signals, for the case of a subject standing at 50 cm is reported in Fig. 3, where the reference signal (acquired with LDVi) is depicted on the left, while the signal measured with the proposed EM system is reported on the right.

Signals are acquired simultaneously and it is possible to observe that for each peak detected by the LDVi (measuring the velocity of the thorax surface), it is possible to record a corresponding peak in the EM system trace.

In Fig. 4, three examples of time signals and relative power spectrum densities are reported for the subject standing at: 25 cm (top), 150 cm (center) and 250 cm (down).

It is possible to observe that, for the time signals the amplitude of the phase variations is reduced with the increase of the distance D. The 74 % of signal reduction has been reported when the subject was measured at 1 m, with respect to the case of the subject standing at 0.5 m. While for longer distances the signal amplitude (phase variation) is always $<5^\circ$ [6, 9]. More in detail, we have explored the dependence of

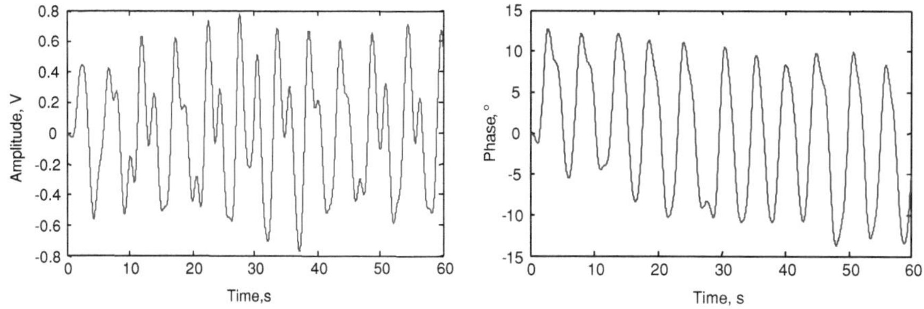

Fig. 3 Example of time signals acquired during the test (subject standing@50 cm): LDVi (reference) signal (*left*) and EM phase signal (*right*)

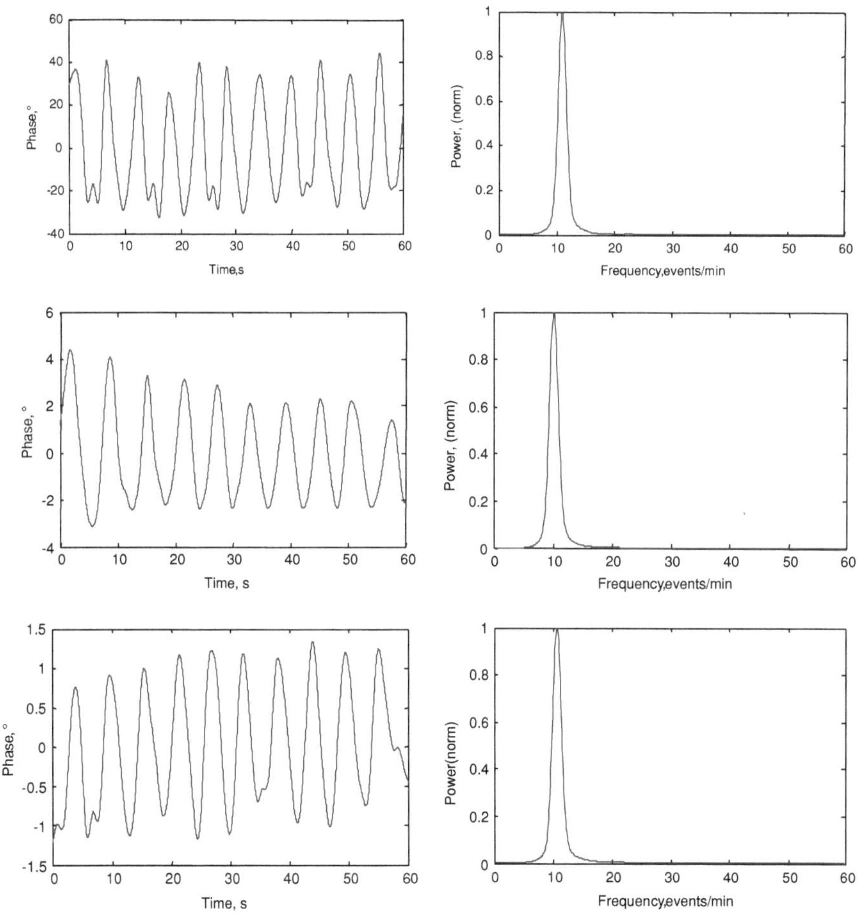

Fig. 4 Example of time signals (*left*) and power spectrum densities (*right*) measured with the subject standing at: 25, 150 and 250 cm

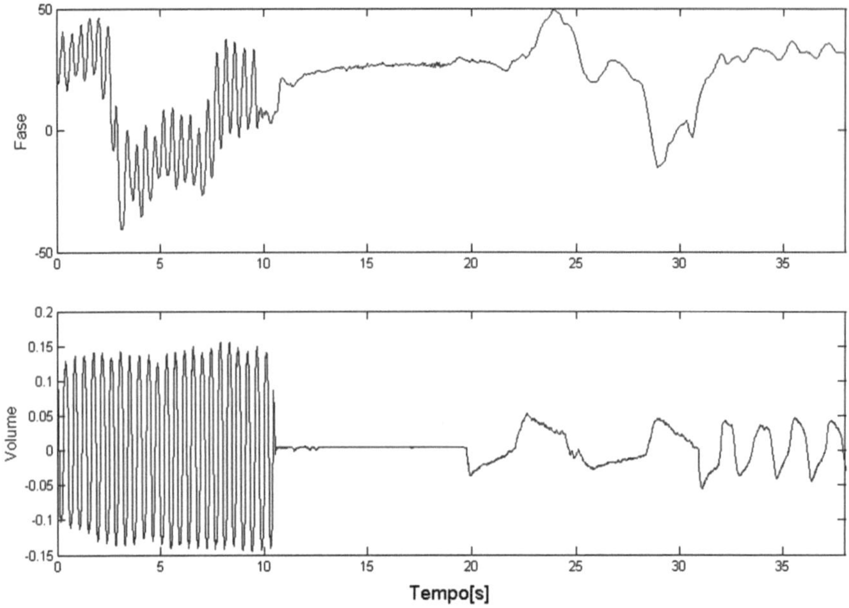

Fig. 5 Example of a respiration rate voluntary variation (high rate, apnea and normal rate) measured with the EM system (*top*) and the spirometer (*down*)

the quality of the measured (in particular the signal SNR) on the distance between antenna and subject, D. A reduction in the SNR with distance D has been reported (in the order of −0.11 dB/cm). From our tests it is shown that despite the fact that at 2.5 m the SNR is reduced (about the 50 %) respect to the signal measured at 25 cm, it is still sufficiently high to allow a correct detection of the respiratory activity. In the case where a minimum operating SNR is requested, a possible solution could be to improve the emitted power (1 mW in our case) or to use a narrower radiating beam for the antenna. Although the quality of the signal can be reduced in some conditions (longer distances), the SNR is still sufficiently high (>8 dB) to provide reliable data [6, 9]. Finally, it can also be observed that the mean respiration rate can also be easily measured as the peak of the power spectrum density (PSD) of the time signals (Fig. 4). The measurement system proposed presents also low sensitivity to the subject postures. Good correlations (R = 0.97) with LDVi are also reported for the measured mean respiration frequencies [6].

Change of the respiration frequencies as well as apneas can also be clearly detected as it can be observed in Fig. 5 where the subject (sitting, D = 20 cm) was asked to start the test with an high respiration rate for the first 10 s, to suspend the respiration (simulating an apnea) for 10 s and to end the test with his natural respiration rate. Also in this case, it is evident how the signals are well related and the phase of apnea is clearly individuated.

Very good correlations between the measurement of the time duration of the respiration acts (distances between two consecutive peaks), measured by the pro-

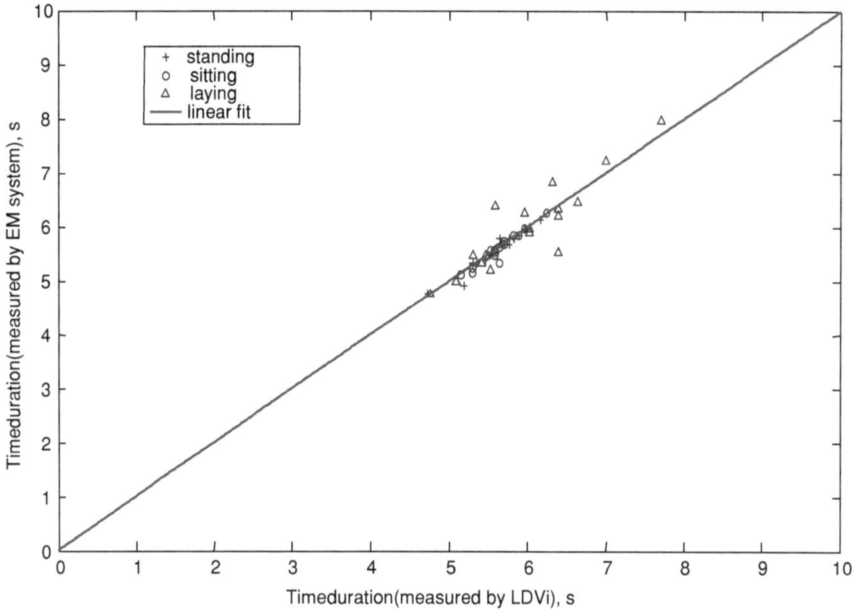

Fig. 6 Scatter plot of the mean time duration of the respiration acts for the three postures (subject posture: standing, sitting and laying): LDVi versus EM system

posed EM system and the LDVi (reference) system are reported for all the tests carried out (N = 50): R = 0.97 for standing posture, R = 0.96 for sitting posture, R = 0.90 for standing. In Fig. 6, the scatter plot of the mean time period (over a time window of 60 s) measured for all the tests and all the postures is reported; the correlation coefficient in this case is R = 0.89.

4 Conclusions

A non-contact approach to the problem of the measuring of the respiratory activity of human subjects is reported, using an electromagnetic sensing system. This method is based on the measurement of the phase variations of the reflection coefficient S_{11} caused by the displacement of the thorax surface.

The solution proposed is feasible for brief as well as for prolonged observation time as it is requested by remote monitoring system to be used at home and for AAL applications. The monitoring systems for the respiration activity available at present are limited by the fact to require a direct contact with the patient (this is the case of the respiration belt or the spirometer) and consequently to be not suited for prolonged observations.

The possible implementation of the proposed EM measurement approach could make use of market ready technologies which are not designed for laboratory use and are available at lower costs such as signal generator and electronic hardware. While specific design for the radiating elements could be studied in order to better integrate them into the furniture or into the building elements.

It is also important to take in consideration the possible interfering inputs such as the clutter noise generated by the indoor environment and caused by the furniture, the metallic elements, etc. To this aim, the authors are exploring the possibility to subtract the static clutter from the received signal in order to improve the signal-to-noise ratio. Moreover for long monitoring (in the order of some hours), automatic re-calibration of the system could also be introduced to reduce the effect of electronic circuit drift.

Another possible cause of interference could be the presence of more than one subject, in fact if two or more subjects will be simultaneously illuminated by the antenna, the sensor will detect a composition of their signals. This can be a limitation of the method investigated, and attempts to resolve this problem have been recently proposed by placing more than one sensor and therefore using spatial diversity.

References

1. World Health Organization: Chronic respiratory diseases. http://www.who.int/gard/publications/chronic_respiratory_diseases.pdf (2012). Accessed 25 Sept 2013
2. Dei, D., Grazzini, G., Luzi, G., et al.: Non-contact detection of breathing using a microwave sensor. Sensors **9**(4), 2574–85 (2009)
3. Folke, M., Cernerud, L., Ekstrom, M., Hok, B.: Critical review of non-invasive respiratory monitoring in medical care. Med. Biol. Eng. Comp. **41**(4), 377–83 (2003)
4. Lai, J.C.Y., Xu, Y., Gunawan, E., Chua, E.C.P., Maskooki, A., Guan, Y.L., Low, K.S., Soh, C.B., Poh, C.L.: Wireless sensing of human respiratory parameters by low-power ultrawideband impulse radio radar. IEEE Trans. Instr. Meas. **60**(3), 928–8 (2011)
5. Lin, J.C.: Non-invasive microwave measurement of respiration. Proc. IEEE **63**(10), 1530 (1975)
6. Scalise L., De Leo, A., Mariani Primiani, V., Russo, P., Shahu, D., Cerri, G.: Non contact monitoring of the respiration activity by electromagnetic sensing. In: Proceedings of the IEEE International Workshop on Medical Measurement and Applications (MeMeA11), pp. 418–422 (2011)
7. Scalise, L., Marchionni, P., Ercoli, I.: Optical method for measurement of respiration rate. in: Proceedings of the IEEE International Workshop on Medical Measurements and Applications (MeMeA '10), pp. 19–22, May 2010
8. Scalise, L.,Marchionni, P., Ercoli, I.: Anon contact, optical procedure for precisemeasurement 175 of respiration rate and flow. In: Biophotonics: Photonic Solutions for Better Health Care II, 77150G of Proc. SPIE 7715, May 14, 2010. doi:10.1117/12.854482
9. Scalise L., Mariani Primiani, V., Russo, P., De Leo, A., Shahu, D., Cerri, G.: Wireless sensing for the respiratory activity of human beings: measurements and wide-band numerical analysis. Int. J. Antennas and Propag. **2013**, Article ID 396459 (2013)
10. International Commission on Non-Ionizing Radiation Protection: ICNIRP Guidelines for limiting exposure to time-varying electric, magnetic and electromagnetic fields (up to 300GHz). www.icnirp.de/PubEMF.htm (1998)
11. Computer Simulation Technology: http://www.cst.com
12. Gazzetta Ufficiale n. 199, 28-8-2003: http://www.iss.it/binary/elet/cont/DPCM_1.120238 3676.pdf

Part II
Robotic Assistance for the Elderly

Robot Interface Design of Domestic and Condominium Robot for Ageing Population

Niccolò Casiddu, Filippo Cavallo, Alessandro Divano, Irene Mannari, Emanuele Micheli, Claudia Porfirione, Matteo Zallio, Michela Aquilano and Paolo Dario

Abstract This chapter presents a preliminary study to design a couple of robots, conceived to assist senior citizens 65+ in domestic and public space. The design and development of these two robots, named Domestic and Condominium, concerned, from one hand, appropriate criteria of acceptability, usability, aesthetic and safety, and on the other hand specific functionalities to satisfy users' needs.

1 Introduction

European population projections have highlighted that the number of elderly people 65+ in the world will quickly increase in the coming years [1]. Nowadays the society sustains elderly people to age well by means of medical cures, socio-medical services, and other social activities but these requests of support and assistance expected to become so high that it is becoming difficult to manage and sustain [2]. In addition the society is also facing the sustainability of the pension system that leads to an extension of the working life of senior citizens [3]. In this societal landscape, advanced robotic systems have the potential to drive tangible benefits in quality of life, improving citizens outcomes and reducing health and social costs.

"Robot Companions" is a concept that refers to a new generation of machines that will primarily help and assist citizens in daily activities at home, in their workplace and in other environments [4]. Robot Companions could be implemented as a plurality of complete advanced robotic services, integrated in intelligent environ-

N. Casiddu (✉) · A. Divano · E. Micheli · C. Porfirione · M. Zallio
Dipartimento di Scienze per l'Architettura, Scuola Politecnica dell'Università di Genova, Genova, Italy
e-mail: casiddu@arch.unige.it

F. Cavallo · I. Mannari · M. Aquilano · P. Dario
The BioRobotics Institute, Scuola Superiore Sant'Anna, viale Rinaldo Piaggio 34, 56025 Pontedera, PI, Italy

S. Longhi et al. (eds.), *Ambient Assisted Living*, DOI: 10.1007/978-3-319-01119-6_6,

ments, which will actively work in real conditions and cooperate with real people and between them to provide favourable independent living, improving the quality of life and the efficiency of care for elderly people. The innovation of this concept is to significantly enhance the performance and acceptability of the current ICT services for aging well to a new level of quality, provided by the cooperation of a plurality of robots and with the support of an Ambient Intelligence (AmI) infrastructure. Different robotic systems could be installed and integrated to cooperate and operate in domestic, condominium and outdoor environments (homes, hospitals, towns, public spaces), appropriately equipped with AmI infrastructure, and could implement an ecosystem that connects different stakeholders, such as elderly people and caregivers, social services, medical centres, municipalities, shops, pharmacies, etc.

This idea of "Robot Companions" is under investigation and development in the context of the Robot-Era Project (FP7-ICT-2011.7) that, sustained by a multidisciplinary teams from different technological, medical and industrial design disciplines, aims to design and implement social service robotics in a user and town centred design approach and characterized by a high level of technology and acceptability. Particularly the Robot-Era Project aims to develop, implement and demonstrate the general feasibility, scientific/technical effectiveness, social/legal plausibility and acceptability by end-users of advanced robotic services integrated with intelligent environments to provide help facilitate independent living, improving the quality of life and the efficiency of care for senior citizens.

This chapter faces the design aspects of a couple of robots, conceived to assist senior citizens 65+ in domestic and public space, and particularly focuses on the criteria of acceptability, usability, aesthetic and safety. The study highlights those properties that favour the human robot interaction, i.e. the affordance (immediately show its functionalities), safety (perceived safe), aesthetics (perceived familiar with an opportune combination of colours and soft/rigid materials), friendliness (emotionally accepted), usefulness (concretely useful in daily activities) and dependability (robust, effective and reliable).

2 Robot Interface Design

Interaction, Acceptability, Design process, Interface design, Usability.

Through the introduction of electronics and informatics the technological development, started in 1980s, had a profound impact on mass products, even in those fields that were previously exclusive only to mechanics. Smartphones are an example of how the shapes of these modern devices are not bound to one univocal purpose or use, but this innovation in design can come at a cost. These devices are often difficult for "the non-alphabetized" in both their physical usability and their cognitive approach based on unusual and less intuitive behavioural codes. Hence, the "digital-divide" concept becomes a paradox, when taking into account the elderly population, that would potentially profit the most from a smart environment [5]. As a discipline, design has always been interested in physical interfaces based on specific ergonomic relations aimed to improve their use. During the mechanic era designing a product was a straightforward process: measurements were dictated by anthropometry, while functions determined the product shape. Unlike recent times, when mono-directional approaches were used solely in designing an interface, nowadays the AAL, the ICTs [6] and robotics in general are leading the way to new multidirectional scenarios: instead of focusing on machines themselves, research is showing more and more interest on the attention machines should foster towards human beings. In other words, this rapidly-advancing technological evolution leaves behind the dialogic interaction based on a dual pattern, in which the user acts and the calculator answers. Not to mention that thanks to these new scenarios, "friendly" and "usable" [7] products can be achieved and this means that they meet specific acceptance requirements such as safety, affordance (i.e. a sum of shape, dimension, weight, texture, ...), aesthetics and congeniality. Therefore it should be highlighted that interfaces cannot only be described as a means to connect human beings and products—interfaces are an open, dynamic and interactive system, able to ensure full accessibility to those advanced technologies, which effectively support an active ageing. Interface and object design also falls within the "project-making method": the most recent method is to move from centre to periphery or vice versa, by gradually approaching what will be recognized as the designed object, i.e. the final result. If designing means giving a shape, a function and a sense to ideas, then clearly there cannot be a design without a shape. In the field of Human Centred Robotics Design [8] a project is defined as a process on which designers, computer technicians and engineers work together as a team, sharing their planning expertise in order to develop a final product that is the result of all their synergetic inputs. Within this framework, the project aims to combine software and hardware components so that the device features may be controlled and programmed. Making a project today means facing this complexity and organizing it in an ordered system of different competences, and at the moment our part is a mere tile that will fall into place with other tiles to create a much wider mosaic.

3 End-users' Needs and Service Analysis

The current design and development of the Robot-Era system is consequence of a substantial study, carried out with elderly people and caregivers with an User Centred Design (UCD) approach [9], that allowed to identify the primary end-user needs, possible structures of services to satisfy such needs and the scientific and technological requirements to develop the relative service robotic solutions. That study highlighted that a high willingness to use robotic systems in case of need was detected for transporting/manipulating objects at home, cleaning, garbage collection, surveillance, outdoor walking support, indoor escort at night and reminding events, laundry support, communicating with persons, for food delivery and shopping/drug delivery [10].

Table 1 The Robot-Era services identified during the tests

	Robot-Era Services	Environments
(1)	Communication	Domestic environments
(2)	Indoor escort at night	
(3)	Reminding	
(4)	Cleaning	
(5)	Objects transportation and manipulation	
(6)	Outdoor walking support	Public spaces: squares, gardens, walking lanes, etc.
(7)	Laundry support	Condominium, residential spaces, etc.
(8)	Food delivery	
(9)	Surveillance	
(10)	Drug and shopping delivery	Outdoor environments, shopping centers, drug stores, etc.
(11)	Garbage collection	

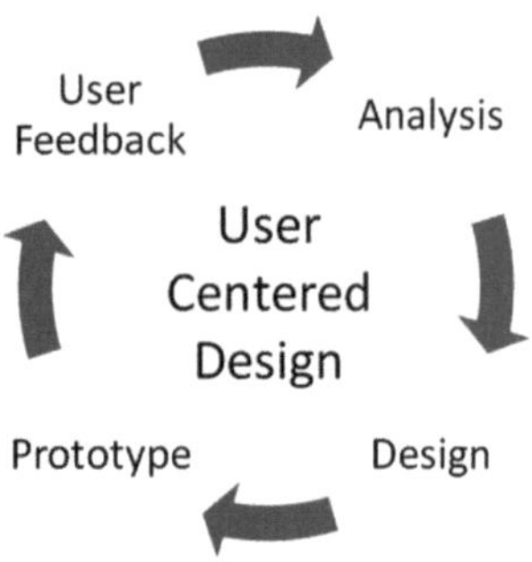

Fig. 1 Description of the UCD approach to design and develop the Robot-Era robotic services

The main phases of the UCD methodology were carefully scheduled and included the (1) end-user needs analysis, (2) the design of services related to needs and the (3) production and implementation of robots for the testing phase (Fig. 1).

The first two phases were performed in three different European countries (Italy, Germany and Sweden), involving about 67 senior citizens 65+ and 17 caregivers. The study highlighted different end-users' needs that could be provided not only inside the domestic environment, but also outdoor in the public spaces. Table 1 describes the Robot-Era Services and the places where they are performed.

4 Shapes for Interaction

Human Interaction, Machine Interaction, Safety.

People, machines, environments. They are the three main factors interacting within a complex ecosystem in constant evolution. People may be defined as "biological machines" gifted with thought, emotions and cultural heritage. In this context they communicate and interact with "new machines", which are now able to act and decide autonomously. Additionally, other types of interaction (besides people-machine interaction) take place within relatively confined environments, *smart* environments and in the urban ecosystem. The study on man-machine interaction has developed inside an experimental context involving the potential users, in order to improve the quality of autonomous life and the efficiency in the care of the elderly [11]. Based on a previous study, four fundamental acceptance parameters have been brought to light in man-machine relation:

- Safety: robot level of danger as perceived by the user.
- Aesthetics: robot capability to trigger familiar reactions in the user through its form, colours and materials used to build it.
- Recognisability: robot capability to highlight its function.
- Friendliness: robot capability to establish a positive emotional relation with the user.

Starting from the robot fundamental design characteristics, the present study focuses on volumes, shapes, the anthropometric relation, colours and material to be systemized for the development of the robot external shell. Moreover, this study analyses the main aspects regarding human perception of service robots with anthropomorphic features and the human acceptance of these features, which can be understood through shape symmetry, proportions and aspects specifically connected with those humans.

4.1 Human–Robot Interaction

The new general asset of the robots discussed in this study is based on unity of shape derived from functions, coherence and recognisability of these shapes [12]. The aim of this design process is to boost the technical aspect of the robot family and its performance, thus enabling it to provide a more efficient service. However, it should not be forgotten that final users are the main focus, together with their needs on various levels: ergonomic, functional, cognitive, psychological and relational.

The geometry utilized for the head of the Domestic and Condominium robots conforms with the pattern used for the DustCart, the third robot of the fleet. Furthermore, research was carried out for each machine to give the Robot-Era fleet a common identity, that would translate in the greatest recognisability and friendliness. The elements characterising the exterior aspect of each robot are the large dimension of their eyes, to enhance perception and interaction; the personalized headgear, to stress their different functions in a household and external environments; the proportions between the cover components for a precise harmony of colours and materials, to facilitate the product acceptance [13].

The Domestic and Condominium robot head covers stem from the archetypical concept of "clothing-uniform". It means that each robot resembles a specific

Fig. 2 Concept sketches of the Robot-Era robots: in *blue* the Domestic Robot with the manipulator and in *orange* the Condominium robot

assistant: a housemaid inside the household and a janitor for public spaces. This last characteristic is conveyed through different colours and different hats for each robot.

The Domestic robot is equipped with a cover that aesthetically resembles a housemaid, while the geometry of the Condominium robot cover uses an actual hat, which typically characterizes a janitor uniform. It is important to add that in both covers tele-transparency and recognition systems have been installed, such as cameras in the robot eye zone and a system to track people and environments thanks to optical tools in the area above the eyes (two video cameras and a supplementary infrared detector) for visual recognition of bodies in motion [14].

4.2 Robot–Robot Interaction

In the interaction between robots exchanging objects is of pivotal importance. However, in this study several other aspects are also taken into account: from object manipulation performed by the robotic arm (a part of the domestic robot), to the mechanical alignment of the robots when they exchange objects.

The torso of the robots is comprised of a shell with a tablet on its anterior part and this works as a communicative interface between user and robot. In the posterior part of the robot a foldaway support system has been installed so to assist elderly users with reduced mobility in those particularly difficult actions, like getting up from a chair or from the bed.

Objects may be easily transported thanks to a horizontal loading platform positioned in the anterior part of the robot. This device works through motorized rollers that allow an optimal sliding movement. It is worth mentioning that the geometry of the anterior loading platform is designed so that two robots are mechanically aligned. This is important because it means that objects may be safely transferred from one platform to the other, with the aid of the mechanical arm of the Domestic robot (Fig. 2).

4.3 Robot Environment Interactions

Sensors placed in the environment together with sensors in the robot allow machines to move accurately around the given space and this is important because it ensures greater safety for the user.

These robots are programmed to communicate both with their users and the environment where they perform their functions. For this reason they are denominated "intelligent machines". They can recognise obstacles inside the household, such as architectural features, furniture and other static objects. Additionally, they are also able to recognise moving objects, thus enabling them to better interact with elderly users, both in their daily routines or in situations of need or danger [15].

Acknowledgments The research leading to these results has received funding from the European Community's Seventh Framework Programme (FP7/2007–2013) under grant agreement num. 288899 (Robot-Era Project)

References

1. World Population 2012. United Nations publications. USA (2012)
2. van der Broek, G., Cavallo, F., Wehrmann, C.: AALIANCE Ambient Assisted Living Roadmap, vol. 6. IOS Press, Amsterdam (2010)
3. Eurostat Pocketbooks. European Social Statistics—edition 2013. Publications office of European Union. Luxembourg (2013)
4. Dario, P., Verschure, P.F.M.J., Prescott, T.: Robot companions for citizens. Procedia Comput. Sci. **7**, 4751 (2011)
5. Bennett, K.B., Flach, J.M.: Display and Interface Design: Subtle Science, Exact Art. CRC Press, London (2011)
6. Lazakidou, A., Siassiakos, K., Konstantinos, I.: Wireless Technologies for Ambient Assisted Living and Healthcare—Systems and Applications. IGI Global, Hersey (2010)
7. Albert, W., Tedesco, D., Tullis, T.: Beyond the Usability Lab: Conducting Large-Scale Online User Experience Studies. Morgan Kaufmann, Boston (2009)

8. Casiddu, N., Micheli, E.: Human Centered Robotic Design, pp. 1–176. Alinea Editrice, Firenze (2011)
9. Cavallo, F., Aquilano, M., Bonaccorsi, M., Limosani, R., Manzi, A., Carrozza, M.C., Dario, P.: On the design, development and experimentation of the ASTRO assistive robot integrated in smart environments. In: IEEE International Conference on Robotics and Automation (ICRA 2013), Karlsruhe, Germany, 6–10 May 2013
10. Bevilacqua, R., Felici, E., Marcellini, F., Klemcke, S., Nedopil, C., Glende, S., Filippo, C., Aquilano, M., Carrozza, M.C., Dario, P.: Robot-Era project: from the end-user's perspective to robotics. Preliminary findings. In: Ambient Assisted Living Forum, Eindhoven (2012)
11. Lupacchini, A.: Design olistico. Progettare secondo i principi del DfA. Alinea, Firenze (2010)
12. Del Curto, B., Fiorani, E., Passaro, C.: La pelle del design. Progettare la sensorialità. Lupetti, Milano (2010)
13. Karwowski, W., Rahimi, M.: Human-Robot Interaction. Taylor and Francis, London (2003)
14. Accolla A.: Design for All. Il progetto per l'individuo reale, FrancoAngeli, Milano (2009)
15. Cavallo, F., Aquilano, M., Bonaccorsi, M., Mannari, I., Carrozza, M.C., Dario, P.: Multidisciplinary approach for developing a new robotic system for domiciliary assistance to elderly people. In: 33rd Annual International Conference of the IEEE EMBS Boston, Massachusetts, USA, 30 Aug–3 Sept 2011

Autonomous Assistive Robot for Respiratory Rate Detection and Tracking

Flavia Benetazzo, Alessandro Freddi, Andrea Monteriú, Panu Harmo, Ville Kyrki and Sauro Longhi

Abstract Assistive robotics has the objective to improve the quality of life of people in daily living, with a special aim to those who suffer of physical disabilities or cognitive impairment, which may be caused by an accident, disease or the natural process of ageing. Autonomous mobile robots can be adapted to work as assistive robots, because of their capabilities to navigate in unstructured environments and react to changes in it. In order to turn a mobile robot into an assistive robot, the contest of use and the set of functionalities to perform should be clearly identified, while limiting at the same time hardware complexity and costs, which are key factors against the application of assistive technologies into the real world. The present paper deals with the development of a home robot companion, an assistive robot, which is being realized at Università Politecnica delle Marche: its main contribution is the implementation of both a stance detection and respiratory rate algorithm by using the information provided by non-invasive sensors, and their integration into a low

F. Benetazzo (✉) · A. Freddi · A. Monteriú · S. Longhi
Dipartimento di Ingegneria dell'Informazione, Università Politecnica delle Marche,
60131 Ancona, Italy
e-mail: f.benetazzo@univpm.it

A. Freddi
e-mail: a.freddi@univpm.it

A. Monteriú
e-mail: a.monteriu@univpm.it

S. Longhi
e-mail: s.longhi@univpm.it

P. Harmo · V. Kyrki
Department of Automation and Systems Technology, School of Electrical Engineering,
Aalto University, Esbo, Finland
e-mail: panu.harmo@aalto.fi

V. Kyrki
e-mail: ville.kyrki@aalto.fi

S. Longhi et al. (eds.), *Ambient Assisted Living*, DOI: 10.1007/978-3-319-01119-6_7,

cost robotic platform which can perform navigation, user identification and tracking. The overall result is an assistive mobile robot which can increase safety at home for the user.

1 Introduction

The use of technology to improve the quality of life of people is becoming a common feature of the modern society. Quality of Life (QoL), however, is not easy to define, since it is an elusive construct connoting a multidimensional appraisal of a variety of important aspects of life [1]. The wider this variety is, the more difficult is to give a single definition which includes all the possible technological solutions related to it. For this reason Quality of Life Technology (QoLT) is generally defined as any technology which impacts the quality of life of individuals who are using it. When addressing people with special needs, the definition of QoLT usually becomes more specific, referring to intelligent systems that augment body and mind functions for self-determination of older adults and people with disabilities [2], which is basically an alternative definition of Ambient Assisted Living (AAL) technology.

AAL is typically classified according to the functional domain targeted, and might include safety systems, medical devices, telemedicine platforms, assistive robots and many others. Among them, during the last years, assistive robotics for domestic use has been a rapidly increasing field of research and development [3]. In the literature, several are the papers addressing the development of mobile robots which can autonomously navigate, detect and interact with the user, usually called home robot companions. In [4] a good overview of mobile robots with manipulation skills is presented, which includes the well-known MOVAID, Hermes, ARMAR and Care-OBot. A recent mobile robot with assistive capabilities has being developed at the Carnagie Mellon University [5–7]: it can perform localization, navigation and user detection, moreover it is well known for its ability to finely interact with the user through a robotic arm. These systems, however, require an extremely complex hardware, are expensive and their possible integration inside real homes in a near future is not easy. The recent European FP7 COMPANIONABLE project [8] is trying to overcome these problems using a low-cost solution for the robot hardware, a friendly appearance and an adequate size and weight for adoption in common houses, showing an increasing interest of simple and affordable assistive robots.

Navigation and user tracking are functions typically available in robotic companions. Few researchers, however, have focussed on the assistive functionalities that such robots should possess in order to be distinguished from classical mobile robot solutions.

The present paper deals with a home robot companion actually under development at Università Politecnica delle Marche, which can perform navigation and tracking by using a low cost platform, to implement two assistive functions: respiratory rate measurement and stance detection of a person being monitored. Breathing is an important physiological tasks in living organisms, and there are many respiratory diseases which require attentive care and respiratory training. So it is very important to monitor the respiratory activity. Among the possible respiratory rate measurement techniques, a non invasive method has been adopted. In the literature several such approaches have been investigated: CCD camera (see [9]), Structured Light Plethysmography (SLP, see [10]), Slit Light Projection Pattern (see [11]) or Ultra Wide Band (UWB) sensors (see [12]). The use of stereocameras for the detection of breathing is a quite recent technique and is described only in a few papers in the literature (see [13–16]). The present paper presents a RGB-D (Red, Green, Blue-Depth) camera system for both respiratory rate measurement and stance detection. The information about the stance of the person can both be used for detecting possible anomalies in his/her behaviour or for deciding a proper way to interact with him/her.

The robot can move inside an indoor environment and look for the person to monitor: when a person is found, then the procedure to determine his/her stance is activated. If the person is standing, the tracking algorithm starts. If the person is sitting, the respiratory rate algorithm starts. The robot can detect if the person is lying on the ground, in which case it can ask for assistance. These procedures can improve the quality of life of people at home during their daily activities.

The paper is organized as follows. Section 2 describes the hardware setup of the autonomous mobile base and the software layers which are adopted to programme and pilot it. Section 3 gives a short overview of the navigation problem and the algorithms adopted to solve it. Section 4, which represents the core of the paper, details the problem of detecting a person, finding his/her stance, tracking him/her and detecting his/her respiratory rate through the use of the robot sensors. Section 5 finally summarizes the main points of the paper and gives an overview of the future developments of the home robot companion.

2 Robot Configuration

The robot companion under development is based on the TurtleBot kit: a commercial, low-cost mobile robot which can be equipped with sensing capabilities and programmed through an open-source software, see [17]. The TurtleBot represents a solid starting point for developing a robot companion, since it can perform simple tasks with a minimal set of sensors and, at the same time, its intelligence can be enhanced by integrating additional sensors and developing custom software algorithms, while keeping development costs low.

2.1 Hardware

The hardware configuration of the robot companion includes a mobile support, a vision sensor, a gyroscope sensor, a set of mounting plates and a laptop for data processing (see Fig. 1).

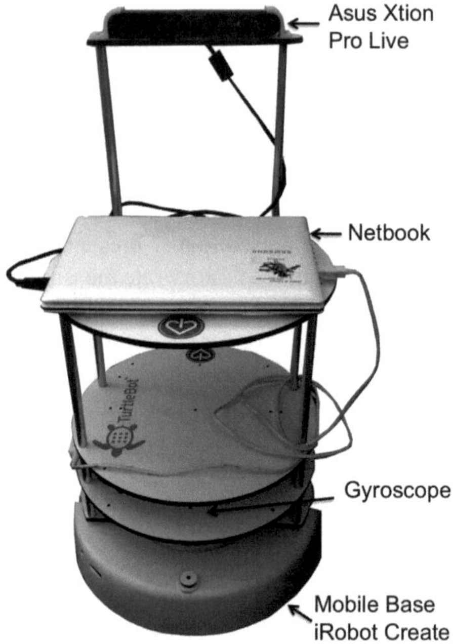

Fig. 1 Robot companion: hardware configuration of the autonomous mobile base

The robot mobile base is the iRobot Create, a mobile base equipped with two wheels driven by DC motors, two support wheels and a microcontroller for differential guide. A gyroscope sensor measuring the robot orientation is mounted directly inside the mobile base in order to minimize the effect of vibrations. Three round plates are fixed on the mobile base at different heights, and are used to carry the laptop, the vision sensor and possible additional sensors.

A RGB-D camera has been used as vision sensor, that is to say a camera which can also measure the distance of objects within its field of view. A RGB-D camera is effective for the identification of persons and objects, even if the background and the person or the object have the same color. RGB-D cameras can recognize overlapped objects by calculating the distance for each of them. Structured Light sensors proved to be effective for breath rate measurement as well (see [13–16]): the sensors resolution is usually adequate to sense small movements like those performed by the thorax during the respiratory phase (see [18]).

More in detail, an Asus Xtion Pro Live camera has been adopted, which falls within the category of RGB-D Structured Light (SL) cameras. With a range of 3.5 meters, the Asus Xtion Pro Live is suitable for navigation, user tracking and breath rate measurement in a home environment (see [19] for more information).

2.2 *Software*

The algorithms of the robot companion are programmed using the C++ language, and are integrated with the Robot Operating System (ROS), a meta-open source operating system which provides hardware abstraction, device drivers, libraries, visualizers, message-passing, package management, and more. Open Natural Interaction (OpenNI), instead, is used to implement further functionalities of the vision sensor.

3 The Navigation Problem

The robot companion must be capable of moving inside an indoor environment, whose map is assumed to be known, looking for a user to track. The knowledge of the map is not a strict requirement: the first time the robot is required to operate in a new environment, the user has the task to guide the robot through it, while a mapping algorithm is activated. In this way a map of the surrounding is always available, however for best results maps generated through laser scanners or surveys should always be preferred.

The navigation problem can thus be decomposed into:

localization, the robot must localize itself within the map;
target definition, the robot must select the next point where to navigate;
path planning, the robot must calculate the trajectory for reaching the next target.

3.1 *Localization Algorithm*

The robot must be able to estimate its position on the map in order to navigate autonomously: this problem is named localization. Solving the problem of localization means to identify global coordinates in the 2D plane and the orientation of the robot (i.e. the robot pose). It is a key problem in mobile robotics, especially in dynamic environments, e.g. if robots operate in the proximity of people who corrupt the robot's sensor measurements. Addressing the problem within a probabilistic framework, it is then possible to use a Monte Carlo Localization (MCL) approach.

The dynamic system is represented by the mobile robot together with its surroundings. The states to estimate are the robot pose, while the measurements are the distances acquired through the Asus camera infrared sensor and the indirect odometry readings. The localization algorithm is able to find the position of the robot within the known map: this information is then used to determine the new initial pose of the robot after each movement, for further information about the *localization algorithm* see [20, 21]. The algorithm is already present in ROS, in the package named AMCL.

3.2 Target Definition Algorithm

The objective of the robot is to move and search for a person to monitor. If he/she is found, then the detection procedure is activated, otherwise the robot returns to its initial position and starts again from the beginning.

We developed a *target definition* algorithm which selects a precise set of points within the map. Each point is identified by three coordinates: the x-position, the y-position and the orientation that the robot must have when reaching that point. At each step the robot has the information on the initial pose, provided by the localization algorithm, and on the next goal that must be reached, provided by the target definition algorithm. This information is fed into the path planning algorithm which calculates the path to the goal.

3.3 Path Planning Algorithm

Path-planning is the process of choosing a course of actions to reach a goal, given the current position. To perform this activity we developed a *path planning algorithm* based on the potential field method. The main idea of this method is to model a virtual potential field in which the robot, modelled as a particle of defined radius, moves following the gradient field until reaching the objective point (see [22] for a description of the theory behind the developed algorithm).

4 User–Robot Interaction

When the robot finds an obstacle during the navigation, it tries to identify it. If the identification succeeds, i.e. the obstacle is identified as a person, then the robot detects his/her stance. If the person is standing, the robot starts to follow him/her. At the present state, the robot can distinguish between three different stances: standing, sitting and lying.

4.1 Stance Identification Algorithm

The identification procedure initially requires an operation called calibration. The *calibration algorithm* recognizes different parts of the person's body, associating a point (joint) to each of them. In particular, after the calibration, the positions of 15 joints are estimated. The calibration operation is required by the vision sensor to track the person and is activated every time the camera finds a moving object within its field of view (see Fig. 2).

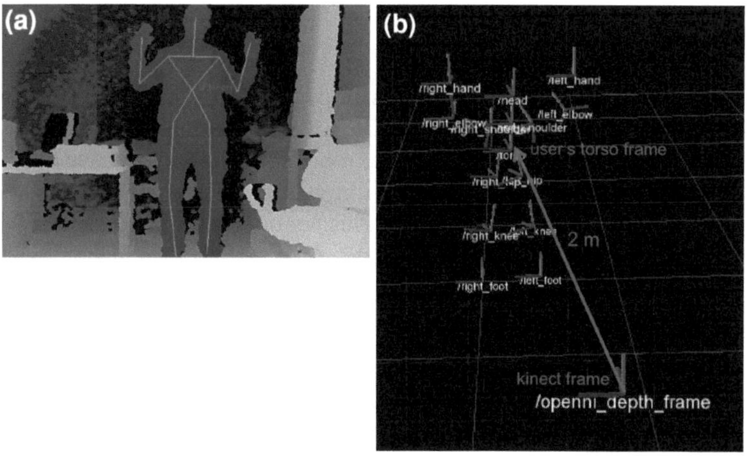

Fig. 2 **a** User calibration and **b** user tracking **c** the robot remains at a fixed distance from the joint torso of the calibrated user

Table 1 Joints values (in *mm*) that allow to determine the stance of the user

	Standing	Sitting	Lying
TORSO	≥−700	≥−700	<−700
HIP	≥−700	≥−700	<−700
FEET	≥−700	≥−700	<−700
TORSO-HIP	≥100	≥100	
KNEE-FOOT	≥150	≥150	
HIP-KNEE (y)	≥200		
HIP-KNEE (z)		≥150	
TORSO-FOOT	≥750	≤750, <500	

Once the calibration data are obtained, they are stored for future use. In this way the calibration is performed only once per person. The robot companion is supposed to be a personal assistive device, thus even if it can recognize multiple persons, calibration can be performed for only one person at a time (i.e. the user).

The *stance identification algorithm* identifies whether the person is standing, sitting or lying on the ground. To do this, the algorithm analyses the relative position of the joints of the torso, hips, knees and feet. The values to define the stance of the user have to be found experimentally according to the person which is going to be assisted by the personal companion. The parameters experimentally calculated for a specific case of use within the development lab are reported in Table 1 for completeness.

If the person is standing, the *tracking algorithm* starts (see Sect. 4). If the person is sitting, the *respiratory rate algorithm* starts (see Sect. 4). The robot can recognize when the person is lying on the ground: this information can be used to understand if the person is in need of assistance (e.g. fallen), in this case the robot can send a message asking for human intervention.

4.2 Tracking Algorithm

The developed *tracking algorithm* exploits the potential field theory already described in Sect. 3. Once a person has been detected in the standing position, the goal is calculated at each iteration step so that the robot keeps a fixed distance of 2 m from the joint associated to the torso, and has a front alignment with respect to the same joint (see Fig. 2). If the person is lost (e.g. his/her orientation w.r.t. camera does not allow to compute the joint position) then the robot performs navigation looking for the person to calibrate again.

4.3 Respiratory Rate Detection Algorithm

The *respiratory rate algorithm* counts the number of breaths per minute of a person. When the robot identifies a person in the sitting position, the algorithm for the detection of the respiratory rate begins. Using the depth information provided by the camera, the algorithm identifies the person's chest and calculates the mean value of the depth of the chest at each time:

$$\bar{D}(k) = \frac{\sum_{i=1}^{N} D_i(k)}{N} \tag{1}$$

where $D_i(k)$ is the information of the depth of the ith point associated to the chest at the sampling instant k and N is the number of points of the chest. The mean value $\bar{D}(k)$ is calculated using data sampled at the frequency of $1/T_c = 7\,\text{Hz}$, where T_c is the sampling time. The initial position of the chest of the person monitored is used as the reference value. The subsequent measurements are used to identify the number of breaths. During the time of the measurement, if the person moves, the algorithm recalculates the position of the chest and it uses this information as the new reference value. Once the time of the measurement is finished, the algorithm calculates the weighted average of the mean values of the depth. This weighted average $WA(k)$ is calculated over a window of 4 samples with the following formula:

$$WA(k) = \sum_{i=0}^{3} w_{(k-i)} \bar{D}(k-i) \tag{2}$$

where $\bar{D}$ is calculated according to Eq. (1) and $w_{(k-i)}$ is the weight associated to the mean value $(k-i)$. After calculating the weighted average, the algorithm calculates the derivative of the weighted average:

$$dWA(k) = \frac{WA(k) - WA(k-1)}{T_c} \tag{3}$$

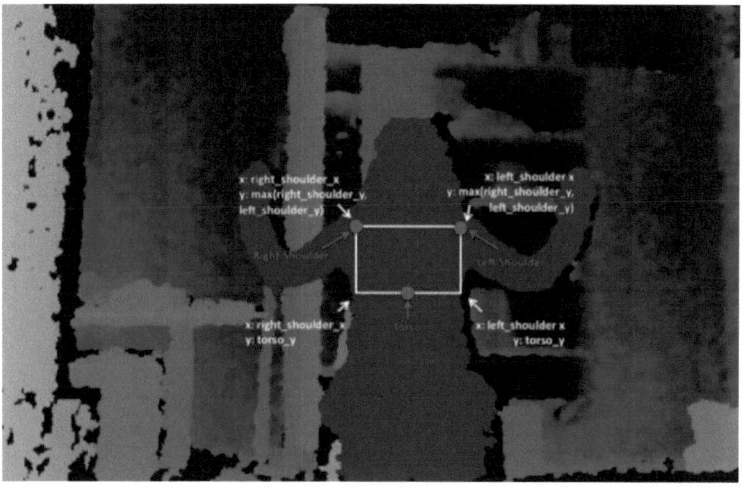

Fig. 3 The Region of Interest. The *white arrows* indicate the vertices of the box and their coordinates. The *red arrows* indicate the Torso and Shoulders joints found in step 1

The derivative was chosen because through its analysis it is possible to identify the maxima and the minima of the average value. The analysis of the derivative also allows to eliminate irregularities in breathing. In fact, in case of irregularities, the analysis of the weighted average is not sufficient to calculate the number of breaths. The algorithm analyses the derivative and counts the number of times that it becomes positive to calculate the respiratory rate. It is possible to extract further information from the weighted average and the derivative:

time of exhalation, $\triangle TE_i = TE_i - TI_{(i-1)}$
time of inhalation, $\triangle TI_i = TI_i - TE_{(i-1)}$
depth if exhalation, $\triangle DE_i = WA(TE_i/T_c) - WA(TI_{(i-1)}/T_c)$
depth of inhalation, $\triangle DI_i = WA(TI_i/T_c) - WA(TE_{(i-1)}/T_c)$

where TE_i is the instant of time in which the exhalation of the ith breath ends, TI_i is the instant of time in which the inhalation of the ith breath ends, WA is the average value of the mean values of the depth of the chest at the sampling instant in which the exhalation or the inhalation of the considered breath ends.

The algorithm introduced above can then be described by the following steps:

1. Detection of torso and shoulder
2. Definition of the Region of Interest (ROI), see Fig. 3
3. Depth measurements
4. Calculation of the mean value of the depth
5. Calculation of the weighted average on a window of 4 samples
6. Calculation of the derivative of the weighted average
7. Calculation of the respiratory rate, the depth and the time of inhalation and exhalation.

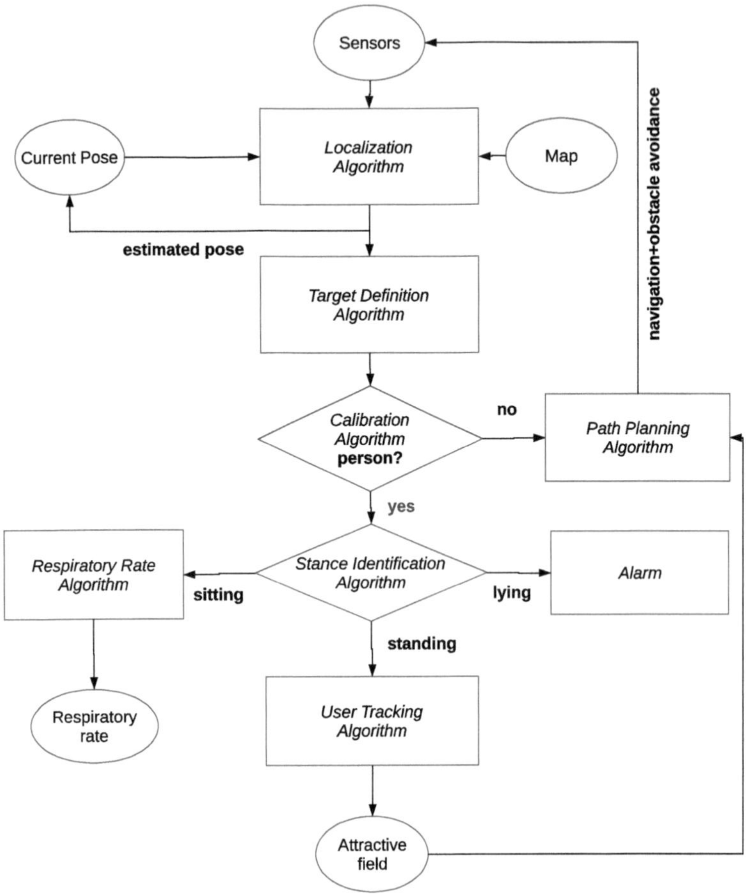

Fig. 4 Flowchart of the navigation and user interaction algorithms implemented in the home robot companion

A flowchart describing both the navigation algorithm and the user interaction algorithm is provided in Fig. 4.

5 Conclusion

In this paper a robot companion capable of performing navigation, user identification, tracking and respiratory rate measurement is presented. The robot can navigate inside an indoor environment, look for a person, identify his/her stance, follow him/her if he/she is standing, calculate his/her respiratory rate if he/she is sitting and send a message if he/she is lying on the ground. The robot companion is thus capable of performing a set of assistive functions to improve the safety of people at home.

The following aspects are actually under investigation. The first one regards the hardware configuration: in order to improve the estimation of the robot pose (especially the orientation) the robot should be equipped with an inertial measurement unit. Several low-cost units are available on the market, thus the robot sensors can be improved without a major impact on the costs. The second one is about the use of the OpenNI library for part of the calibration, stance detection, tracking and respiratory rate algorithms. The library has a limitation: the person must be in frontal position with respect to the robot. When this does not happen, then two or more body joints overlaps and the system has not enough information to detect the target anymore. In order to avoid that, the algorithm should be able to identify a higher number of joints: in this way the system should be able to keep track of the person even when several joints overlap. The third one regards validation of the respiratory rate algorithm assuming different scenarios: the authors are currently evaluating the performances of the algorithm w.r.t. light conditions variation, different clothes of the user and different angles of view of the camera.

References

1. Schulz, R., et al.: Designing and evaluating quality of life technologies: an interdisciplinary approach. Proc. IEEE **100**(8), 2397–409 (2012)
2. Kanade, T.: Quality of life technology [scanning the issue]. Proc. IEEE **100**(8), 2394–6 (2012)
3. Tapus, A., Mataric, M., Scasselati, B.: Socially assistive robotics [grand challenges of robotics]. Rob. Autom. Mag. **14**(1), 35–42 (2007)
4. Graf, B., Staab, H.: Service Robots and Automation for the Disabled/Limited, pp. 1485–1500. Springer Handbook of Automation, Berlin (2009)
5. Srinivasa, S., et al.: The robotic busboy: steps towards developing a mobile robotic home assistant. In: International Conference on Intelligent Autonomous Systems, pp. 651–656 (2008)
6. Srinivasa, S., et al.: Herb: a home exploring robotic butler. Springer Auton. Rob. **28**(1), 5–20 (2010)
7. Srinivasa, S., et al.: Herb 2.0: lessons learned from developing a mobile manipulator for the home. Proc. IEEE **100**(8), 2410–28 (2012)
8. Gross H.M., et al. (2011) Ill keep an eye on you: home robot companion for elderly people with cognitive impairment. In: IEEE International Conference on Systems, Man, and Cybernetics (SMC), pp. 2481–2488
9. Nakajima, K., Matsumoto, Y., Tamura, T.: Development of real-time image sequence analysis for evaluating posture change and respiratory rate of a subject in bed. Physiol. Meas. **22**(3), N21 (2001)
10. Wareham, R., Lasenby, J.J., Cameron, P.D.B., Iles, R.: Structured light plethysmography (slp) compared to spirometry: a pilot study. In: European Respiratory Society Annual Congress (2009)
11. Hirooki, A., Kohji, K.: Non-contact respiration monitoring method for screening sleep respiratory disturbance using slit light pattern projection. World Congr. Med. Phys. Biomed. Eng. **14**, 680–3 (2007)
12. Staderini, E.: Uwb radars in medicine. IEEE Aerosp. Electron. Syst. Mag. **17**(1), 13–8 (2002)

13. Burba, N., et al.: Unobtrusive measurement of subtle nonverbal behaviors with the Microsoft Kinect. In: Proceedings of the 2012 IEEE Virtual Reality (VR '12), pp. 1–4. DC, USA, Washington (2012)
14. Martinez, M., Stiefelhagen, R.: Breath rate monitoring during sleep using near-ir imagery and PCA. In: 21st International Conference on Pattern Recognition (ICPR), pp. 3472–3475, Tsukaba, Japan (2012)
15. Yu, M.C., et al.: Noncontact respiratory measurement of volume change using depth camera. In: Annual International Conference of the IEEE on Engineering in Medicine and Biology Society (EMBC), pp. 2371–2374. Osaka, Japan (2012)
16. Yu, M.C., et al.: Breath and position monitoring during sleeping with a depth camera. In: International Conference on Health Informatics, pp. 12–22. Vilamoura, Portugal (2012)
17. Gerkey, B., Conley, K.: Robot developer kits. IEEE Robot. Autom. Mag. **18**(3), 16 (2011)
18. Khoshelham, K., Elberink, S.O.: Accuracy and resolution of kinect depth data for indoor mapping applications. Sensors **12**(2), 1437–54 (2012)
19. Mrazovac, B., Bjelica, M.Z., Papp, I., Teslic, N.: Smart audio/video playback control based on presence detection and user localization in home environment. In: Proceedings of the 2011 Second Eastern European Regional Conference on the Engineering of Computer Based Systems (ECBS-EERC '11), pp. 44–53. DC, USA, Washington (2011)
20. Fox, D.: Kld-sampling: adaptive particle filters and mobile robot localization. Adv. Neural Inf. Process. Syst. (NIPS) **14**(1), 26–32 (2001)
21. Fox, D., et al.: Monte Carlo localization: efficient position estimation for mobile robots. Proceedings of the 16th National Conference on Artificial intelligence (AAAI '99), pp. 343–9. Menlo Park, CA, USA (1999)
22. Khatib, O.: Real-time obstacle avoidance for manipulators and mobile robots. Int. J. Robot. Res. **5**(1), 90–8 (1986)

Telepresence Robot at Home: A Long-Term Case Study

Roberta Bevilacqua, Amedeo Cesta, Gabriella Cortellessa, Alessandro Macchione, Andrea Orlandini and Lorenza Tiberio

Abstract The paper presents the current findings of an almost one-year case study of fielding the GIRAFF robot in a real home. The testing phase, which is still running, involves an older user, affected by Parkinsons disease, and his family, in the daily usage of a telepresence robot. Initial evidence show that GIRAFF has become part of the daily routine of the user being judged as a valuable tool for communication and close connection with the remote user. In general, the attitude of the user towards the aid remains stable being constantly positive over time. Nevertheless we also noted an impact of the technology robustness on user's acceptance that will be further investigated in this continuation of the study.

R. Bevilacqua
INRCA, Direzione Scientifica, Polo Scientifico-Tecnologico, Ancona, Italy
e-mail: r.bevilacqua@inrca.it

A. Cesta (✉) · G. Cortellessa · A. Orlandini · L. Tiberio
CNR—Consiglio Nazionale delle Ricerche, ISTC—Istituto di Scienze e Tecnologie della Cognizione, Rome, Italy
e-mail: amedeo.cesta@istc.cnr.it

G. Cortellessa
e-mail: gabriella.cortellessa@istc.cnr.it

A. Orlandini
e-mail: andrea.orlandini@istc.cnr.it

L. Tiberio
e-mail: lorenza.tiberio@istc.cnr.it

A. Macchione
Universitá Politecnica delle Marche, Dipartimento di Ingegneria Industriale e Scienze Matematiche, Ancona, Italy
e-mail: a.macchione@univpm.it

S. Longhi et al. (eds.), *Ambient Assisted Living*, DOI: 10.1007/978-3-319-01119-6_8,

1 Introduction

In the past decade, the Human Robot Interaction (HRI) was approached in many different ways, mainly focusing on the human perception of robots appearance, behaviours and personality [16]. Analysis were mostly conducted in controlled settings, despite the overall aim of creating robots for assisting people in performing daily activities, from entertainment to cooking, for example. This paradoxical situation is due to the challenging and unpredictable nature of the real setting characteristics, that does not allow a deep control over all the key variables as the lab condition, as well as the short time reserved to the in-home trials that does not permit the real engagement of the users with the robot. The interest in longitudinal studies on human-robot interaction has mainly originated from non-residential environments (such as offices, schools and hospitals) as ecological setting of study, emphasizing that the prolonged interaction may result in a change in attitude and behavior of the person, influencing the user experience and its relationship with the adopted aid [4–7, 9, 10]. The objective of our research is to deepen the knowledge of what really happens when a social assistive robot is introduced in the home of an elderly person and his/her caregiver, becoming part of their daily lives for a long period of period time. To understand how the effect of a long-term interaction influences the behaviour and attitudes of end-users can indeed be fundamental for the design of an aid that hopefully remains useful and acceptable in the long term.

In this perspective, the ExCITE project (http://excite.project.eu), part of the AAL EU initiative (http://www.aal-europe.eu), is assessing the robustness and validity of a telepresence robotic platform, named GIRAFF produced by the Swedish company Giraff AB (http://www.giraff.org), as a means to support elderly and to foster their social interaction and participation. The whole project is based on the idea of deploying several GIRAFF for long periods of time (at least 3 months, and possibly 1 year) in three different countries (Italy, Spain and Sweden) in real contexts of use. Feedback obtained from the users (both older users having the robot at home and the clients, who is people connecting and visiting the older) is used to both improve the robot and gather results on aspects related to the long-term human-robot interaction. The GIRAFF case study represents a good example of a long-term analysis where a robot interacts with an older user in a real setting, giving the opportunity of deeply investigating the effects of a long term interaction on the person's behaviour, attitude, technology acceptance and general experience of use.

To find and set up such long terms studies turned out to be a rather difficult and demanding activity. In this light a specific CNR effort within ExCITE was devoted to establish relationship with possible user groups representative of different realities (e.g., healthcare professional, healthy and not healthy elderly, nurses, relatives, friends, etc.).

In this paper we specifically focus on one of the on-going case studies performed in Italy derived by the cooperation of CNR participants to the ExCITE project as responsible for the evaluation activities, with colleagues of INRCA (Istituto Nazionale di Riposo e Cura per Anziani) and UNIVPM (Universitá Politecnica delle Marche) who have collaborated in the execution and monitoring of this specific test site. It is worth highlighting how the specific test sites has presented specific challenges for its peculiarity of being installed in a distant city entailing a continuous collaboration between the CNR group with the team in Ancona who followed the instantiation and maintenance of the test sites more closely. We here report some results of the work illustrating the experience related to the user interaction findings of the long-term usage. More specifically the paper is organized as follows: Sect. 2 describes the motivation of the study and the technological steps needed to perform the long term test case within the ExCITE project; Sect. 3 introduces in details the specific long-term case study dwelling on its peculiarity and challenges and showing the current results on the long-term user experience; Sect. 4 ends the paper.

2 Assessing a Telepresence Robot in the Field

When a robotic aid is to be introduced and evaluated in the home, to promote and sustain a long-term interaction becomes a fundamental aspect of use. The home is becoming one of the contexts of use more attractive for robotic applications especially for the elderly care. However, evaluating a long-term interaction in a domestic context is still an open challenge.

Most of the successful examples of longitudinal researches on human-robot interaction has mainly originated from non-residential environments. Of particular importance in the domain of elder care are the studies on Paro, a seal-shaped robot equipped with touch, light, auditory and balance sensors, capable of responding to external stimuli and designed for therapeutic interventions with older people and children affected by genetic syndromes and developmental disorders. The effects of time in the use of Paro were assessed for example, in a group of 12 elderly residents in a nursing facility, who have interacted with the robot for a continuous period of 30 days. The results of systematic ethological observations by therapists of the structure have shown that Paro was able to improve and enhance the ability to relate in older people with a positive effect on their levels of physical and psychological stress [15]. Another study, [9], confirmed an improvement of the patients' social skills in response to a therapeutic intervention based on prolonged use of Paro, as well as a positive effect on their emotional sphere.

More recently, [12] have described the results of an ethnographic study in which a humanoid social assistive robot (called Robovie) interacted for a period of 3 months and a half, with 55 elderly residents in a care center. The robot was operated remotely to interact verbally with patients. The analysis of the interviews and the results of the observations, performed by a therapist of the structure, suggested that the robot had been well accepted among residents, especially for its ability to socially engage patients through verbal reminders and simple conversations. However, the effects of long-term interaction between a human and a robot does not always have positive implications, but they can reveal changes in the behavior of people who may be indicative of an experience of use and interaction not completely satisfactory. It may happen for instance that the initial effect of novelty and curiosity, towards a robotic aid and its use, quickly fades resulting in a reduction in the interest and attitude of the person towards it [2, 5, 8, 13, 14]. For this reason, longitudinal studies, although they may be difficult to organize especially in ecological contexts, are gaining significant relevance to understand how to promote long-term interactions, considering the effect of time on the behaviour and attitudes of the human user toward the robot. However, the literature on interactions and long-term deployment of robotic platforms in a domestic setting, it is still rather incomplete and this opens new possibility of studies. In fact, our research is motivated by the interest in understanding how the continuous use and the sharing of physical and social spaces in the home, influences the experience, the use and attitude of seniors people and their caregivers (family and/or professional) of the telepresence robot. In so doing we explored the challenge of moving laboratory experiments to real people life settings by delivering robots into the house of old person and assessing several aspects of the long-term usage and interaction.

The next section describes the effort done for one these long-term studies. Specifically we illustrate the evaluation plan, which has been slightly updated with respect to the general procedure conceived for ExCITE due to the particular case we were able to set up. We also present the results we currently have after several months of usage.

3 A Long-Term Case Study

For conducting this specific analysis, an older end-user was recruited in Ancona and he was asked to take part in the ExCITE long-term case study. The beginning of the testing phase was characterized by technical issues that needed to be solved and that have also affected the testing timing delaying the actual start of the evaluation plan. In the next subsections we describe in details the used approach and illustrate the results so far.

3.1 Method

The method used for this assessment is based on longitudinal research and specifically on the single case study, which essentially allows in depth observation within a real-life context, collecting a lot of details that would not normally be easily obtained by other research designs. The present article focuses on long-term experience of usage of the telepresence robotic platform on behalf of the older user at home. Specifically, we aimed at investigating acceptance, attitude toward the robotic aid, appearance, impact on the home environment and usability. It is worth highlighting how we here report some of the results of an on-going test site; hence also the material reported is the one we have used for the specific period of experimentation we are describing here. For the complete list of the questionnaires foreseen for the whole experiment the reader can refer to [1].

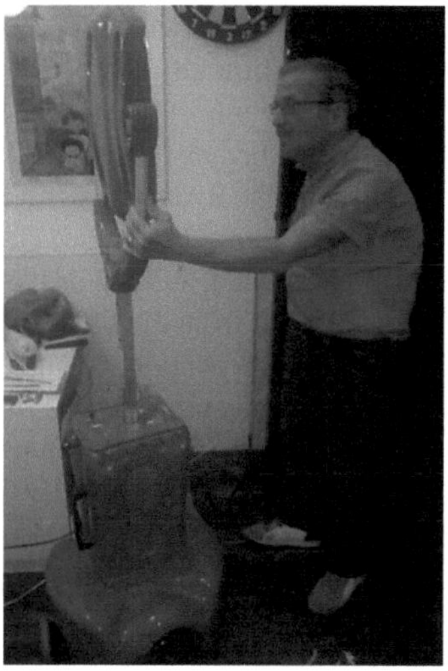

Fig. 1 The primary user interacting with the GIRAFF robot

3.1.1 Material

The material used for this long term assessment is composed of both standardized measures and specific questionnaires we developed on purpose to assess the aspects mentioned above. Specifically at this time, for the primary user we used: a consent form describing the aim and procedure of the study to be signed by participant; a socio-demographic data form to gather some relevant information on the user like sex, age, gender, familiarity with technology, etc.; the UCLA Loneliness Scale [11] a 20-item Likert scale (0–4) designed to measure ones subjective feelings of loneliness as well as feelings of social isolation. The MSPSS Multidimensional Scale of Perceived Social Support [19] which consists of 12 items to identify the social support factors perceived by the individuals; the GDS (Geriatric Depression Scale—[18]) is a 30-item self-report assessment used to identify depression in the elderly; SF-12, a short version of the SF-36 [17] that allows to investigate the perception of the state of physical and mental health; an ad hoc questionnaire based on Almere model [3] that allows assessing dimensions of technology acceptance; a questionnaire we developed on purpose to assess the attitude and general evaluation of older persons toward such technology.

3.1.2 Participant

The older user is affected by Parkinson's disease since 2005. Both his walking ability and the speech were deeply compromised by the severe stiffness in movement as well as the muscle rigidity of the face. In particular, due to the speech restriction, the user often refuses the communication with the others and prefers to stay alone. The fear of feeling bad outside home and the difficulty in speech have led to a progressive departure from the community life. The interruption of his social life has caused him relevant mood alterations: sometimes, he has reported to be very depressed and feel that anyone could help him. Currently, he lives with his wife, in the centre of a small town. He can reach by feet all the most important places in the town and the sea-side too. His wife is a very active woman instead, and she takes care of him constantly. His son (the Client) and his daughter live far from the town (respectively 30 and 60 Km), so he does not see them very much. He is currently engaged in physical rehabilitation and speech therapy. The user has showed some basic skills with technology, mainly motivated by his interests: he is able to use Internet connection, to play online games. In addition, he is able to use the mobile phone basically, just for calling but not for sending text messages or photos, for example, while he uses daily the digital TV for looking at sport events.

3.1.3 Procedure

The method used for this test site is taken from the general evaluation plan conceived for ExCITE and described in [1]. It entails a period of N months (with $3 \leq N \leq 12$) during which the primary user has the robot at home and the secondary users (family members, formal caregivers or health professionals) can visit him/her through the robotic platform. Assessment happens across different steps S_i. Specifically, after an initial assessment at S0 at the beginning of the experimentation (baseline), different potential psycho-sociological variables of interest are measured at regular intervals (S1–S3) to observe changes over time. At the last month the GIRAFF is removed from the end user apartment and the same variables are assessed again after 1–2 months from this removal (S4). A researcher and a technician conducted the training on the robot use, in order to exhaustively answer all kind of questions. At this purpose, verbal explanations and practical examples were mainly given to the user, making him free to ask all the needed information. Finally, the team, who intervened when necessary, both from remote and at home, always supported the user. After the training period, the standard evaluation plan started. We have currently completed the S2 phase, following the schema listed below :

Primary user

- S0: At time S0 we asked the primary user to fill in the consent form and the Socio-Demographics Data Form, the UCLA, the SF-12, the GDS, the MSPSS, and the ad hoc questionnaire based on Almere Model.
- S1: After a period of usage at time S1 the primary users filled in the questionnaire on the Attitude and General Evaluation towards the robot, the UCLA, the GDS, the MSPSS.
- S2: At time S2 we asked the primary users to fill in ad hoc questionnaire based on Almere Model.

For the subsequent phases the users will follow the evaluation plan foreseen in [1].

To explore how primary users perceive and think about the robot and which are the dynamics in their real context of use, we decided, for this test site, to also collect data in the form of ethno-graphic notes through a Participant Observation (PO). The participant observation has been possible taking advantages form the collaboration between CNR and the colleague of INRCA who has specific competence in ethnographic research methodology. The choice of adding a participant observation, was mostly motivated by the setting characteristics of this specific case study as well as the opportunity of drawing up data and concepts to be matched with the past assessment analysis. Some field notes were written during the PO but they were definitely elaborated after each session of observation, which corresponded to the evaluation step S0, S1 and S2. Besides the assessment of the validity of the answers and the user cooperation, self-reflexive notes were also taken, on what the researcher considered important.

3.2 Results

This section describes the results of the long-term experience of use of the system on behalf of the primary user. Results are related to both the questionnaires and participant observation analysis. Findings from questionnaires are related to the instruments for collecting data of general evaluation plan conceived for ExCITE and mainly report to the current status of the test sites, which will be completed toward the end of this year.

3.2.1 General Evaluation Plan Results

Psychological measures. The results of Geriatric Depression Scale (GDS) at time S0 and S1 show scores substantially similar, indicative of the presence of mild depressive symptoms (scores between 10–19). Similarly, the scores on the perceived loneliness scale (UCLA) indicate an experience of strong loneliness (UCLA score = 52 at S0, UCLA score = 47 at S1, range score 20–80). The score of the Short Form-12 Health Survey (SF-12) show low mean meta scores of the Physical Component Summary ($M = 23.5$) and Mental Component Summary ($M = 38.4$) of the score denote a perceived health condition characterized by significant limitations in self-care and in physical, social and personal.

In the pre-adoption phase (S0), the scores of the Multidimensional Scale of Perceived Social Support Scale (MSPSS) indicate that the end user perceives a higher social support from family (sum of the scores = 28) with respect to that received from the friendships network (sum of the scores = 8) and /or from significant other persons close to the user (sum of the scores = 4) (see Fig. 2). After the adoption and the first period usage of the robot (S1), the family continues to be the main source of perceived social support (sum of the scores = 23), while the perceived support from

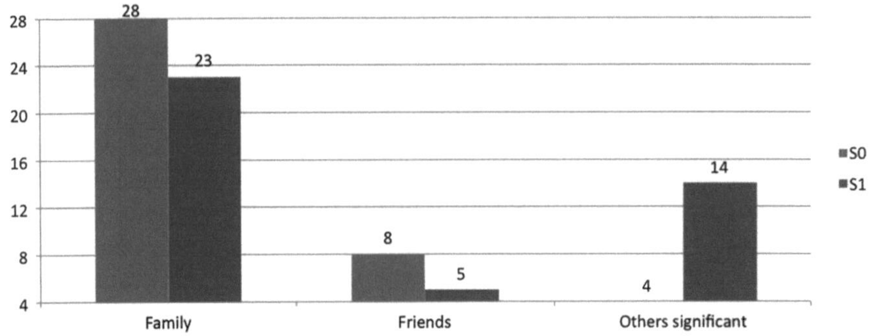

Fig. 2 Total scores for each dimensions of MSPSS obtained from the sum of the responses from the items in each of the three dimensions (range score 4–28)

the friendships network (sum of the scores = 5) further decreases, and the feeling of social support from significant other persons increases to some extent respect (sum of the scores = 14) (see Fig. 2).

Attitude and general evaluation towards the robot. Based on the dimensions of attitude and evaluation described in [1], Fig. 3 shows the results of the questionnaire related to significant dimension of evaluation from the primary users perspective. Specifically at time S1, after the adoption and the first period of usage, the user does not perceive a high level of intrusion into his privacy ($M = 0.67$), does not show distrust in terms of psychological distance between himself and the adopted aid ($M = 0.60$), and recognizes the benefits and advantages ($M = 2.79$), of the system related to his needs (e.g., GIRAFF simplifies the management of daily life or relieves the workload of people who "take care of me"). Nevertheless he shows a partial satisfaction of the GIRAFFs functionalities and features ($M = 2.47$), (e.g., the quality of the video and the movement of the robot are not very satisfactory) and some notable apprehension related to the difficulty of maintenance of the robot ($M = 3.25$), maybe justified also by the technical difficulties that emerged.

Furthermore, the user would like additional features of the robot like for instance the possibility to have a direct link with his doctor via GIRAFF. The emotional reaction of elderly user to the robot is very good, scoring high on the positive adjectives useful, interesting, stimulating, and funny, and very low on the negative adjectives scary, overwhelming, gloomy, dangerous, and uncontrollable.

Robots Acceptance. The mean scores of each Almere model acceptance constructs (see Fig. 4) show an increase on Intention to Use (ITU) the robot between S0 and S2 ($Ms0 = 4$ vs $Ms2 = 5$) and of Social Influence (SI) ($Ms0 = 3$ vs $Ms2 = 4$). Similarly, the Facilitating Conditions (FC) increase ($Ms0 = 3$ vs $Ms2 = 4$). However there is a slight increase in the perceived Anxiety (ANX) towards the use of the robot ($Ms0 = 2.5$ vs $Ms2 = 3$). The Perceived Easy of Use (PEOU) decreases ($Ms0 = 4.8$ vs $Ms2 = 3.8$). as well as the Perceived Usefulness (PU) decrease ($Ms0 = 4$ vs

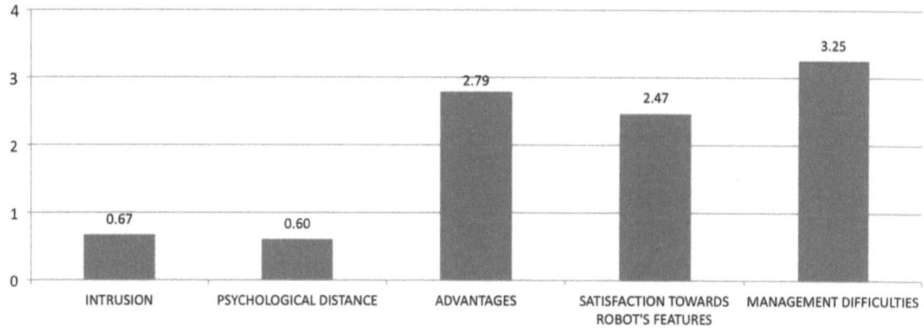

Fig. 3 General Attitude of users toward the GIRAFF system after a period of usage. Mean scores for each dimension explored by questionnaire ad hoc (5 points Likert scale, from 0= completely disagree to 4= completely agree)

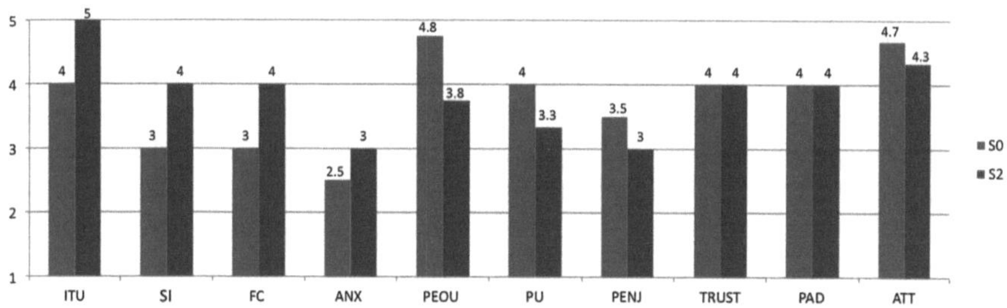

Fig. 4 End User Acceptance over time (differences between S0 and S2). Mean scores for each Almere Model construct (5 points Likert scale, from 1= completely disagree to 5= completely agree)

$Ms2 = 3.3$), accompanied with a lower Perceived Enjoyment (PENJ) of the user ($Ms0 = 3,5$ vs $Ms2 = 3$). The levels of Trust, Perceived Adaptiveness (PAD) of the robot to the individual needs and the Attitude (ATT) towards GIRAFF remain high and stable.

3.2.2 Participant Observation Results

Participant observation results focusing on understanding GIRAFF functioning and perception from the primary users point of view reveal important data on behavior, attitude and reactions of the user towards the introduction and adoption of the robot in his domestic life environment. We discuss and integrated the collected observational notes into the three main categories below.

Appearance. The participant observation showed a positive judgment of the user on this dimension. Since the first home visit (S0), the end-user has liked GIRAFF, considering it was funny and not too "technical" to be used. The animal shape reminds him more to an electronic toy than to an assistive tool for the daily communication.

The familiar shape have also transmitted him the feeling of an easy-to-use tool, as the shape was the metaphor of its functioning. The positive evaluation of the robot appearance was maintained all along the experience with the robot, as well as the absence of stigmatization perception.

Attitude. Results of the participant observation highlight that at the beginning, the end-user has underestimated the complexity of the robot: he felt to be able to use it very quickly and without the support of a technician. In addition, the restricted number of operations that can be performed has given him the sense of mastery in use, as well as the presence of clear commands for answering a call, controlling the volume, and turning on/off the robot. It could be said that, during the first home visit (S0), the perception of easiness of use was the dimension that mostly influenced the positive attitude towards the robot, as well as its use. In addition to the robot characteristics, the reason of this feeling could be partially justified by the technological skills of the user: he is able to use Internet to play online games, send emails, do home banking operations, looking for events and stuff of his own interest. After a period of usage (S1), some technical obstacles that emerged influenced the positive attitude, and specifically a decreasing in the sense of self-efficacy and competence emerged, together with a feeling of frustration. After the resolution of the technical problems, the user partially recovered the positive attitude towards the robot during S2 phase.

The participant observation highlighted the following aspect. The acceptance of the robot seems to be gained after some time of daily use: from the initial underestimation of the robot complexity (S0) to the achieved comfort in use (S1), it seems that the representation of GIRAFF was passed through a "normalization" path (S2). The weak points of the "path to the robot acceptability" could be found in two different phases:

- at the beginning, due to the overestimation of the own technological skills and/or the underestimation of the robot complexity (S0);
- during the first period of use, in which the user considered himself not able to use the robot, instead of researching the reason of the failure in the technical obstacles (S1). In this case, it could be said that there was an erroneous attribution of the events to the internal locus of control.

On the observer side, it seems that the creation of a daily routine with the robot was the key driver for the robot acceptance (S2): before dinner time, they knew they would be in touch with their son and his family, taking part to their domestic life. The affection to the robot was mainly mediated by the successful experiences with it and the overcoming of the technical obstacles (S2). For the end-user, in particular, who presents a severe mobility restriction, as well as difficulty in speech, the robot represents a valuable mean for "feeling connected", not just for communicating.

Usability. Despite the interest and the initial positive approach to the robot usage, the end-user did not learn easily how to use the robot: the turn on/off command and the red/green buttons for answering to a call were often confused (S0). Also the use of the remote control was not successful: the user still preferred to answer directly from the robot. It seems that the end-users initial representation of the robot did not match the real condition of use, at the beginning (S0). Starting with the daily use, the user has gained a good level of autonomy in operating as well as the achieved competence in referring all the robot activities to the technical staff. Once the technical issues were solved, the end-user felt more comfortable and the robot became soon the preferred mean for talking with the client (S1). The key benefit for the user was of course the opportunity of seeing the client as well as the availability of the service as free. At the same time, the user tried to understand how to cope with the robot when system errors occurred, in order to be able to solve them by himself. Moreover, he asked for additional functionalities, in order to ameliorate the communication also with professionals and physician from remote (S2).

4 Conclusions

The profile of the primary user, with respect to the psychological variables, reflects the clinical picture of Parkinson's disease. The presence of depressive symptoms, joint to the sense of loneliness and to a state of health perceived by the user as rather fragile, are to be interpreted as the common psychological reaction of a person suffering from a debilitating chronic disease, such as the Parkinson's disease.

The family is the main source of support (not only physical but also social) perceived by primary. The function of the family in terms of social support is confirmed over time. However, in recent months the use of GIRAFF seems to have also contributed to increase the sense of the perceived social support from other persons significant for the secondary user. Presumably, the presence of personnel for technical support, may have contributed to an increase in the user 's perception of being able to rely on other people outside his family network, in relation to the management and operation of adopted aid.

GIRAFF has become part of the daily routine of the user being judged as a valuable tool for communication and close connection with the secondary user. The function of visual contact of the telepresence system helps to validate the user's primary beliefs about the effectiveness of the robot as a communication aid. In general, the attitude of the user towards the aid remains stable being constantly positive over time. It is plausible to think that the attractiveness of GIRAFF, the users familiarity with the modern information and communication technologies and the adequacy of aid with respect to personal needs, have contributed to the a positive attitude of the user. This attitude has affected his intention to use and acceptance of the robot over time.

Obviously we cannot overlook the role that the technical problems have had on the user's attitude so far. Results show indeed an increase over time of the worries with respect of maintenance of the robot and a decrease of his satisfaction of the system's functionality. The malfunction may be responsible for the perceived anxiety over time related to the use of GIRAFF. Moreover, the presence of feelings of frustration, accompanied with a decrease of the sense of self-efficacy reported during the participant observation, highlight the need of reliable and robust assistive robot to ensure and interaction with the user which is effective and safe.

Acknowledgments CNR authors are partially supported by the EU under the Ambient Assisted Living Joint Program—ExCITE Project (AAL-2009-2-125).

References

1. Cesta, A., Cortellessa, G., Orlandini, A., Tiberio, L.: Evaluating telepresence robots in the field. Commun. Comput. Inf. Sci. **358**, 433–448 (2013)
2. Gockley, R., Bruce, A., Forlizzi, J., Michalowski, M., Mundell, A., Rosenthal, S., Sellner, B., Simmons, R., Snipes, K., Schultz, A., Wang, J.: Designing robots for long-term social interaction. In: Intelligent Robots and Systems. IEEE/RSJ International Conference on. IROS 2005. pp. 1338–1343 (2005)
3. Heerink, M., Kröse, B., Evers, V., Wielinga, B.: Assessing acceptance of assistive social agent technology byolder adults: the almere model. Int. J. Soc. Robot. **2**(4), 361–75 (2010)
4. Huttenrauch, H., Severinson Eklundh, K.: Fetch-and-carry with CERO: Observations from a long-term user study with a service robot. In: Proceedings of 11th IEEE International Workshop. Robot and Human Interactive Communication, 2002. ROMAN. pp. 158–163 (2002)
5. Kanda, T., Sato, R., Saiwaki, N., Ishiguro, H.: A two-month field trial in an elementary school for long-term human-robot interaction. IEEE Trans. Robot. **23**(5), 962–71 (2007)
6. Kim, H., Lee, H., Chung, S., Kim, C.: User-centered approach to path planning of cleaning robots: Analyzing user's cleaning behavior. In: Proceedings of the ACM/IEEE International Conference on Human-robot Interaction. HRI '07. pp. 373–380. ACM (2007)
7. Lee, H., Kim, H.J., Kim, C.: Autonomous behavior design for robotic appliance. In: Proceedings of the ACM/IEEE International Conference on Human-robot Interaction. HRI '07. pp. 201–208. ACM (2007)
8. Leite, I., Martinho, C., Paiva, A.: Social robots for long-term interaction: a survey. Int. J. Soc. Robot. **5**(2), 291–308 (2013)
9. Marti, P., Pollini, A., Rullo, A., Shibata, T.: Engaging with Artificial Pets. In: Proceedings of the 2005 Annual Conference on European Association of Cognitive Ergonomics. EACE '05. pp. 99–106. University of Athens (2005), http://dl.acm.org/citation.cfm?id=1124666.1124680
10. Mutlu, B., Forlizzi, J.: Robots in organizations: the role of workflow, social, and environmental factors in human-robot interaction. In: Proceedings of the 3rd ACM/IEEE International Conference on Human Robot Interaction. HRI '08. pp. 287–294. ACM (2008)
11. Russell, D., Peplau, L., Cutrona, C.: The revised ucla loneliness scale: concurrent and discriminant validity evidence. J. Pers. Soc. Psychol. **39**, 472–80 (1980)
12. Sabelli, A.M., Kanda, T., Hagita, N.: A conversational robot in an elderly care center: An ethnographic study. In: Proceedings of the 6th International Conference on Human-robot Interaction. HRI '11. pp. 37–44. ACM (2011)
13. Salter, T., Dautenhahn, K., Bockhorst, R.: Robots moving out of the laboratory - detecting interaction levels and human contact in noisy school environments. In: 13th IEEE International Workshop on Robot and Human Interactive Communication, 2004. ROMAN. pp. 563–568 (2004)
14. Sung, J.Y., Grinter, R.E., Christensen, H.I.: Domestic robot ecology—an initial framework to unpack long-term acceptance of robots at home. Int. J. Soc. Robot. **2**(4), 417–429 (2010)
15. Wada, K., Shibata, T.: Living with seal robots: its sociopsychological and physiological influences on the elderly at a care house. IEEE Trans. Robot. **23**(5), 972–80 (2007)

16. Walters, M.L., Syrdal, D.S., Dautenhahn, K., te Boekhorst, I.R.J.A., Koay, K.L.: Avoiding the uncanny valley: robot appearance, personality and consistency of behavior in an attention-seeking home scenario for a robot companion. Auton. Robots **24**(2), 159–78 (2008)
17. Ware, J., Kosinski, M., Keller, S.: A 12-item short-form health survey: Construction of scales and preliminary tests of reliability and validity. Med. Care **34**, 220–233 (1996)
18. Yesavage, J., Brink, T., Rose, T., Lum, O., Huang, V., Adey, M., Leirer, V.: Development and validation of a geriatric depression screening scale: a preliminary report. J Psychiatr. Res. **17**, 37–49 (1983)
19. Zimet, G., Dahlem, N., Zimet, S., Farley, G.: The multidimensional scale of perceived social support. J. Pers. Assess. **52**, 30–41 (1988)

A Multi-Agent Expert System Simulation for Ambient Assisted Living: The Virtual Carer Experience

Paolo Sernani, Andrea Claudi and AldoFranco Dragoni

Abstract A shift in the distribution of population towards older ages is occurring in almost every country of the world and it is becoming a major problem in Europe, Japan and USA, due to the high dependency ratio of these countries. Ambient Assisted Living (AAL) and Ambient Intelligence (AmI) are polarising the attention of the research community, trying to improve the quality of life of elderly people promoting their autonomy, self-confidence and mobility. Artificial Intelligence (AI), developing intelligent and adaptable systems, can play a considerable role in AAL and AmI to cope with the changing needs that characterize the life of people with chronic diseases. To show a possible contribution of AI in these fields, this paper introduces the multi-agent architecture of an expert system for Ambient Assisted Living: the Virtual Carer (VC). Based on the Belief-Desire-Intention (BDI) paradigm, it models the behaviours of a human caregiver. The main goal of the Virtual Carer is to help an elderly patient in his daily activities, while his health conditions are monitored, in order to ensure his security. To show the main capabilities of the system, the paper describes some simulations of the proposed agency, highlighting the architecture of the reasoning component.

P. Sernani (✉) · A. Claudi · A. F. Dragoni
Dipartimento di Ingegneria dell'Informazione (DII), Università Politecnica delle Marche,
Via Brecce Bianche, 60131 Ancona, Italy
e-mail: p.sernani@univpm.it

A. Claudi
e-mail: a.claudi@univpm.it

A. F. Dragoni
e-mail: a.f.dragoni@univpm.it

S. Longhi et al. (eds.), *Ambient Assisted Living*, DOI: 10.1007/978-3-319-01119-6_9,

1 Introduction

Population ageing is occurring in almost every country in the world. The rising of life expectancy along with the decreasing of birth rates determine a higher dependency ratio (i.e. the ratio between the number of people over 65 years old and those of working age) in Europe, USA and Japan [1]. The world is facing a situation without precedent: we soon will have more older people than children and more people at extreme old age than ever before [2]. The consequent rise of the number of chronic diseases results in the increasing of health-related emergencies and of the health care spending [3].

Ambient Assisted Living (AAL) focuses on these themes and aims at extending the time older people can live in their home environment, assisting them with the activities of daily living, promoting the use of intelligent products and Information Technology (IT) tools to provide remote care services [4]. Through the AAL Joint Programme, the European Union aims to foster the emergence of AAL services and systems for ageing well at home, in the community, and at work [5]. The AAL Joint Programme defines these key objectives for AAL:

- to extend the time people can live in their preferred environment by increasing their autonomy, self-confidence and mobility;
- to support maintaining health and functional capability of the elderly individuals;
- to promote a better and healthier lifestyle for individuals at risk;
- to enhance the security, to prevent social isolation and to support maintaining the multifunctional network around the individual;
- to support carers, families and care organizations;
- to increase the efficiency and productivity of used resources in the ageing societies.

Every AAL system is based on pervasive devices typically used in Home Automation systems, and on Ambient Intelligence (AmI) technologies to integrate devices and build a safe environment for the assisted person [4]. The main goal of AmI is to help people in their daily activities, building around them an unobtrusive, interconnected, adaptable, dynamic, embedded, and intelligent environment [6]. Humans can interact with AmI-based systems using natural user interfaces as speech and gestures. One of the goals of AmI is to allow the user to interact with an AmI system as it would do with any other human [3].

AAL and AmI systems have to adapt themselves over time to cope with the changing needs and health conditions of the assisted person. Moreover, they need to perceive variation in habits that can signal health-related problems or stressful situations. Applying Artificial Intelligence techniques on this domain seems to be a promising direction for research [7, 8].

In this paper we present Virtual Carer, a Multi-Agent System (MAS) based on the Belief-Desire-Intention (BDI) paradigm, which models, integrates and manages the typical services of an AAL system. The Virtual Carer manages a distributed sensor network formed by ambient and biometric sensors. To increase modularity and reliability of the system, each sensor is associated with an agent; thus the system is resilient to sensor failures and disconnections. Moreover agents, by their own nature, can be distributed over a network. Depending on the complexity of the system to be managed, one or more BDI agents are used to form the Virtual Carer Agency (VCA). The main goal of VCA is to model logical structures, reasoning and behaviours similar to those of a human caregiver, transforming data from sensor agents in logical predicates representing its knowledge base. To perform actions on the monitored ambient and interact with the assisted person, VCA can collaborate with agents controlling actuators. Thus the system can monitor the health conditions of the assisted person and facilitate his daily activities.

The rest of the paper is organized as follows. Section 2 deals with some related works on MAS and AAL; in Sect. 3 the core of the proposed system is described. Section 4 describes the AgentSpeak implementation of the BDI core of the system, highlighting the reasons behind the proposed simulation scenarios. In the end some conclusions are drawn and future work suggested.

2 Related Works

Several definition of agents can be found in scientific literature. Among these, Wooldridge and Jennings [9] list the main properties of an agent: autonomy, social ability, reactivity and pro-activeness indicate that agents operate without external intervention, interact with other agents using some kind of Agent Communication Language (ACL), perceive the environment answering to changes and are able to autonomously exhibit goal-directed behaviours. BDI agents are rational agents having certain mental attitudes of Belief, Desire and Intention [10]: agent's beliefs are its knowledge about the environment, agent's desires are the goals it should achieve and agent's intentions are the goals it is committed to achieve.

MAS and BDI agents with an appropriate knowledge base can meet the guidelines of AAL and AmI, being capable to autonomously adapt to context changes resulting from user activity, device failures, and the addition or removal of devices and services. Hence, they meet the requirements on modularity and adaptability of AAL systems [11].

MAS applications in the AAL domain are proposed in several scientific contribution. Nefti et al. [12] presents a system to monitor patients suffering from dementia, in which a Risk Assessment Agent, using its knowledge base and information provided by agents responsible for various ambient sensors, applies a method based on fuzzy logic to predict risk situations and to trigger proper alarms. In the system described in [13], an agency provides suitable services analysing user's goals. In [14] the focus is on the interaction with the user rather than on monitoring. The work in [15] is based on Home Automation and describes a small society of BDI agents able to cope with energy efficiency issues when different devices are available.

The evaluation of AAL systems should include several perspectives, as described in [16]: at least the points of view of end users (elderly people and their relatives), medical professionals and developers (software and systems engineers) are essential to assess system quality and effectiveness. These perspectives can be combined using participatory evaluations as proposed in [17].

3 The Virtual Carer Architecture

The Virtual Carer is a MAS modelling a distributed, reliable and modular sensor network composed by biometric and ambient sensors, and integrating a BDI core agent (Virtual Carer Agency, VCA). The Virtual Carer is an IT system able to communicate with an assisted person, to monitor his health conditions and to control the environment around him. To cope with a highly variable environment, and following the BDI paradigm, VCA models reasoning mechanisms and behaviours similar to those of a human being. The large amount of information used by VCA is represented by logical predicates, forming its Knowledge Base (KB). VCA chooses the right actions to perform on the environment with an inference engine applied on its KB. The VCA works as follows:

- it analyses data provided by sensors and devices forming the system (i.e. by agents controlling them) and updates its KB;
- when its KB is updated, VCA generates new knowledge using its inference engine;
- with backward reasoning rules applied on its KB, VCA selects the main goal and chooses (through backtracking) the plan to satisfy it;
- VCA carries out the actions of chosen plan, collaborating with agents responsible for actuators, in order to directly act on the environment.

Figure 1 shows the basic architecture of the system. The system is composed by the BDI Virtual Carer Agent, a Register Agent and a number Actuator Agents and Sensor Agents. The Actuator Agents are responsible for the activation and deactivation of the devices composing the system, as speakers, lights, monitors and so on. They must be able to receive requests from other agents and to execute on/off commands on the devices. The Register Agent is responsible for the database storage of all the information provided by sensors and of the anomalies detected by the VCA.

Sensor Agents can be distinguished in two types: Ambient Agents and Health Agents. Ambient Agents are responsible for reading the value of ambient sensors, for instance the temperature of a specific room. They check if the recorded values are in a predetermined range: if the value is outside of the range, the Ambient Agent communicates the anomaly to the VCA. Presence sensors are managed by Ambient Agents, too. For example, we can consider a Passive Infrared (PIR) sensor: the Ambient Agent controlling it merely informs the VCA about the presence of someone in the monitored room. Health Agents, instead, manage sensors measuring values related to the health conditions of the assisted person. For instance, a Health Agent can control a heartbeat sensor.

The arrows in Fig. 1 highlight the information flow characterizing the system. The VCA receives alarms from Sensor Agents when a value is out of the predetermined

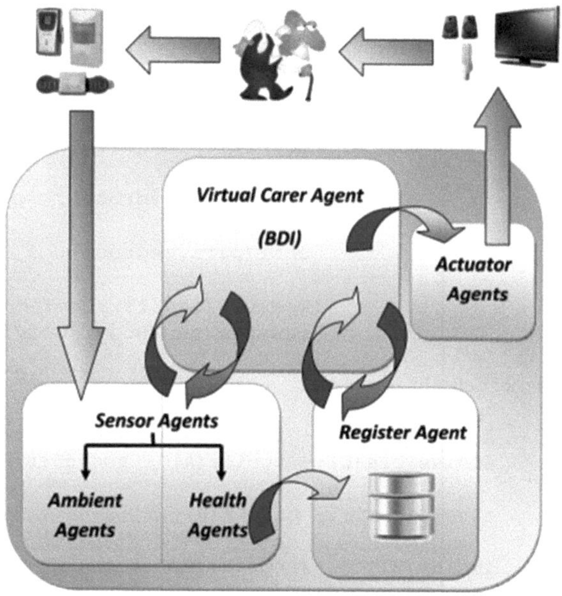

Fig. 1 Virtual Carer basic architecture

range; in addition, the VCA can directly require to read a value, because the chosen plan requires data to satisfy the goal. The Register Agent receives and store in a database values from Sensor Agents and warning and anomalies reports from the VCA. The VCA can require values from the database to the Register Agent. The VCA sends request to the Actuator Agents in order to execute actions corresponding to the adopted plan. For example the VCA can send a request to turn the light on in the living room because the PIR agent detects the presence of the assisted person.

Sensor and Actuator Agents are implemented using the JADE framework [18] whilst the BDI agent representing the VCA and the Register Agent are implemented using JASON and the AgentSpeak language [19]. It is important to notice that the use of JADE framework allows the communication between agents with FIPA-ACL messages, ensuring the modularity of the system: different devices can be added any time simply integrating more agents. JASON is used for the BDI agents composing the VCA, in order to apply the BDI paradigm, and for the Register Agent, allowing to store beliefs directly in the database.

4 Simulation Scenarios

To show the capabilities of the proposed architecture, a preliminary implementation of the Virtual Carer was carried out, simulating its functionalities in a home environment. The main goal of the simulation is to demonstrate how the proposed expert system can generate plans, take decisions and perform actions in order to help

```
partOf(bed,bedroom).
partOf(bedroom,home).

audioDevice(speaker1).

covers(speaker1,bed).

light(all,bedroom).

spir(pir1,bedroom).

measure(sens1T,roomTemp).
measure(sensBodyT,bodyTemp).
```

Fig. 2 Belief base examples

```
admissible(D):- inside(A) & videoDevice(D) & covers(D,A).
admissible(D):- inside(A) & audioDevice(D) & covers(D,A).

covers(D,R1):- partOf(R1,R2) & covers(D,R2).

inside(R1):- partOf(R1,R2) & inside(R2).
```

Fig. 3 An example of the rules in the VCA's knowledge base, indicating when a device is admissible, i.e. available to perform an action

the assisted person in his daily activities. It is important to underline that the Virtual Carer can infer new knowledge and perform actions autonomously, without the direct intervention of the assisted person. However, as one of the goal of AAL (and thus of the Virtual Carer) is to improve the patient's quality of life, the system has to interact with him, being capable of receive requests and commands. For example the patient could query the system about his body temperature or request it to turn a light on. In other words the system has to be *intelligent* both in its autonomy and in the interaction with the assisted.

The system was complemented by a BDI agent, the Elderly Agent, modelling behaviours and activities of the person assisted by the Virtual Carer System. For simulation reasons the Elderly Agent is able to send messages to other agents of the system: as an example of the presence detection in a specific room, the Elderly Agent informs the PIR agent when it moves in the controlled room.

Figure 2 shows some possible facts in the initial belief base of the VCA: the beliefs, easily understandable thanks to the use of the AgentSpeak language, indicate how the environment is structured, available sensors and the areas covered by them and by actuators.

The VCA, in order to perform actions, needs to know information such as the position of the assisted person, the actuators covering that area, the possibility to activate an emergency plans. Thus, beside facts similar to those already listed, the VCA's knowledge base includes logical rules; Fig. 3 contains some of these rules.

Figure 4 shows the AgentSpeak code for one of the emergency plans used in the simulation; it is executed when an Ambient Agent, responsible for a temperature

```
+emergency1(V)[source(A)] : true <-
        .print("Room temperature out of range");
        .print("from sensor ", A);
        .print("value ", V);
        .time(H,M,S);
        .send(register,tell,anomaly(A,V,H,M,S));
        .print("Information sent to Register Agent");
        ?em1(N);
        New = N + 1;
        -+em1(New);
        !notify.
```

Fig. 4 Emergency plan for room temperature: the VCA send to the Register Agent the detected anomaly to store the information in the database, and then informs the elderly

Fig. 5 Plan for the elderly simulating the request for a parameter

```
+!queryP(R)[source(U)] : true <-
        .print("Requiring parameter ");
        .print(R);
        ?measure(Sens,R);
        .send(Sens,askOne,query,Reply);
        .print("The last value is ");
        .print(Reply).
```

sensor, reads a value out of the predetermined range. In this case the VCA communicates the anomaly to the Register Agent and activates the plan "notify" in order to evaluate if the monitored person has to be notified with some alarms. This notification is possible thanks to the rules of Fig. 3: the VCA will be able to send the information of the elderly with the *admissible* devices, i.e. those present in the current position of the assisted. A similar plan is activated when the body temperature of the assisted person is out of a security range.

To simulate the different behaviours of the assisted person, the Elderly Agent was provided with different plans. In a first simulation scenario, we modelled the situation in which the assisted person goes to bed and measures its body temperature (through a plan like the one in Fig. 5) imagining, merely as an example, he is feeling a weak cold. The main reason behind this plan and others similar is that the system has to be ready to receive requests from the assisted person, even at the same moment it is autonomously executing other actions.

The simulation works as follows. At first, the Elderly Agent initializes the VCA Agent, providing the information about its initial position; then, after 5 s (i.e. the time necessary to get to bedroom), it informs the PIR Agent of its presence in the final position. The PIR Agent controlling the bedroom notifies the VCA about the presence of a person in the bedroom; thus VCA has a new belief (i.e. the presence of the assisted person in the bedroom), and activates the plan to turn the light on in the bedroom and to turn it off in the hallway (Fig. 6). Finally, the Elderly Agent sends a message to the VCA to know its body temperature value. VCA activates a plan that includes all the actions necessary to require the data from the proper Health Agent and to communicate it to the Elderly Agent.

Fig. 6 When the VCA has a new belief "enter" sent by a PIR sensor, it updates its belief base with the new position of the elderly and activates the plan "powerOn"

```
@ent[atomic]
+enter[source(A)] : true <-
        ?inside(M);
        ?spir(A,R);
        M \== R;
        -+inside(R);
        !powerOn(R,M);
        .print("light on in ");
        .print(R);
        .print("light off in ");
        .print(M);
        -enter[source(A)]-
```

Beside plans similar to the one just described, we simulated plans without an explicit request by the Elderly Agent, in which the Virtual Carer had to infer the right actions. A case can be when a window remains open in the bedroom and the assisted person is sleeping, so the decreasing of room and body temperature should activate the plan to wake up the person and ask him to close the window.

Of course, the ideal test for the proposed approach should include an AAL scenario with sensors and devices deployed in a real home environment, allowing to fully validate the MAS approach. However, the simulations we carried out are adequate to underline that MAS and BDI paradigm are useful when applied to AAL contexts:

- MAS approach guarantees modularity to the system, in order to cope with the addition or removal of devices, simply wrapping them with agents (the only requirement is that they have to share a common ontology);
- BDI paradigm allows to quickly respond to changes in a dynamic environment, permitting to compose a system able to act in a goal-oriented way.

5 Conclusions

We described a MAS approach to AAL to deal with a dynamic environment, monitoring the health conditions of an elderly or disabled person. The core of the system is a BDI agent representing a Virtual Carer. The Virtual Carer collaborates with other agents modelling various sensors and devices to simplify the daily activities of the assisted person and to trigger alarms when something is wrong with parameters regarding his health conditions.

We tested the system considering several simulation scenarios and representing the assisted person using another BDI agent. Several plans were taken into account, modelling different actions performed by the Elderly Agents (e.g. movements, direct requests, changing of health parameters). Tests highlight that the Virtual Carer can both act autonomously and react to received commands. Tests also confirm that MAS and BDI paradigm improve modularity and adaptability of AAL and AmI systems.

The Multi-Agent approach guarantees the modularity and the distribution of the system, whilst the BDI paradigm permits to react to changes in the environment and in the needs of the assisted person. Thus the Virtual Carer architecture could represent a "sketch" for AmI systems, providing an agent structure to model distributed sensor networks associated to reasoning modules. The carried out simulations highlight that Artificial Intelligence can provide a significant contribution in the development of systems with the aim to improve the quality of life of elderly or disabled people, especially in the proposal of autonomous solutions capable of adapting to dynamic contexts and in the representation of human mental attitudes.

As future work more insightful tests have to be conducted: a real AAL scenario, i.e. a daily-used home environment equipped with devices and sensors, where an assisted person can live, is beyond our possibilities at this stage of the work, but is the only way to fully validate the proposed approach. The next step could be towards the integration of more technologies in the Virtual Carer system: for example those for video surveillance described in [20] could be useful with the purpose of monitoring the conditions of the assisted person, ensuring its security. The integration with Human-Computer-Interaction techniques, especially voice and gesture recognition, will be investigated, trying to improve the communication between the expert system and the assisted person, enabling him to interact in a natural way as he would with a human caregiver. Finally, the Virtual Carer could be integrated in health information systems to simplify the communication with the medical personnel: a multi-agent architecture, as proposed in [21], can simplify the integration of multiple information sources.

References

1. Christensen, K., Doblhammer, G., Rau, R., Vaupel, J.W.: Ageing populations: the challenges ahead. Lancet **374**(9696), 1196–208 (2009)
2. World Health Organization: Global health and ageing. World Health Organization (2011)
3. Kleinberger, T., Becker, M., Ras, E., Holzinger, A., Müller, P.: Ambient intelligence in assisted living: enable elderly people to handle future interfaces. In: Universal access in human-computer interaction. Ambient interaction. Volume 4555 of Lecture Notes in Computer Science. pp. 103–112. Springer, Berlin, (2007)
4. Sun, H., De Florio, V., Gui, N., Blondia, C.: Promises and challenges of ambient assisted living systems. In: Proceedings of the Sixth International Conference on Information Technology: New Generations, 2009, ITNG '09, Las Vegas, pp. 1201–1207. (2009)
5. AAL Europe: Ambient assisted living joint programme. http://www.aal-europe.eu/ (2007). Accessed 3 Apr 2013
6. Sadri, F.: Ambient intelligence: a survey. ACM Comput. Surv.**43**(4), 36:1–36:66 (2011)
7. Cook, D.J., Augusto, J.C., Jakkula, V.R.: Ambient intelligence: technologies, applications, and opportunities. Pervasive Mob. Comput. **5**(4), 277–98 (2009)
8. O'Grady, M.J., Muldoon, C., Dragone, M., Tynan, R., O'Hare, G.M.: Towards evolutionary ambient assisted living systems. J. Ambient Int. Humanized Comput. **1**(1), 15–29 (2010)
9. Wooldridge, M., Jennings, N.R.: Intelligent agents: theory and practice. Knowl. Eng. Rev. **10**(2), 115–52 (1995)
10. Rao, A.S., Georgeff, M.: Bdi agents: from theory to practice. In: Proceedings of the 1st International Conference on Multi-Agent Systems, San Francisco, pp. 312–319. (1995)

11. Ayala, I., Amor, M., Fuentes, L.: Self-configuring agents for ambient assisted living applications. Pers. Ubiquit. Comput. **17**(6), 1159–1169 (2013)
12. Nefti, S., Manzoor, U., Manzoor, S.: Cognitive agent based intelligent warning system to monitor patients suffering from dementia using ambient assisted living. In: Proceedings of the International Conference on Information Society (i-Society), London, pp. 92–97. (2010)
13. Cavone, D., De Carolis, B., Ferilli, S., Novielli, N.: An agent-based approach for adapting the behavior of a smart home environment. In: Proceedings of the 12th Workshop on Objects and Agents, WOA-2011. Volume 741 of CEUR Workshop Proceedings, pp. 105–111. (2011)
14. McNaull, J., Augusto, J., Mulvenna, M., McCullagh, P.: Multi-agent interactions for ambient assisted living. In: Proceedings of the 7th International Conference on Intelligent Environments (IE), Nottingham, pp. 310–313. (2011)
15. Reinisch, C., Kastner, W.: Agent based control in the smart home. In: Proceedings of the 37th Annual Conference on IEEE Industrial Electronics Society, IECON 2011, Melbourne, VIC, pp. 334–339. (2011)
16. Kleinberger, T., Jedlitschka, A., Storf, H., Steinbach-Nordmann, S., Prueckner, S.: An approach to and evaluations of assisted living systems using ambient intelligence for emergency monitoring and prevention. In: Universal Access in Human-Computer Interaction. Intelligent and Ubiquitous Interaction Environments. Volume 5615 of Lecture Notes in Computer Science. Springer, pp. 199–208. (2009)
17. Demiris, G., Oliver, D.P., Dickey, G., Skubic, M., Rantz, M.: Findings from a participatory evaluation of a smart home application for older adults. Technol. Health Care **16**(2), 111–8 (2008)
18. Bellifemine, F.L., Caire, G., Greenwood, D.: Developing Multi-Agent Systems with JADE, vol. 7. Wiley, London (2007)
19. Bordini, R.H., Hübner, J.F., Wooldridge, M.: Programming Multi-Agent Systems in AgentSpeak Using Jason, vol. 8. Wiley, London (2007)
20. Claudi, A., Di Benedetto, F., Dolcini, G., Palazzo, L., Dragoni, A.F.: Marvin: mobile autonomous robot for video surveillance networks. In: Proceedings of the Sixth UKSim/AMSS European Symposium on Computer Modeling and Simulation, EMS 2012, Valetta, pp. 21–26. (2012)
21. Palazzo, L., Rossi, M., Dragoni, A.F., Claudi, A., Dolcini, G., Sernani, P.: A Multi-Agent Architecture for Health Information Systems. In: Advanced Methods and Technologies for Agent and Multi-Agent Systems. Volume 252 of Frontiers in Artificial Intelligence and Applications, pp. 375–384. IOSPress (2013)

Part III
Elderly People Monitoring

Expert System for Wearable Fall Detector

Gabriele Rescio, Alessandro Leone and Pietro Siciliano

Abstract Falling down can cause moderate to severe injuries, increasing the risk of death among elderly. For this reason there is a substantial growth of Ambient Assisted Living technologies, including smart environments, in order to support elderly and fragile people in potentially dangerous situations. The paper describes an expert system based on a wireless wearable low-cost accelerometer able to automatically detect falls, generalizing the detection of critical events in several practical conditions. The algorithmic scheme appears invariant to age, weight, height of people and relative positioning area (even in the upper part of the waist), resulting compliant with many commercial wearable devices. Experimental results show high generalization properties and better performances than well-known threshold-based approaches.

1 Introduction

The main reason for the development of the presented system is to allow partially self-sufficient people to live safely in their own homes as long as possible. The problem of falls in the elderly has become a health care priority due to the related high social and economic costs [9]. Many solutions have been proposed in detection and prevention of falls and some excellent reviews have been presented [9, 10, 13]. The availability of miniaturized low-cost MEMS accelerometers on the market and the new reliable wireless communication technologies allow the realization of affordable wearable systems useful for daily activities monitoring [4]. However, this kind of technology

G. Rescio (✉) · A. Leone · P. Siciliano
CNR – Institute for Microelectronics and Microsystems, Via Monteroni, 73100 Lecce, Italy
e-mail: gabriele.rescio@le.imm.cnr.it

A. Leone
e-mail: alessandro.leone@le.imm.cnr.it

P. Siciliano
e-mail: pietro.siciliano@le.imm.cnr.it

S. Longhi et al. (eds.), *Ambient Assisted Living*, DOI: 10.1007/978-3-319-01119-6_10,

presents some drawbacks: they are prone to be forgotten, worn in a wrong body position or accidentally damaged. Regarding fall detection, an accelerometer-based solution presents some advantages respect to vision or acoustic sensors: just one sensor needs to be used, re-design of the environments is not required and ethical issues (such as privacy) are always satisfied. On the other hand, camera-based or acoustic-based approaches are non-invasive [6] since contactless solutions. In this paper a fall detector through a wearable tri-axial MEMS accelerometer is presented. The proposed solution overcomes the limitation of well-known threshold-based approaches [1] in which several parameters need to be manually estimated according to the specific features of the end-user. In particular, a machine learning scheme has been used and high generalization capabilities in the fall detection discrimination process have been recovered. The expert system uses robust features extracted taking into account important constraints and/or requirements of mobile solutions (workload). The extracted features are (quasi-) invariant both to specific characteristics of the mounting setup (device on chest, on waist, on abdomen) and specific characteristics of the end users in terms of age, weight, height and gender.

2 Framework Overview and Materials

The main computational steps of the software architecture are data acquisition, noise filtering, data pre-processing, system calibration, feature extraction and classification. Each step is described in the following sections. The capability of the system to detects a fall is evaluated by using a state-of-the art supervised classifier, according to the procedure described in [5]. The algorithmic framework has been developed by using the wearable device [12] composed by commercial discrete circuits.

The system integrates a ST LIS3LV02DL tri-axial MEMS accelerometer with digital output, an FPGA for computing functionalities and a Zigbee module for wireless communication up to 30 m, suitable for indoor contexts. The power consumption is about 190 mW in streaming mode and 9 mW in idle. The wearable device can operate in streaming mode (raw data are sent via ZigBee to an external computing platform for data analysis with a 10 Hz frequency) or in standalone mode. In the last release of the device a threshold-based fall detector has been integrated on the on-board FPGA and the implementation has been used as comparison item for the evaluation of the proposed scheme. Raw data are in hexadecimal 16-bit and represent the acceleration values with full scale in the range ± 2 g for higher sensitivity. The accelerometer is DC coupled (it responds up 0 Hz) and it measures both static and dynamic acceleration along the 3 axes, describing the 3D spatial relative position of the person who wears it. Generally speaking, the accelerometer measures the projection of the gravity vector on each sensing axis. Assuming a particular axis, the component of the acceleration (amplitude A, Eq. 1) is defined according to the value of the sin of the angle α between the considered axis and the horizontal plane, which is perpendicular to the Earth's gravity component (g):

$$A = g\sin(\alpha) \quad (1)$$

In this way, if the accelerometer relative orientation is known, the resulting data can be used to determinate angle of the user posture respect to the vertical direction.

3 Self-Calibration System

The acceleration data on three axes (A_x, A_y, A_z) were read out from the device worn by a user during the data collection. Data were stored and converted from hexadecimal to decimal format to be compliant the implemented software. A normalization step has been used in order to represent acceleration data in the range ± 2 g. The samples coming from the device are filtered out by a low pass 8-order, 8 Hz cut-off FIR (Finite Impulse Response) filter to reduce the noise due to electronic components, environment and human tremor. In order to handle correctly pre-processed data, a calibration procedure has been accomplished by recovering the initial conditions after the device mounting. During the step, the correct placing of hardware is verified by checking if two acceleration axes are orthogonal to Earth's gravity g (Fig. 1): the acceleration values measured on the two orthogonal components must be close to zero. The calibration procedure is achieved by the following steps:

1. The user wears the device in a still standing position for 10 s, as shown in Fig. 1.
2. The calibration routine calculates the average of the acceleration (A_{x0}, A_{y0}, A_{z0}) on each axis over this period.
3. If (A_{x0}, A_{y0}, A_{z0}) are close to those expected (-1, 0, 0 according to Eq. 1), the calibration routine finished and (A_{x0}, A_{y0}, A_{z0}) is the reference in the features extraction.

The calibration procedure is concluded if the initial measured values do not differ more than 30 % from the expected ones and the values A_{x0}, A_{y0} and A_{z0} will be recorded and used as references in the feature extraction phase. Otherwise, a routine to compensate the sensed misplacement is enabled and the angles displacements of the sensor axes (αA_{x0}, αA_{y0} and αA_{z0}) are calculated using the following trigonometric Eq. 2:

$$\alpha A_{x0} = arctan\left(\frac{A_{x0}}{\sqrt{(A_{y0})^2 + (A_{z0})^2}}\right), \quad \alpha A_{x0} = arctan\left(\frac{A_{y0}}{\sqrt{(A_{x0})^2 + (A_{z0})^2}}\right),$$
$$\alpha A_{x0} = arctan\left(\frac{A_{z0}}{\sqrt{(A_{x0})^2 + (A_{y0})^2}}\right) \tag{2}$$

These values are stored and will be used for the correction of the misplacement during the feature extraction phase.

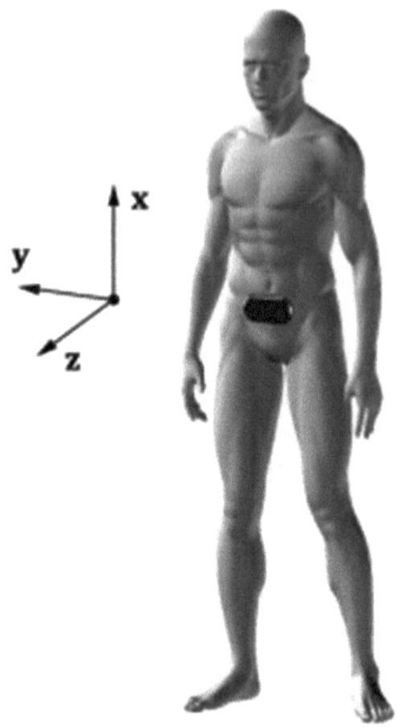

Fig. 1 Mounting position and calibration

4 Robust Feature Extraction

Robust features are extracted in the time domain by considering both quick and relevant acceleration changing along each axis (due to the fall) in a 5 s sliding window and the posture changing registered after the shock. The aim is to produce robust features, taking all the information able to discriminate falls from other events. It is also important that such features have a low dependence on both the position of the sensor (whether it is placed on the waist, on the chest or on the abdomen) and the human body characteristics of the user. For the features extraction process, both critical and post fall phases are of interest and corresponding two kind of features are extracted. In the former, the shock is measured due to the impact toward the floor plane and a dynamic acceleration changing is registered. In the latter (the body is already lying) the static acceleration value records a great change due to the new position of the individual with respect to the calibration phase. The posture changing with respect to the initial condition (A_{x0}, A_{y0}, A_{z0}) can be realized through the static information of acceleration which varies with the inclination of the accelerometer sensing axis. Hence the difference between the value of the 3D-static acceleration after the fall and the one stored in the calibration phase will result in an offset, called Changing Position Offset (CPO), which is proportional to the user displacement. In this way, a study of posture was not made: only the relative varying posture analysis was considered, causing a computational cost reduction and mounting setup invariance.

If the device is not worn correctly, the routine to compensate the device misplacement is enabled (see Self-calibration system section). In this case, for the CPO

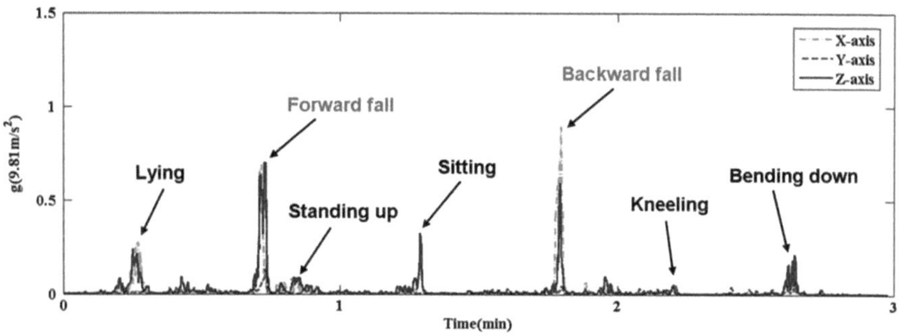

Fig. 2 Extracted features for simulated falls and ADLs (the device is on the waist)

calculation, the axes angles of the sensor in the initial condition (αA_{x0}, αA_{y0} and αA_{z0}) need to be taken into account. The CPO value for x-axis is obtained as:

$$CPO = \left|\sin\left(\alpha\overline{A_x} - \alpha A_{x0}\right)\right|, \tag{3}$$

where $\alpha\bar{A}_x$ is the X-axis angle of the sensor during the post-fall phase; it is calculated considering the static averaged acceleration $\bar{A}$ and the same trigonometric equation used in (2):

$$\alpha\overline{A_x} = arctan\left(\frac{\overline{A_x}}{\sqrt{\left(\overline{A_y}\right)^2 + \left(\overline{A_z}\right)^2}}\right), \tag{4}$$

The same procedure must be done for the two other axes. Thanks to this routine the problems of device positioning are strongly reduced at the expense of an increase in the computational costs.

The feature vector is made up by three parameters, one for each acceleration component. It makes sense to consider the acceleration signal on each axis singularly, because a fall event leads to a change in the value of the static acceleration in at least two of the three acceleration axes (due to the orientation change of sensing axes). In Figs. 2 and 3 the features of a sequence of falls and daily events, when the device is worn on the waist and on the chest, are shown. So it is evident that the features obtained discriminate the falls from typical Activities of Daily Living (ADLs) also when the device is placed in other area of the torso.

5 Supervised Classification and Experimental Results

Once features are extracted, the fall events are detected by a One Class Support Vector Machine (OC-SVM) which is less computationally intensive than other algorithms like neural networks [7]. SVM is a robust classification tool (in presence of outliers

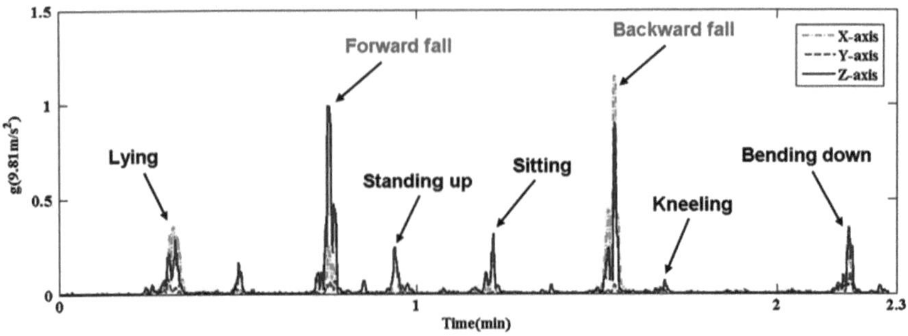

Fig. 3 Extracted features for simulated falls and ADLs (the device is on the chest)

too) with a good generalization ability. OC-SVM divides all samples into objective field and non-objective field and then non linearly maps those sample into high dimensional features space with some efficient functions, called kernel function. The commonly types of kernel found in literature have been tested. They were analyzed using Matlab tool STPRtool [2], changing the values of key parameters as described in [11]. For the extracted features, the optimal kernel are Gaussian Radial Basis Function (GRBF) [14] and polynomial, as better detailed in the following.

The OC-SVM classifier has been trained by using about 40 falls and 50 ADLs belonging to a large dataset in which more than 250 falls and 200 ADLs were performed compliant the specifics proposed in [8]. The remaining 210 falls and 150 ADLs have been used for testing. To validate the implemented algorithmic framework, the previously described dataset has been considered. The performances of the system have been evaluated considering two normally used metrics, sensitivity and specificity [8], respectively. The algorithm is tested when the device is placed on waist, abdomen or chest. GBRF and polynomial kernel functions give the best results in terms of performance, even the polynomial kernel shows a lower computational cost (relative number of support vectors and relative execution time are considered). In particular, GRBF (with $\sigma = 3$ and $C = 2$) and the Polynomial (with $P = 3$ and $C = 2.8$) provide similar values of specificity and sensitivity but the last one works faster and its number of vectors is slightly lower. Misclassifications are for falls presenting slow dynamics or falls with partial recovery. The implemented SVM improves in the specificity and sensitivity respect to the threshold-based approaches detailed in [1, 3], as reported in Table 1. As already discuss, the two fall detection systems have in common hardware, benchmark dataset and training/test sets. Of course, the computational cost of threshold-based is lower than the implemented expert systems, but it suffers in detection rate due to high false positive and true negative. Since the number of features is compact and the computational cost of the extracted features is low, the overall system workload is compatible with an integration in embedded low-power solutions (DSP, FPGA, microcontroller).

Table 1 Comparison of the proposed OC-SVM method and threshold-based algorithm detailed in [3]

	Sensitivity (%)	Specificity (%)	Relative execution time
OC-SVM (Polynomial C = 2.8, P = 3)	97.7	94.8	1x
OC-SVM (GRBF C = 32, $\gamma = 3$)	97.4	95.2	1.5x
Threshold-based	89.5	85.7	0.3x

6 Conclusions

The proposed supervised scheme overcomes the limitation of well-known threshold-based approaches in which a heuristic choice of the parameters is accomplished. A specific study on postures was not made in order to make a low computational power system. A new inferred information has been computed (Changing Position Offset) in order to intrinsically acquire relevant posture changing with possible affection on fall events. The calibration step guarantees the generalization of the approach in terms of invariance to physics characteristic of the end-users, during the fall detection process. High performances in controlled conditions (simulated events) in terms of sensitivity and specificity were obtained using only the 20 % of dataset for training purposes. Performance metrics of different kernels in One Class SVM have been compared and the best results are obtained with polynomial function and Gaussian Radial Basis Function, even the polynomial kernel presents a limited workload. Future works are addressed both to validate the solution in real conditions, to test the methodology with a large set of different MEMS accelerometers and to port the framework on embedded mobile solutions.

References

1. Bagalà, F., Becker, C., Cappello, A., Chiari, L., Aminian, K., Hausdorff, J.M., Zijlstra, W., Klenk, J.: Evaluation of accelerometer-based fall detection algorithms on real-world falls. PLoS One **7**, e37062 (2012)
2. Franc, J., Hlavac, V.: Statistical Pattern Recognition Toolbox for Matlab. http://cmp.felk.cvut.cz/cmp/software/stprtool/ (2004)
3. Grassi, M., Lombardi, A., Rescio, G., Malcovati, P., Malfatti, M., Gonzo, L., Leone, A., Diraco, G., Distante, C., Siciliano, P., Libal, V., Huang, J., Potamianos, G.: A hardware-software framework for high-reliability people fall detection. In: Proceeings of IEEE Sensors 2008, pp. 1328–1331 (2008)
4. Karantonis, D., Narayanan, M.R., Mathie, M., Lovell, N.H., Celler, B.G.: Implementation of a real-time human movement classifier using a triaxial accelerometer for ambulatory monitoring. IEEE Trans. Inf. Technol. Biomed. **10**(1), 156–167 (2006)
5. Liu, S.H., Cheng, W.C.: Fall detection with the support vector machine during scripted and continuous unscripted activities. Sensor **12**, 12301–12316 (2012)
6. Leone, A., Dirago, G., Siciliano, P.: Detecting falls with 3D range camera in ambient assisted living applications: a preliminary study. Med. Eng. Phys. **33**, 770–781 (2011)

7. Manevitz, L.M., Yousef, M.: One-class SVMs for document classification. J. Mach. Learn. Res. **2**(1), 139–154 (2001)
8. Noury, N., Fleury, A., Rumeau, P., Bourke, A.K., Laighin, G.O., Rialle, V., Lundy, J.E.: Fall detection–principles and methods. In: Proceedings of the 29th IEEE EMBS, pp. 1663–1666 (2007)
9. Noury, N., Rumeau, P., Bourke, A.K., Laighin, G.O., Lundy, J.E.: A proposal for the classification and evaluation of fall detectors. IRBM **29**(6), 340–349 (2008)
10. Sadigh, S., Reimers, A., Andersson, R., Laflamme, L.: Falls and fall-related injuries among the elderly: a survey of residential-care facilities in a Swedish municipality. J. Community Health **29**, 129–140 (2004)
11. Sangeetha, R., Kalpana, B.: Performance evaluation of kernels in multiclass support vector machines. Int. J. Soft Comput. Eng. **1**(5), 2231–2307 (2011)
12. Siciliano, P., Leone, A., Diraco, G., Distante, C., Malfatti, M., Gonzo, L., Grassi, M., Lombardi, A., Rescio, G., Malcovati, P.: A networked multisensor system for ambient assisted living application. In: 3rd International, Workshop on Advances in Sensors and Interfaces, IWASI 2009, pp. 139–143 (2009)
13. Yu, X.: Approaches and principles of fall detection for elderly and patient. In: Proceedings of the 10th IEEE HealthCom, pp. 42–47 (2008)
14. Zhang, T., Wang, J., Xu, L., Liu, P.: Fall Detection by Wearable Sensor and One-Class SVM Algorithm. Lecture Notes in Control and Information Science, vol. 345, pp. 858–863 (2006)

Quality of Kinect Depth Information for Passive Posture Monitoring

Enea Cippitelli, Samuele Gasparrini, Ennio Gambi and Susanna Spinsante

Abstract The availability of a low cost device incorporating RGB and depth sensors, such as the Microsoft Kinect device, has enabled a plethora of applications and solutions aiming at automatic monitoring of gait and posture, in order to assess the subject's behavior and possibly prevent future health problems. This paper discusses the quality of the information provided by the depth sensor the Kinect device is equipped with, in order to assess the precision of such information, in different operational contexts, and properly identify the performance that may be reasonably expected from the adoption of such device, in the field of passive monitoring of gait and posture. Some techniques borrowed from the image processing field are also suggested, in order to improve the depth information in specific conditions, by means of a low complexity post-processing of the raw depth data.

1 Introduction

The availability of a relatively low cost device featuring both an image and a depth sensor, such as the Microsoft Kinect sensor device, has enabled a plethora of applications aiming at adopting such a device for posture or gait monitoring and evaluation.

E. Cippitelli · S. Gasparrini · E. Gambi · S. Spinsante (✉)
Università Politecnica delle Marche, via Brecce Bianche 12, I-60131 Ancona, Italy
e-mail: s.spinsante@univpm.it

E. Cippitelli
e-mail: e.cippitelli@univpm.it

S. Gasparrini
e-mail: s.gasparrini@univpm.it

E. Gambi
e-mail: e.gambi@univpm.it

S. Longhi et al. (eds.), *Ambient Assisted Living*, DOI: 10.1007/978-3-319-01119-6_11,

Several studies have shown the importance of long term monitoring and measuring of a person's gait and posture, in order to implement automatic solutions for gait assessment and falls prevention [1, 2]. As long as a passive monitoring system may be used, the results obtainable are considered more reliable and adherent to the real conditions of the subject observed [3, 9]. As a matter of fact, a noninvasive monitoring system does not affect the natural and genuine subject's behavior and movements, with respect to a system adopting wearable sensors or accelerometer based devices. In the framework of Ambient Assisted Living (AAL) technologies, this could lead to the availability of datasets related to long-term observations of subject's behaviors, useful for different kinds of analyses and evaluations. Further, a device that integrates depth and camera sensors can be even adopted as a low-cost alternative to more complex systems, for real-time motion capture and body tracking in health applications, such as physical exercise of elderly population [5]. In [7], Stone et al. discuss the initial results provided by a gait monitoring system based on the Kinect sensor, in the real life scenario of an independent living facility for older adults. The system is able to monitor the subjects, and to identify walking sequences, so that habitual and in-home gait parameter estimates may be automatically generated. In the proposed study, the focus is not on the precise and exact evaluation of the movements performed by the monitored subjects, but on the proper identification of the types of movements. Other experiments, such as the ones discussed by Ren et al. in [6], and by Wan et al. in [8], highlight the need of estimating the precision of the Kinect sensor in providing depth information, as that specific information is the one used to enable Human Computer Interaction (HCI), and gesture recognition, that is preparatory for issuing proper commands to a given system. The geometric accuracy of the depth data provided by Kinect is also analyzed by Khoshelham [4], with a specific reference to the field of mapping and 3D modelling. The author presents a theoretical model of the error affecting the depth estimation, which exhibits a quadratic trend with respect to the distance between the sensor and the target. The same trend is confirmed by the experimental evaluation presented in this paper, in which the focus of the study is specifically devoted to indoor environments that may host AAL systems. As a matter of fact, continuous, ongoing assessments of physical function would help older adults live more safely in independent settings, while also facilitating targeted medical interventions when needed. The Kinect sensor, as a passive and noninvasive means to monitor the behaviors of a subject in his living environment, may consequently have a relevant role in enabling truly effective AAL solutions, for real life adoption.

The paper is organized as follows: Sect. 2 describes the setup installation, and the Kinect performance in gathering depth information, with respect to relative distance and direction between the target and the sensor. Depth resolution and edge detection capability are discussed in Sect. 3, whereas Sect. 4 suggests a possible low complexity post processing to improve the edge information obtained from depth data collection. Finally, Sect. 5 concludes the paper.

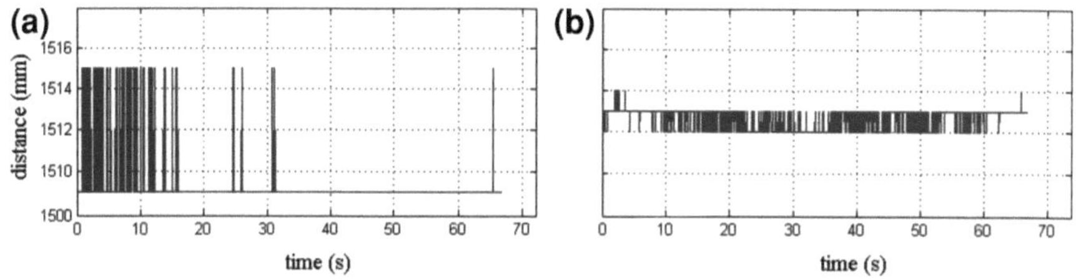

Fig. 1 Transient behavior of the sensor in depth information acquisition: **a** single pixel, **b** area of 400 pixels (average), at a distance of 1,500 mm

2 Experimental Setup and Preliminary Evaluations

The performance provided by Kinect in gathering depth information are analyzed, with respect to two main parameters that may affect the acquisition process, i.e. relative distance, and direction between the sensor and the target. Situations typical of videosurveillance applications, in which the sensor is usually located in elevated position with respect to the ground (on the ceiling), with a quite relevant tilting angle, are not considered in this paper, and should deserve specific evaluation.

The device used is a Kinect for Xbox sensor, operated through the Microsoft v1.5 SDK. In order to ensure that data variations may be ascribed to the sensor itself, and not to the target (for example, due to uncontrolled movements), tests are run by using a fixed object as a target: a wood block of dimensions $39.6 \times 35.1 \times 1$ cm. Such a block is located at a distance of 1,000, 1,500, 2,500, and 3,500 mm from the sensor, at three possible directions ($0°$, $\pm 20°$). The Kinect sensor is elevated at 77.8 cm from the ground, whereas the target object is at an height of 87.4 cm. At each test, 25 depth frames are collected; the sensor works at a rate of 30 frames per second, with a frame format of 640×480 pixels.

As a preliminary evaluation, in order to correctly perform the acquisition process, we are interested in recognizing possible transient effects due to the sensor. To this aim, 2,000 depth frames of dimensions 320×240 are collected at 30 fps, starting from the time instant at which the sensor is switched on. Depth information is acquired, referred to a single pixel, and to a set of 400 pixels located in the central area of the target object. Figure 1a, b show the quite different transient behavior of the depth information referred to a single pixel, and to pixels in the central area of the target object, respectively. When a single pixel is considered, the depth information may vary in a range of $6 \div 7$ mm, whereas, if the average behavior of a set of pixels is analyzed, the variation range amounts to 2 mm only. The depth value provided by the Kinect sensor for a single pixel may vary only by integer multiple of the resolution range; when a 20×20 pixels area is considered, the average depth value is less affected by time variability, and the amount of depth variation is reduced. In fact, adjacent pixels of the same area do not exhibit correlated depth variations over different frames, and the average operation reduces the impact of single pixel depth variability. Transient variations in the depth information are still present after 60 s

Table 1 Estimation of depth variability range (Δ) with respect to relative distance (d) and direction (φ) between the target and the sensor

Distance d (mm)	$\varphi = -20°$ (mm)	$\varphi = 0°$ (mm)	$\varphi = +20°$ (mm)
1,000	$\Delta = 12$ [1004, 1007, 1010, 1013, 1016]	$\Delta = 12$ [995, 998, 1001, 1004, 1007]	$\Delta = 12$ [1013, 1016, 1019, 1022, 1025]
1,500	$\Delta = 27$ [1502, 1509, 1515, 1522, 1529]	$\Delta = 20$ [1502, 1509, 1515, 1522]	$\Delta = 27$ [1515, 1522, 1529, 1536, 1542]
2,500	$\Delta = 75$ [2505, 2523, 2542, 2560, 2580]	$\Delta = 55$ [2505, 2523, 2542, 2560]	$\Delta = 73$ [2487, 2505, 2523, 2542, 2560]
3,500	$\Delta = 147$ [3493, 3528, 3565, 3602, 3640]	$\Delta = 147$ [3493, 3528, 3565, 3602, 3640]	$\Delta = 147$ [3493, 3528, 3565, 3602, 3640]

from the sensor switch on time instant, but they are negligible in the case of a whole pixel area. In any case, in order to avoid additional noisy effects on the data acquired, due to the intrinsic features of the sensor, depth data will be processed following an initial time interval of 60 s acquisition.

In a second evaluation step, the performance of the Kinect sensor in depth estimation are analyzed, with respect to the relative distance and direction between the sensor and the target object. Table 1 shows how the range of depth values assigned by the sensor to pixels belonging to the same area of the target object changes, according to the distance at which the target is located, and its direction with respect to the sensor. Ideally, all the pixels belonging to the same area of the target object should be recognized as located at the same depth from the sensor; Table 1 shows that, on the contrary, the sensor performance are affected by both distance and direction. At the shortest distance of 1,000 mm, the depth information is provided with the highest precision: the variability range is only 12 mm. This value differs from the one obtained by analyzing the behavior of the sensor on a single pixel. This is due to the fact that each pixel in the area 20×20 exhibits uncorrelated depth variations with respect to adjacent pixels, and the range of variation is evaluated over all of them. Looking at Table 1 for a distance of 1,500 mm, similarly to the results shown for a single pixel, the discrepancy between adjacent depth samples amounts to $6 \div 7$ mm, but, within the 20×20 pixels area, up to 5 different depth values are provided by the sensor, for a total variation range equal to 27 mm. When looking at a single pixel, such a behavior cannot be evidenced, as its depth may take only up to two different values, differing by an amount of 7 mm. In the test setup adopted, the target object is located on a cart, in order to have an acceptable ground elevation. The depth image captured by Kinect shows a noisy contribution in the bottom area, near to the floor, where light reflections are able to reduce the sensor precision. In any case, the target object is properly identified, at any possible direction ($\varphi = 0°, +20°, -20°$), and its edges clearly defined. As long as the relative distance between the sensor and the target object increases, the precision of edge detection decreases, and pixels belonging to the edges of the object exhibit a greater variance around an average value. When the relative distance between the sensor and the target object increases,

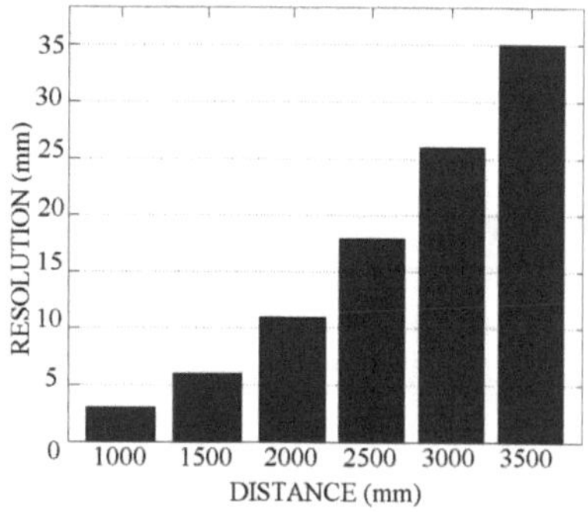

Fig. 2 Depth resolution provided by Kinect for different values of its distance from the target object

up to 3,500 mm, it is clear that the precision in the definition of the object edges is reduced and the variance of the depth information grows. The behavior described by the tests discussed above leads to the definition of specific requirements, according to which a range of acceptable distances and directions between the sensor and the target is defined, in order to ensure the correct performance in depth estimation. The suitability of the Kinect sensor for a specific posture monitoring application shall be taken into account in order to ensure the expected performance in terms of precise estimation.

3 Depth Resolution and Edge Detection Capability

A second issue that is interesting to evaluate deals with the depth resolution supported by Kinect. In this context, the term resolution is intended as the shortest depth range the Kinect is able to detect.

3.1 Depth Resolution in the Case of Generic Target Object

The same tests discussed in Sect. 2 are considered, and depth frames analyzed individually. A central area in the target object is selected, of 40×40 pixels; the pixels are ordered in an increasing fashion, and their multiple occurrences rejected. By this way, the smallest gap between two consecutive values represents the depth resolution provided by the sensor. As shown in Fig. 2, when the distance between the sensor and the target object increases, the depth resolution grows: it goes from 3 @ 1 m from the sensor, to 35 @ 3.5 m distance from the sensor, thus featuring a reduced precision in depth estimation. Tests performed by varying the relative direction between the sensor and the target object have shown that this parameter does not affect the depth resolution supported by Kinect, that does not vary according with φ.

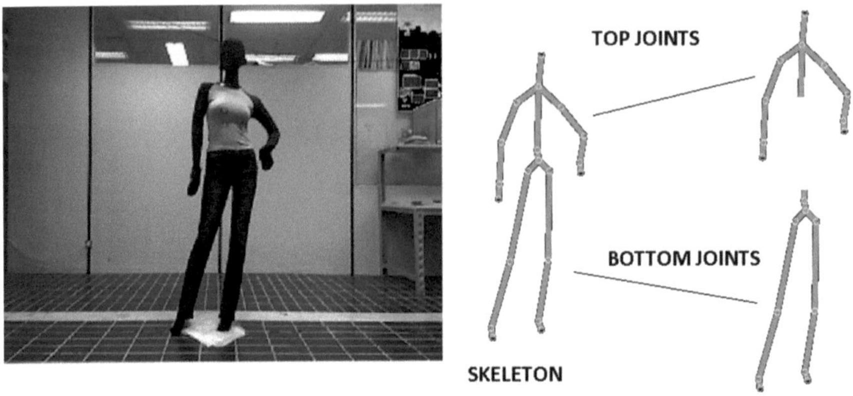

Fig. 3 Kinect joints definition and classification (dummy case)

3.2 Depth Resolution in the Case of Human Subject

The evaluation of the depth resolution provided by Kinect is important when dealing with human subjects, that feature more complex shapes to be observed by the sensor. In applications adopting the Kinect sensor for posture and gait monitoring, the depth information detected on human shapes is important in order to understand the subject's conditions (if standing, or sitting, or fallen), and even to evaluate the accuracy of movements and gestures performed for rehabilitation purposes. When focusing a human shape, the Kinect is capable to extract so called joints, i.e. specific points in the shape corresponding to so called *repere points*, such as: left and right hand, wrist, elbow, shoulder, hip, knee, ankle, foot, and so on. For easiness of discussion, we consider two sets, the top joints and the bottom joints, as depicted in Fig. 3, for the case of a dummy.

To detect all the joints in a human shape, it is necessary to ensure a distance from the sensor in a specific range; two distances are considered, i.e. 2.5 and 3.5 m. Depth variations are estimated for either a dummy and a real human subject focused by the sensor. The idea is to highlight possible variations in depth estimation that are due to the unavoidable and accidental body movements featured by a human being, apparently standing fixed in front of the sensor.

Figure 4a, b show the variance in depth estimation for the top and bottom joints of a dummy and a human being, at a distance of 2.5, and 3.5 m from the sensor, respectively. Irrespective of the distance at which the target is located, the depth information of the joints extracted by Kinect from the human target features greater variance than the corresponding one provided by the dummy. It is possible to ascribe such a behavior to the accidental but unavoidable small movements performed by a human being, even when it is apparently standing fixed in front of the sensor. In fact, the top joints of the dummy have an almost null variance, thanks to the truly static and fixed position of the dummy in front of the sensor. Moving from a distance

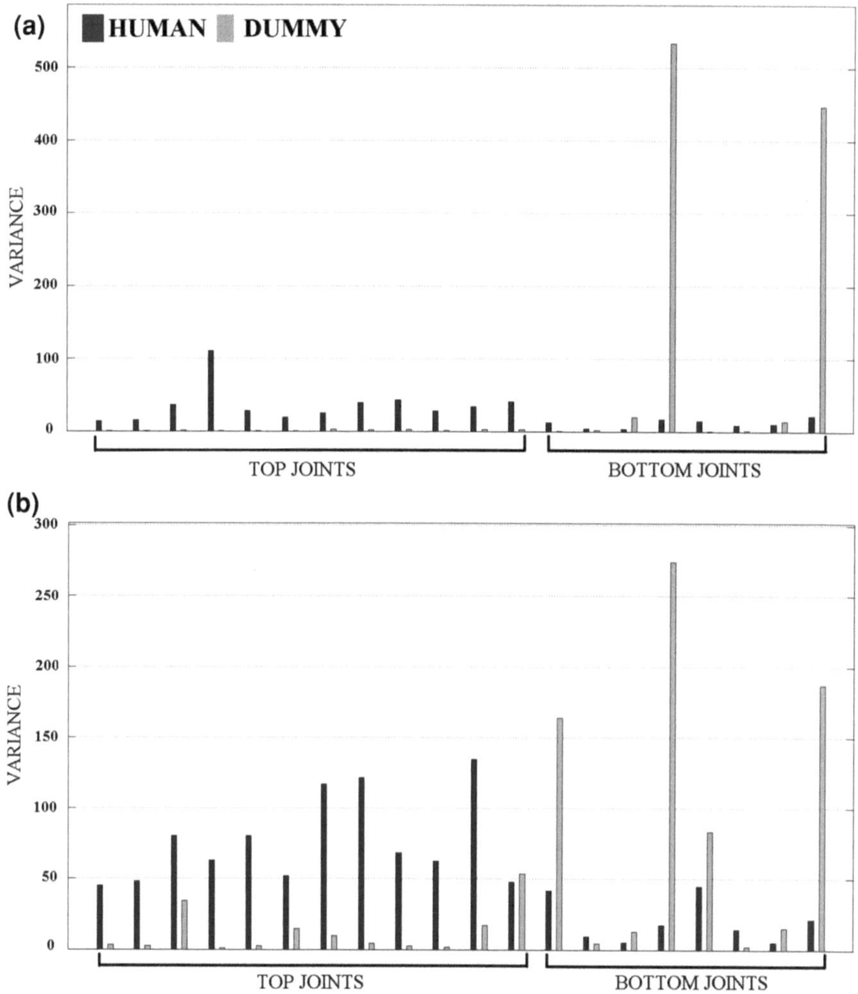

Fig. 4 Variance (in mm^2) in depth estimation for: **a** $d = 2.5$ m, and **b** $d = 3.5$ m, in the case of human subject and dummy (please note that different scales are used in the graphs)

of 2.5–3.5 m, the number of noisy joints increases; especially the top joints get more affected by uncertainty on their real depth value. Bottom joints are basically affected by disturbances in depth estimation due to light reflections from the ground, given the test setup depicted in Fig. 3. Bottom joints in the dummy, expecially those corresponding to the feet, exhibit greater noise effects than in the human subject: looking at Fig. 5, it is clearly shown that the Kinect sensor is not able to correctly understand the position of the dummy feet, which is quite unnatural. In fact, the coordinates of the feet joints provided by Kinect show great variability during the acquisition process, as the sensor is not able to match the position of the dummy feet to any of its embedded body models.

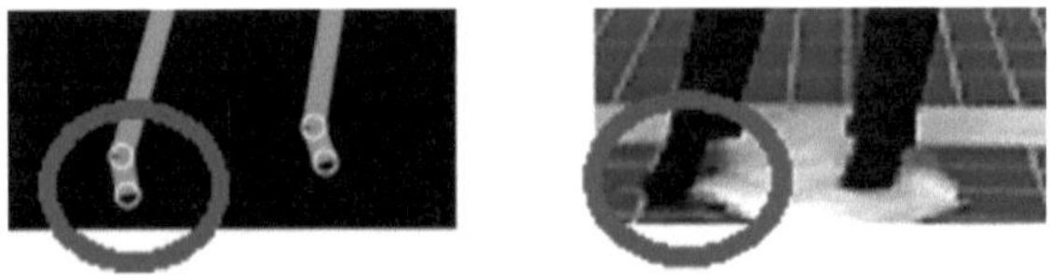

Fig. 5 Detail of feet joints location by Kinect, dummy case

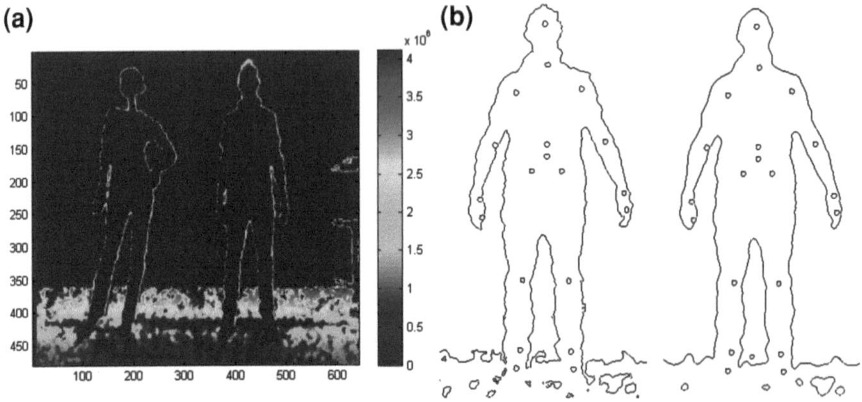

Fig. 6 **a** Depth map referred to two subjects focused by Kinect (*left* dummy, *right* human subject), **b** Edge detection on original depth frame (*left*), and median filtered depth frame (*right*), for the case of a human subject at a distance of 2.5 m

3.3 Depth Resolution in the Presence of Two Targets

The performance provided by Kinect in depth resolution are affected by the number of subjects focused by the sensor, at the same time. To check this effect, we tested the depth resolution capabilities of the sensor when the dummy and a real human subject are monitored by the device, at the same time.

By comparing the depth evaluation results in the presence of two subjects, to the previous tests executed at a distance of 2.5 and 3.5 m, it is possible to evidence a decrease in performance, particularly in the case of the dummy. As a matter of fact, the dummy exhibits an unnatural posture (see the position of the feet), that is hardly understood by the sensor, thus providing a less precise estimation of the joints position. The depth map shown in Fig. 6a, referred to both the subjects (dummy and human), has been obtained by computing, pixel by pixel, the variance of their depth value over 25 consecutive frames. The numerical values corresponding to the plot herein provided show a greater amount of noise in the dummy shape, with respect to the human one, and confirm the strongest impact of disturbances on the bottom joints.

4 Edge Detection Based on Depth Information

The variance in the depth estimation featured by Kinect when dealing with a human subject, as discussed in Sect. 3.2, affects the correct detection of the edges in the human shape. On the contrary, an accurate edge detection can be important when the goal of an application is to monitor and check the precision of the movements performed by the subject, e.g. during a rehabilitation session remotely guided by a health operator. In the case of the generic object (not human), the variations in the edge estimation due to different distances and relative directions between the sensor and the target are limited to a few pixels, and do not affect the sensor capability of correctly locating the shape edges on the depth map.

Depth frames are generated on board the Kinect sensor, and it is not possible to act on the sensor itself in order to improve the quality and precision of the information generated. However, it is possible to apply post-processing algorithms to improve, for example, the precision of the edges extracted from the depth maps. Edge detection and extraction play an important role in any application devoted to posture and gait analysis, as the subject segmentation allowed through edge detection may dramatically decrease the amount of data that are to be analyzed in order to identify and track the subject's movements. It has been shown above that as long as the distance between the subject and the sensor increases, pixels located along the target edges are affected by greater variance. The same pixels are also used to perform edge detection; as a consequence, we aim at applying post-processing algorithms to the depth frames provided by Kinect, in order to improve the obtainable performance in edge detection.

A first simple algorithm that is applied to improve edge detection is a median filter, i.e. a filter of dimension $N \times N$ pixels, that is ideally shifted over the depth frame, to replace the value of the central pixel with the median value of the neighboring ones. Once the median filter has been applied, a Sobel operator on the filtered depth image provides the edge detection results. Fig. 6b shows the different outcomes of the edge detection step in the case of the original and the median filtered depth frame, for a human subject. The median filter acts on a 7×7 pixels basis; the improved edge detection is quite evident, and confirmed at a distance of 2.5 and 3.5 m.

5 Conclusion

This paper presented a set of preliminary results about the evaluation of the precision and resolution features of the depth information provided by the Kinect sensor, in different operational conditions. Tests have been performed by considering either a generic static object, a dummy, and a human subject, in order to evidence possible differences in the behavior of the sensor, with respect to the focused target. The analyses performed allow to say that after a few seconds from the switch on, the Kinect provides quite stable data about the target location; the depth resolution featured by the sensor decreases when the relative distance from the target increases. Accidental and unavoidable movements by a human subject, even when standing apparently fixed in front of the sensor, reduce the precision in joints detection and depth estimation. Current research is focused on establishing specific guidelines for the correct use of Kinect in posture and gait monitoring and analysis applications, according to the performance provided by the sensor, and partly presented in this paper.

References

1. Camicioli, R., Howieson, D., Oken, B., Sexton, G., Kaye, J.: Motor slowing precedes cognitive impairment in the oldest old. Neurology **50**, 1496–1498 (1998)
2. Hausdorff, J.M., Rios, D.A., Edelberg, H.K.: Gait variability and fall risk in community-living older adults: a I-year prospective study. Arch. Phys. Med. Rehabil. **82**, 1050–1056 (2001)
3. Jia, W., Yi, W.J., Saniie, J., Oruklu, E.: 3D image reconstruction and human body tracking using stereo vision and kinect technology. In: IEEE International Conference on Electro/Information Technology (EIT), pp. 1–4 (2012)
4. Khoshelham K.: Accuracy analysis of kinect depth data. In: International Archives of the Photogrammetry, Remote Sensing and Spatial, Information Sciences, vol. XXXVIII-5/W12 (2011)
5. Obdrzalek, S., Kurillo, G., Ofli, F., Bajcsy, R., Seto, E., Jimison, H., Pavel, M.: Accuracy and robustness of Kinect pose estimation in the context of coaching of elderly population. In: 2012 Annual International Conference of the IEEE Engineering in Medicine and Biology Society (EMBC), pp. 1188–1193 (2012)
6. Ren, Z., Meng, J., Yuan, J.: Depth camera based hand gesture recognition and its applications in human-computer-interaction. In: 8th International Conference on Information, Communications and Signal Processing, pp. 1–5 (2011)
7. Stone, E., Skubic, M.: Passive, in-home gait measurement using an inexpensive depth camera: initial results. In: 6th International Conference on Pervasive Computing Technologies for Healthcare, pp. 183–186 (2012)
8. Wan, T., Wang, Y., Li, J.: Hand gesture recognition system using depth data. In: 2nd International Conference on Consumer Electronics, Communications and Networks, pp. 1063–1066 (2012)
9. Yeh, S.C., Hwang, W.Y., Huang, T.C., Liu, W.K., Chen, Y.T., Hung, Y.P.: A study for the application of body sensing in assisted rehabilitation training. In: International Symposium on Computer, Consumer and Control (IS3C), pp. 922–925 (2012)

A Wearable Multi-sensors Device for AAL Environment

Francesco Montalto, Valentina Bianchi, Ilaria De Munari and Paolo Ciampolini

Abstract MuSA (Multi Sensor Assistant) is a wearable multi-sensor device designed for elderly people monitoring. The system features healthcare services equipping a fall detector, a user alarm button, heartbeat, breathing rate and body temperature evaluation. By integrating MuSA in an ambient-assisted living framework, CARDEA, data fusion approaches can be implemented to obtain a behavioral profile which can be significant for the caregivers. The objective is to provide a set of features that integrated all together may foster a safe, independent and autonomous life to elderly in their home in accordance to the AAL (Ambient Assisted Living) paradigm. This paper describes the main concepts of MuSA and the details of the single functionalities. The suitable low-cost approach and the adequate quality of the system response, at the same time, have been proved by field tests of the device.

1 Introduction

The population is increasingly becoming older and a growing amount of people is facing weakness and disability problems. In 1999, the World Health Organization (WHO) launched a new campaign called "Active Ageing" [1]. With this initiative the WHO aims to improve the quality of life as people age, allowing people to realize

F. Montalto (✉) · V. Bianchi · I. D. Munari · P. Ciampolini
Dipartimento di Ingegneria dell'Informazione, Università degli Studi di Parma,
Parco Area delle Scienze 181/A, 43124 Parma, Italy
e-mail: francesco.montalto@unipr.it

V. Bianchi
e-mail: valentina.bianchi@unipr.it

I. D. Munari
e-mail: ilaria.demunari@unipr.it

P. Ciampolini
e-mail: paolo.ciampolini@unipr.it

S. Longhi et al. (eds.), *Ambient Assisted Living*, DOI: 10.1007/978-3-319-01119-6_12,

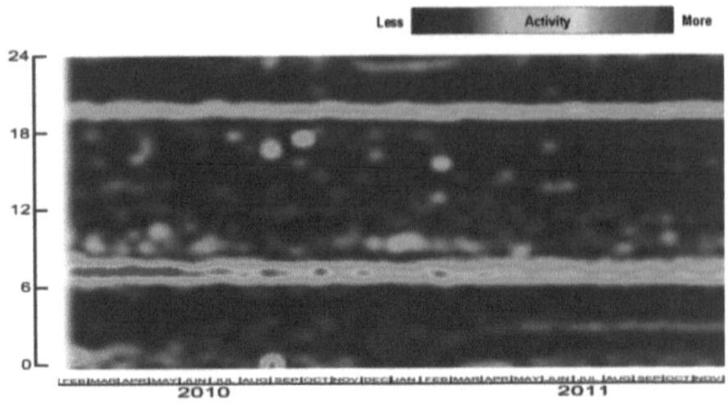

Fig. 1 Use of PIR sensors to evaluate quantity of activity

their potential for physical, social, and mental well-being throughout the life course and to participate in society, while providing them with adequate protection, security and care.

In this context the technology may play an important role both through the Assistive Technologies (AT) used to increase, maintain, or improve functional capabilities of individuals with disabilities and the Ambient Assisted Living (AAL) for an independent, autonomous and safety life at home. Exploiting the fusion of these two technologies new assistance and monitoring services can be carry out.

For example it is possible to collect the information from AT and AAL systems and process them to evaluate a person wellness and the evaluation of his health status, doing a sort of behavioral analysis.

A simple behavioral analysis can already be done using some simple information from environmental sensors used in AAL systems. Our project among the AAL environments is called CARDEA [2]. It is based on TCP/IP communication, i.e., all the system components are connected to the same Local Area Network. It features the capability of controlling a wide range of environmental sensors and home appliances and includes "assistive" interface modules (vocal, brain-computer); it also enables full web-based monitoring and control through a simple user interface. A wireless sensor network (WSN) contributes to the system architecture as well, dealing with mobile devices. Here, we shall refer to a WSN based on the IEEE-802.15.4 (ZigBee) standard protocol, exploited for the implementation of the wearable sensor described in the following.

An example of behavioral analysis obtained from environmental sensors is depicted in [3]: by tracking the activity of the PIR sensors we are able to quantify the activity of the person who lives in the room, and to outline a trend in his habits (Fig. 1).

Although this approach is simple to implement, environmental sensors have some limitations. First they are not able to distinguish which person has been detected and second they are strictly related to fixed positions.

Our purpose is to extend the behavioral analysis to wearable sensors in order to evaluate more information about people status. Wearable sensors are useful for early risk detection and continuous health monitoring that play an important role among

Fig. 2 MuSA

the assistive functions in the home environment. Many technological solutions have been developed in order to monitor personal activities and vital signs. In this area, Wearable Monitoring Health Systems (WHMS) realize consumer operated personal prevention and early risk detection [4]. A variety of prototypes and commercial products have been designed with the aim of providing real-time information about a person's health, either to the user himself or to a specific caregiver.

In this paper, MuSA, Multi Sensor Assistant, will be described (Fig. 2). It has been conceived for personal activity and vital parameters continuous supervision [5]. MuSA has been specifically developed to be embedded in CARDEA even if it is able to communicate with any Ambient Assisted Living (AAL) systems. Our aim is to extract useful information from the raw data obtained from the sensors implemented on MuSA. By processing these data on-board we send only significant information to CARDEA. Finally, from the vast availability of information coming from wearable sensors and environmental sensors, data fusion can be performed in order to obtain reliable behavioral analysis and hence a good awareness of people status.

2 MuSA

MuSA is a wearable multisensor platform, specifically designed with assistive purposes. It is compliant with ZigBee 2007 PRO standard protocol, and the Home automation standard profile thanks to a CC2531 SoC [6]. MuSA is designed to be worn at belt or at chest: it is quite small (78 × 48 × 20 mm), and lightweight (about 70 g, Li-Ion battery included). Different functions can be implemented on the same platform, basic configuration of MuSA includes a call button and automatic fall detection. MuSA can be extended with further functions, hosted by the same hardware platform: a single-lead ECG system is used to evaluate heart rate and respiration rate using EDR technique [7] and a NTC thermistor is included to evaluate body temperature. All of the signal acquisition and processing is carried out by MuSA on-board circuitry: detection of abnormal behaviors or deviation of vital signs from their "normal" range is carried out by this device. Radio communication is hence kept at a bare minimum (alarm messages and network management), saving battery energy.

Two basic building blocks can be identified: a IEEE 802.15.4 radio transceiver, and a microcontroller taking care of ZigBee stack management. The same microcontroller is exploited for digital signal processing. The board also includes sensors and analog front-end circuitry needed to acquire vital signs.

2.1 Fall Detection

Fall movements can be quite different, depending heavily on the actual situation. In literature three different kinds of most common falls for an older adult can be found: fall during sleep, from the seated position and from standing up to lying on the floor. Whereas the first two can be somehow monitored by means of bed- or chair-occupancy sensors the last one (also being the most frequent kind) requires a smarter automatic detectors. Basic fall detection algorithms exploit threshold comparison [8]: since falls are often associated to acceleration peaks, current acceleration (the Euclidean norm of the acceleration vector, actually) is checked against a given threshold, which depends on personal physical features (height, weight, etc.). However, tuning such a threshold is critical: hence a more reliable detection strategy is exploited: MuSA correlates acceleration pattern with postural information.

A tri-axial MEMS accelerometer, LIS331DLH [9], has been chosen here to evaluate human body position and orientation. The implemented algorithm detects, at first, specific acceleration peaks and then evaluates the eventual variation of the body orientation. Observing Fig. 3, the CC2531 reads the acceleration components from the accelerometer every 16 ms, then MuSA computes the acceleration norm and compares it to a particular threshold. If this condition takes place, it is necessary to look for the tilt angle computed just before and after the acceleration peak. The scalar product of the static acceleration components, before and after the acceleration peak, is used to evaluate the orientation by comparing this value to a suitable threshold, in order to detect an actual fall.

2.2 Heart Rate

Heart activity monitoring is necessary to promptly identify a person's abnormal heart rate rhythms (i.e. arrhythmia, tachycardia, etc.). Moreover this analysis can be useful to combine heart information with motion data, in order to classify human activities [10, 11].

In order to provide a continuous monitoring and a non-obtrusive design at the same time, a simple electrocardiogram (ECG) has been implemented on MuSA by using a single-lead (I Einthoven) derivation using an elastic strap belt. An analog front-end circuit has been implemented on-board. It mainly consists of a differential instrumentation amplifier and an analog low-pass filter; moreover a feedback op-amp integrator has been used to keep to a constant value the DC component, in order

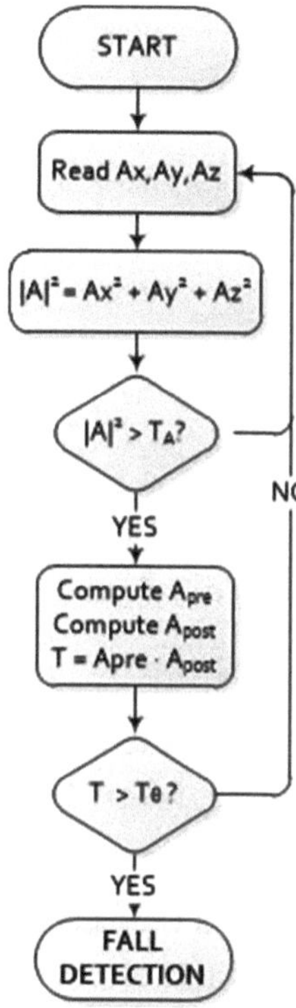

Fig. 3 Fall detection algorithm

to avoid skin-electrode resistance dependence. The acquired ECG signal is then properly filtered by the CC2531 for the purpose of emphasizing the QRS complexes.

Heart rate is measured detecting and counting every R-peaks in the ECG filtered signal. It is done by detecting the local maximum, observing when the ECG derivative passes through zero (Fig. 4).

2.3 Respiration Rate

Healthy people's respiration depends on both age and physical activity. Monitoring breathing activity allows the evaluation of abnormalities such as apnea, hyperpnea, tachypnea, etc.

In a first time, respiration signal was obtained from a piezoelectric sensor attached to the elastic chest belt. Unfortunately, the non-perfect signal periodicity and even more the motion-induce artifacts, makes the algorithm unreliable.

A sensorless solution, based on the fact that the respiration is highly correlated to heart activity, has been studied and implemented. Hence, the EDR (ECG Derived Respiration) method has been chosen [7]: small ECG morphology variations during

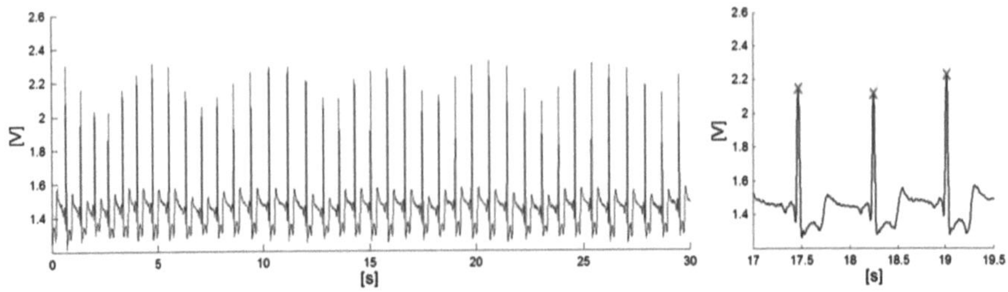

Fig. 4 ECG signal and R-wave peaks detection

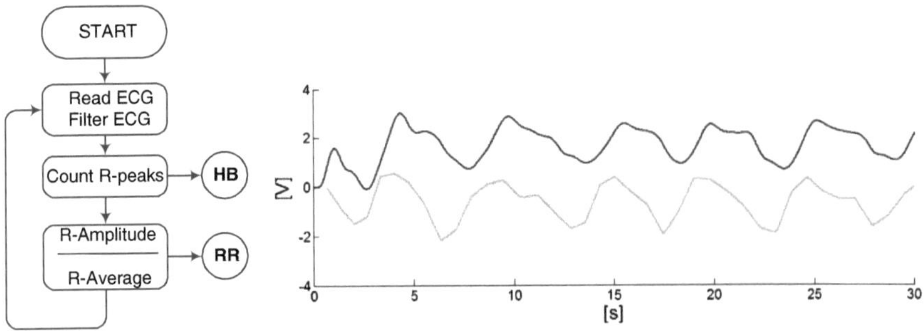

Fig. 5 ECG and EDR routine (**a**); respiration waveforms: piezo (*blue*) versus EDR (*green*) (**b**)

the breathing cycle happen due to the movement of the heart position in relation to the chest electrodes and the mutation of the lung volume that causes a dielectric changes. Here, the EDR method focus on the amplitude modulation of the ECG signal: heart's apex move towards the abdomen while inspiring, and it relaxes during expiration.

The implemented algorithm used to estimate the respiration signal from the ECG, described in detail in [12], has its main advantage in being a real-time operation, suitable for our low-power CC2531 SoC. Shortly, we can describe the ECG signal as a composite R-wave amplitude, consisting of the sum of the amplitude modulation due to breathing activity of the chest and various noise components. After the ECG is properly filtered, the algorithm identifies every R-wave peak, stores the value of its amplitude and computes the current and previous amplitudes running average. By evaluating the ratio of the current amplitude and the running average we obtain the estimation of the respiration signal.

This method has been validated comparing a piezo-sensor outcome and an EDR signal of the same respiration activity (Fig. 5b). It is important to emphasize that the EDR technique does not aim to provide an highly accurate respiration waveform, but rather to give a precise respiration rate measure. Moreover this solution allows to reduce the overall cost and increase the user's comfort by removing the piezo-sensor from the system. From the EDR-inferred signal has been possible to realize the respiration cycles count algorithm by simply counting waveform maximums.

2.4 Body Temperature

Body temperature is another important indicator for health status monitoring. Since changes during the day can be related to actual activity, a combination of the above information and the body temperature might be relevant. A value around 37 °C is considered to represent healthy status and the temperature setpoint is defined as the level which the body try to maintain, if this value rises we speak of fever status. However this value varies during the day for each person.

Here a temperature probe capable of measuring in a range between 35–45 °C, with a 0.1 °C accuracy, has been implemented on MuSA. An inexpensive NTC thermistor has been chosen here as the sensing element. From its temperature-resistance relationship, a multi-stage analog circuit has been designed to provide a linear characterization of the voltage-temperature relationship in the desired range. Power consumption issues have been taken into account to prevent the thermistor self-heating. As it was done before, the successful validation of the temperature system has been conducted by comparison with a mercury thermometer, as reference instrument.

2.5 Behavioral Analysis

As stated in the beginning of this article, our vision is to extend the process of the information coming from the previous algorithms in order to obtain meaningful analysis.

A first example of what we could do about is showed in Fig. 6. From the filtered accelerations, we already use for the fall detection routine, we examine a person quantity of movement. In particular by calculating the discrete norm of the three accelerations component and comparing this value with proper thresholds, it is possible to distinguish static from active situations and, at the same time, still detect fall events.

Furthermore, from the same acceleration components, we are able to measure the body orientation (tilt angle) and so discriminate is a person is sitting/standing or lying.

Although this approach needs improvements, this is a first interesting step towards behavioral analysis.

3 Results and Discussion

In this article we presented an AT solution called MuSA. It was at first conceived as a fall detector. MuSA has been later enhanced with more health monitoring functionalities. Our vision is to fuse the most of the data available in order to obtain reliable behavioural analysis. This can be done exploiting MuSA itself but even more by correlating MuSA and CARDEA's environmental sensors information.

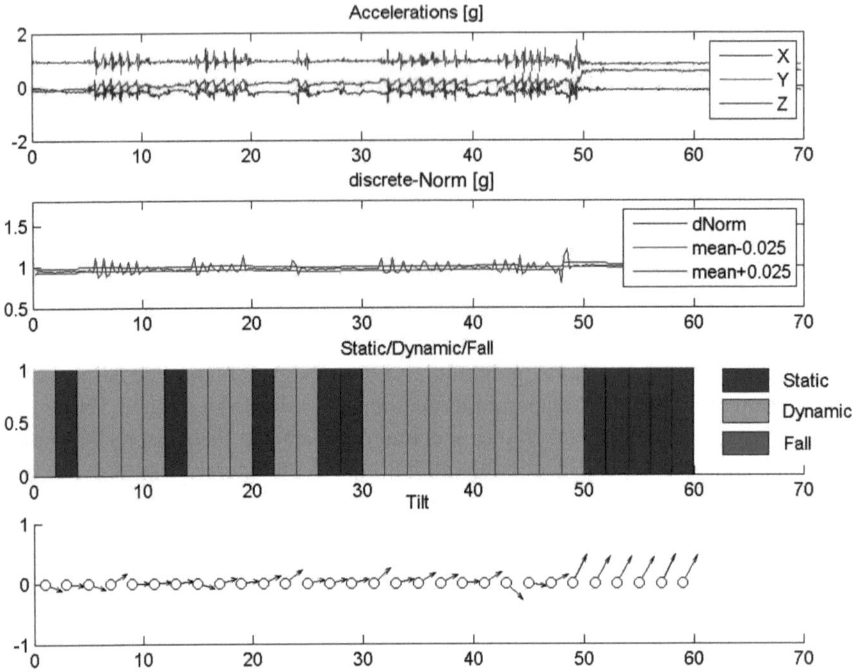

Fig. 6 Acceleration components (**a**); discrete Norm and thresholds (**b**); Quantity of movement analysis, static (*green*), dynamic (*blue*), fall (*red*) (**c**); body orientation evaluation (**d**)

Concerning the singular functionalities of our wearable device, preliminary tests on people have been conducted in laboratory. For the fall detection, by considering the number of True Positives and the number of False Negatives, the algorithm assesses 99 % of sensitivity and 97.8 % of specificity [5]. Heartbeat and respiration rate have been tested together on ten healthy volunteers. The relative error of both algorithms has been estimated as the difference between the counted rates by the user and the ones by the algorithms themselves in a one-minute window. First results state a mean relative error about 0.56 % for the heart-rate and about 12.5 % for the respiration rate. It is clear that if the ECG has a typical generic waveform for every person, the respiration signal changes due to different body characteristics and respiration mechanism. Nevertheless, the attained accuracy is adequate to the application at hand. As mentioned before, the body temperature system provides an accuracy lower than 0.1 °C.

In this work a multi-sensor wearable device has been expanded toward low-cost healthcare functionalities. The medium-term target is to combine all the kinetic and physiological parameters in order to lead data-fusion strategies and obtain user's behavioral analysis and health-status profile. The integration of this device into the AAL system CARDEA provides an innovative solution in the assistive technologies.

References

1. World Health Organization. Active ageing, a policy framework. http://www.who.int/ageing/active_ageing/en/index.html (2002)
2. Grossi, F., Bianchi, V., Matrella, G., De Munari, I., Ciampolini, P.: An assistive home automation and monitoring system. In: ICCE 2008 Digest of Technical Papers, pp. 341–342 (2008)
3. Losardo, A., Grossi, F., Matrella, G., De Munari, I., Ciampolini, P.: Exploiting AAL environment for behavioral analysis. In: Assistive Technology: From Research to Practice. Assistive Technology Research Series, vol. 33, pp. 1121–1125 (2013)
4. Pantelopoulos, A., Bourbakis, N.G.: A survey on wearable sensor-based systems for health monitoring and prognosis. IEEE Trans. Syst. Man Cybern. Part C Appl. Rev. **40**, 1–12 (2010)
5. Bianchi, V., Grossi, F., De Munari, I., Ciampolini, P.: Multi sensor assistant: a multisensor wearable device for ambient assisted living. J. Med. Imaging Health Inf. **2**, 70–75 (2012)
6. Texas Instruments. www.ti.com
7. Mazzanti, B., Lamberti, C., de Bie, J.: Validation of an ECG-derived respiration monitoring method. IEEE Proc. Comput. Cardiol. (2003). doi:10.1109/CIC.2003.1291230
8. Chen, J., Kwong, K., Chang, D., Luk, J., Bajcsy, R.: Wearable sensors for reliable fall detection. In: Proceeding 27th Annual International Conference of the Engineering in Medicine and Biology Society (2005)
9. STMicroelectronics. http://www.st.com/
10. Fujimoto, T., Nakajima, H., Tsuchiya, N., Marukawa, H., Hata, Y.: Human activity classification by ECG and accelerometers aided by fuzzy logic. In: IEEE International Conference on Systems, Man, and Cybernetics, Seoul, 14–17 Oct 2012
11. Tapia, E.M., Intille, S.S., Haskell, W.: Real-time recognition of physical activities and their intensities using wireless accelerometers and a heart rate monitor. In: Proceedings of the 2007 IEEE International Symposium on Wearable Computers, pp. 37–40, 11–13 Oct 2007
12. Arunachalam, S.P., Brown, L.F.: Real-time estimation of the ECG-derived respiration (EDR) signal using a new algorithm for baseline wander noise removal. In: Proceedings of IEEE Engineering in Medicine and Biology Society Conference, pp. 5681–5684 (2009)

RGBD Sensors for Human Activity Detection in AAL Environments

Emanuele Frontoni, Adriano Mancini and Primo Zingaretti

Abstract This paper aims to propose a novel idea of an embedded intelligent system where a low cost embedded vision system can analyze human behaviors to obtain interactivity and statistical data, mainly devoted to human behavior analysis in Ambient Assisted Living environments. We construct a system that uses vertical RGBD sensor for people tracking and interaction analysis, where the depth information has been used to remove the affect of the appearance variation and to evaluate users activities inside the home and in front of the fixtures. Also group interactions are monitored and analyzed with the main goal of having a better knowledge of the users activities, using real data in real time. All information coming from this human behavior analysis tool can be used to provide basic data gathered in real time for an Ambient Assisted Living environment. Even if preliminary, the results are convincing and most of all the general architecture is affordable in this specific application, robust, easy to install and to maintain and low cost.

1 Introduction

The Ambient Assisted Living (AAL) has the main goal to promote a healthy, safe and functional living environment for old people and people with disabilities that limit their activities, movements and sense, but who want autonomously live in their home. For this purpose, AAL environments use Information and Communication Technolo-

E. Frontoni (✉) · A. Mancini · P. Zingaretti
Dipartimento di Ingegneria dell'Informazione,
Università Politecnica delle Marche, Ancona, Italy
e-mail: frontoni@dii.univpm.it

A. Mancini
e-mail: mancini@dii.univpm.it

P. Zingaretti
e-mail: zinga@dii.univpm.it

S. Longhi et al. (eds.), *Ambient Assisted Living*, DOI: 10.1007/978-3-319-01119-6_13,

gies (ICT) and Ambient Intelligence paradigm, to provide solutions that ensure the livability of a home [1–3]. This is a serious problem because, unfortunately, in all the countries, old age people and people with disabilities have a limited support for their daily activities. Moreover the increase of senior population has major consequences for all aspects of people everyday life. So, the number of people that will need a form of institutionalized help is increasing. It is necessary make easier for people with age and other problems to live a satisfying and independent life in their own home environment.

AAL has recently attracted a growing attention in scientific community since it involves innovative and emerging technological solutions providing integral solutions in the area of home environment, independent living that will increase the quality of life and lower costs.

In this ambit, this project focuses on developing automated RGBD video analysis techniques for tracking users in AAL environments with the goal of recognizing different user activities that may be relevant for assistance, with a particular focus on the interactions between users and home fixtures or objects in the environment. The approach uses low-level segmentation and tracking methods, mapping the users to the floor plan, identifying events such as picking up objects and interacting with the home environment.

The system provide many statistical data that, opportunely elaborated, can give much information about users, customizing the home for their needs and adapting home design to their behaviors. Some data are: number of interactions, average of interaction time, direction left-right or right-left and so on. Data extracted in this way are objective, while often this kind of decision is made looking to subjective data. Some data can also generate alarms together with a high-level decision support system able to analyze user activity data.

The processing framework of visual surveillance in dynamic scenes involves the following stages: modeling of environments, detection of motion, classification of moving objects, tracking, understanding and description of behaviors, human identification, and fusion of data from multiple cameras [4–6]. As a result, more researches have been conducted on this field involving a great quantity of different vision based approaches. Automatic recognition of human actions and interactions remains a very challenging problem. The key difficulty stems from the fact that the appearance of a person who performs a particular action can significantly vary due to many factors such as camera viewpoint, clothing of a person, occlusions, variation of body position, object appearance and layout of the scene. In addition, motion cues often used to disambiguate actions in video [3, 7, 8] are not available in vertical acquired RGBD images. A number of previous works [9–12] focus on exploiting body pose as a cue for action recognition. In particular, several methods address joint modeling of human positions, objects and relations among them [13–18].

Reliable estimation of body configurations for people in arbitrary poses, however, remains a very challenging research problem. Less structured representations, e.g. [19, 20] have recently emerged as a promising alternative demonstrating state-of-the-art results for action recognition in static images.

In this paper we introduce the general framework of our system and we discuss how vision and, in particular, people and object tracking, are involved in different tasks of the whole system. We show successful applications for AAL environment management, showing the visual tracking methodology and results of real scenarios with real time performances. Software analysis, tests and experiments developed are presented to show the feasibility of the proposed approach in large-scale applications, using only RGBD vision sensors.

The paper is more focused on the application of vision to AAL environment and on the application in general in visual tracking as a main information source; main novelties are in the original and wide application of the tracking system and activity analysis focused on user-environment interactions.

The paper is organized as follows: Sect. 2 introduces the general concept of the system and describes the peculiarity of the particular methodology described in this paper. The experimental results are provided in Sect. 3. Conclusions are given in the last Section.

2 Methodology

2.1 Users Tracking and Behaviors

The goal of the system is the use of vertical mounted RGBD camera (Fig. 1) to detect people activities and in particular their interactions with everything that surrounds them. The first part of the activity recognition method is based on the water filling algorithm to identify people from a vertical depth image even when they are moving in group inside the home. Details of the tracking algorithm can be found in [21] and other aspects and applications of this approach in different scenarios are detailed in [22–24]. This last aspect of group behaviors will be better investigated in future tests in real scenario and the results on this side are not reported in this paper even if the methodology will be the same described here following.

The system needs to identify every person moving in the scene (Fig. 1) to track their activities and in particular to analyze their interaction with the environment. Section 2.2 will give more details on this second aspect.

We take the depth image as a function f, where f(x, y) stands for the depth information of pixel (x, y). Due to the noise of RGBD sensor, f(x, y) can be non-derivable or even discontinuous. Finding people in depth image equals to finding local minimum regions in f. Mathematically, the problem can be defined as finding the region A and N that satisfy the following constraint:

$$E_A\,(\mathrm{f(x,\,y)}) + \eta \leq E_{N_A}\,(\mathrm{f(x,\,y)}) \tag{1}$$

where $A \in N$, A is the local region and N is its neighborhood; E(.) is an operation to pool the depth information in the region to a real value that reflects the total depth

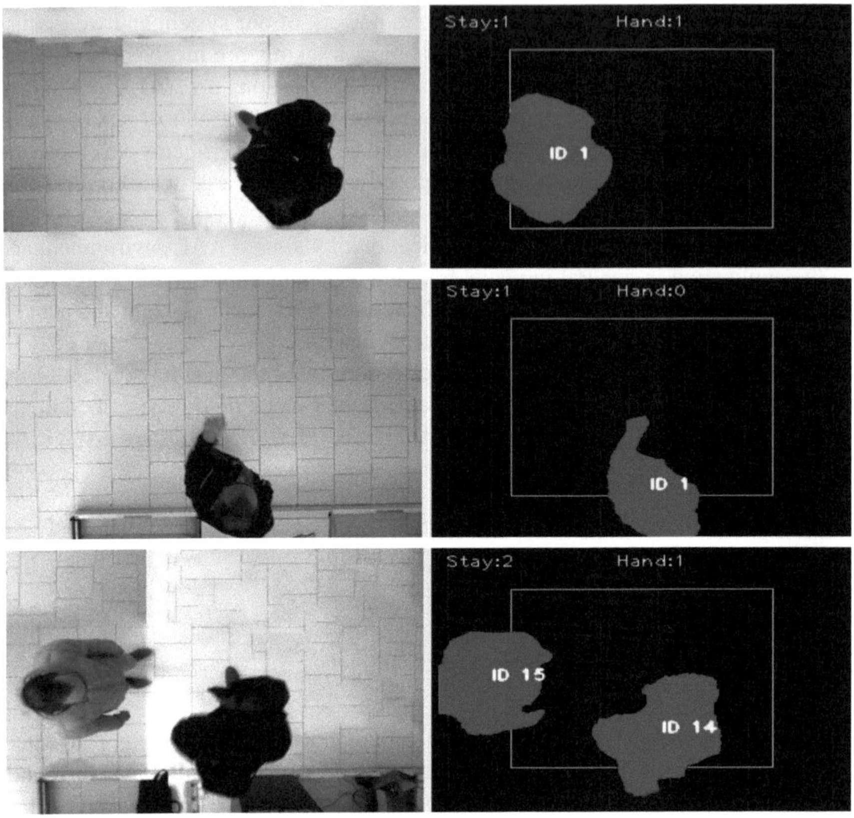

Fig. 1 Example of the water filling algorithm detecting user with its relative ID together with the interaction line estimated by the depth sensor and the reference system. Three different cases with different lighting conditions and number of people are shown

information in the region; η is a pre-defined threshold to ensure that depth in A should lower than N_A with a margin.

Note that A and N can be of arbitrary shape, and finding all the regions in image can be very time consuming.

2.2 Users Interaction Aspects

The second methodological aspect of this work is the idea to have an estimation of the interaction with environment and with objects and a classification of the interaction type. In particular we define *positive* interaction when the object is picked up and *negative* one when the object is put back after a pickup. This is very useful information when we try to identify the activities of users in front of shelves, fixtures in general,

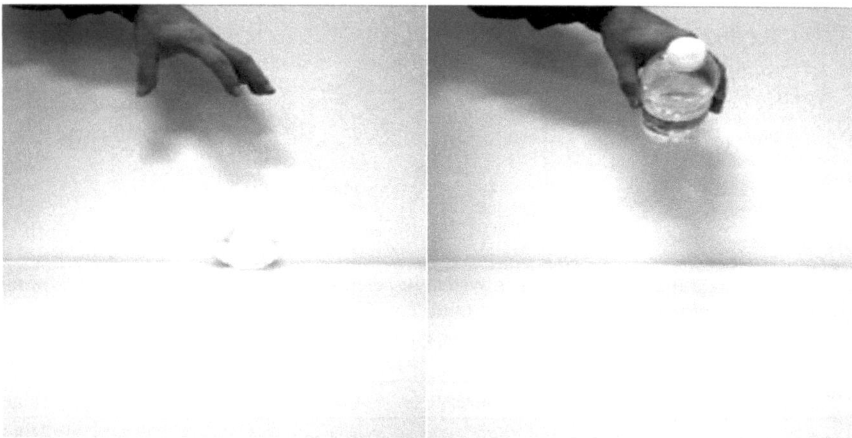

Fig. 2 Example of hand template and hand/object template, the *first* one entering the shelf line, the *second* exiting with object

windows and doors. This allows the description of the activity also with an interaction map that will be discussed in the result section.

To solve the hand problem we designed a template matching method dealing with hand and objects templates. A sample of two of these templates is reported in Fig. 2. The template frame is identified when the hand is entering or exiting a shelf line identified by the depth sensor.

In general template matching requires similarity measures between the features of a template and the query image. Image intensity patches are often compared by normalized cross-correlation whereas Hausdorff and chamfer measures are popular with edge-based features. A common approach is to have a number of prototype shape templates and search for them in the image. Templates used for hand with objects and hands without were collected in real scenarios. Chamfer score and a closely related measure Hausdorff score have been used in many shape matching schemes. We use truncated chamfer score, which makes it more robust to outliers [12, 13].

3 Results

The system was installed in a real home environment using 3 sensors mounted in the ceiling of particular locations of the house (main entrance, high interest areas such us dinner room and bathroom, etc.).

The hardware was selected respect to the low cost aspect of the project and in particular the idea to use commercial components that are suitable for low maintenance and low cost requirements. The Raspberry processing unit was used to compute the

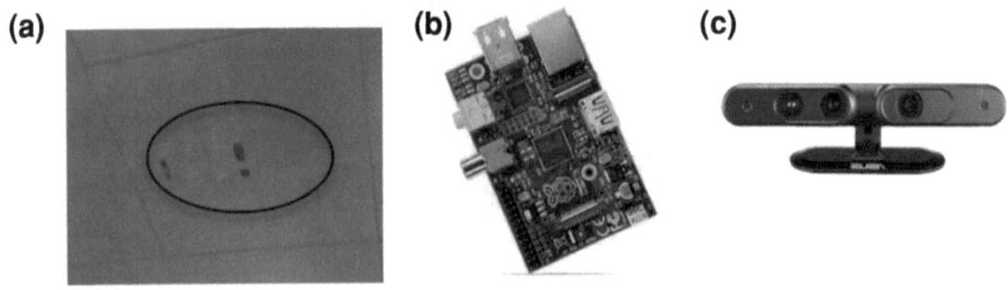

Fig. 3 Showing different components **a** system installed in the ceiling of a real house (3 of this unit was installed for test in particular areas of the house). **b** Raspberry used as processing unit (one for each RGBD camera with WiFi modules for synthetic data transmission) **c** Asus RGBD sensor

Table 1 List of parameters provided by the system for activity analysis in a day of test in a real scenario

Parameter	Value
1. Total number of user presences	46.5
2. Average interaction time	36 s
3. User passing by (no interaction)	43 %
4. Average group number	1.2
5. Number of interactions	59.0
6. Interaction per presence	2.1
7. Average X, Y (to evaluate interaction)	134.20
8. Direction Left-Right	72 %
9. Direction Right-Left	28 %

RGBD video streaming at 10 frames/second using OpenCV and our software with drivers for the Asus Pro Live RGDB sensor. The components are showed in Fig. 3.

The processing unit is devoted to all image-processing elaborations and only synthetic data are transmitted over the WiFi network of the house to a web server that collect numerical data in respect to privacy limitation.

The system on the web server collects several data suitable for user activity analysis together with the interaction map of the shelf. The following Table 1 presents a list of parameters that are grabbed from the real environments that was on line for 30 days. The table is referred to a complete day analysis and is reported to give an idea of possible data comparisons that can be performed both between different areas (locations inside the home) and different periods.

Figure 4 shows an example of the interaction map. This particular result, that is one of the innovations of this paper, is performed on the web application using data of hand interaction and their x, y, z labeled with the object/no-object features and with the id of the person that is performing the interaction. In particular the sequence of the interactions can be used to represent over the image an intensity map that shows knowledge about the interaction sequence: red areas identify that the object has been picked up and put id back on the shelf (*negative interaction*); green areas identify that users retains the object (*positive interaction*).

Fig. 4 Example of user interaction map. *Red circle* are negative interactions; *green circle* are positive interactions

Further results analysis are currently being developed mainly integrating data coming from 3 sensors. Until today, the system demonstrated to be inexpensive and robust and it collected data for 30 days with no evidence of particular errors. Even if preliminary and open to huge further investigations, results are convincing and also the discussion with marketing people involved in the project demonstrate a lot of interest for the application here described.

4 Conclusion and Future Works

In this paper we presented a novel application of visual activity recognition, using a combination of techniques to better understand users behaviors also in the ability to interact with the home environment. The proposed approach was tested in a real environment with interesting results in the field of AAL behavior analysis.

The paper presented also an integrated architecture for mixing together different kind of vision based applications such as people and interaction tracking. Future work about this project is, first of all, the optimization of these systems in terms of stability, performances and robustness to environmental inconveniences, considering that in a home there is no special worker able to reset or modify these systems in case of trouble.

The final step of this project is to provide the home with a fully automated user interaction model that can also give information about human behavior analysis for AAL systems. The goal is also to have a fast system able also to evaluate user interactions patterns and behaviors.

Acknowledgments Authors would like to thank Daniele Liciotti for his support. This work is partially supported by INRCA and Marche Region in the HDOMO 2.0 project.

References

1. Ascani, A., Frontoni, E., Mancini, A., Zingaretti, P.: Feature group matching for appearance-based localization. In: IEEE/RSJ 2008 International Conference on Intelligent RObots and Systems—IROS 2008, Nice, pp. 3933–3938 (2008)
2. Desai, C., Ramanan, D., Fowlkes, C.: Discriminative models for multi-class object layout. In: International Conference on Computer Vision IEEE/ICCV, pp. 229–236 (2009)
3. Ferrari, V., Marin-Jimenez, M., Zisserman, A.: Pose search: retrieving people using their pose. In: International Conference on Computer Vision and Pattern recognition IEEE/CVPR, pp. 1–8 (2009)
4. Brox, T., Bourdev, L., Maji, S., Malik, J.: Object segmentation by alignment of poselet activations to image contours. In: International Conference on Computer Vision and Pattern recognition IEEE/CVPR, pp. 2225–2232 (2011)
5. Desai, C., Ramanan, D., Fowlkes., C.: Discriminative models for static human-object interactions. In: Computer Vision and Pattern Recognition Workshops (IEEE/CVPRW), pp. 9–16 (2010)
6. Farhadi, A., Endres, I., Hoiem, D., Forsyth, D.: Describing objects by their attributes. In: International Computer on Vision and Pattern Recognition, IEEE/CVPR, pp. 1778–1785 (2009)
7. Moeslund, T.B., Hilton, A., Kruger, V.: A survey of advances in vision-based human motion capture and analysis. Comput. Vis. Image Underst. **104**(2–3), 90–126 (2006)
8. Zhang, J., Marszalek, M., Lazebnik, S., Schmid, C.: Local features and kernels for classification of texture and object categories: a comprehensive study. In: International Conference on Computer Vision and Pattern Recognition Workshop, pp. 13–20 (2006)
9. Bourdev, L., Malik., J.: Poselets: body part detectors trained using 3D human pose annotations. In: International Conference on Computer Vision IEEE/ICCV, pp. 1365–1372 (2009)
10. Csurka, G., Bray, C., Dance, C., Fan, L.: Visual categorization with bags of keypoints. In: Workshop on Statistical Learning in Computer Vision, ECCV, pp. 1–22 (2004)
11. Felzenszwalb, P., Girshick, R., McAllester, D., Ramanan, D.: Object detection with discriminatively trained part based models. IEEE Trans. Pattern Anal. Mach. Intell. **32**(9), 1627–1645 (2010)
12. Gupta, A., Kembhavi, A., Davis, L.: Observing human-object interactions: using spatial and functional compatibility for recognition. IEEE Trans. Pattern Anal. Mach. Intell. **31**(10), 1775–1789 (2009)
13. Johnson, S., Everingham, M.: Learning effective human pose estimation from inaccurate annotation. In: International Conference on Computer Vision and Pattern recognition IEEE/CVPR, pp. 1465–1472 (2011)
14. Lampert, C., Nickisch, H., Harmeling, S.: Learning to detect unseen object classes by between-class attribute transfer. In: International Conference on Computer Vision and Pattern recognition IEEE/CVPR, pp. 951–958 (2009)
15. Laptev, I., Marszałek, M., Schmid, C., Rozenfeld, B.: Learning realistic human actions from movies. In: International Conference on Computer Vision and Pattern recognition IEEE/CVPR, pp. 1–8 (2008)
16. Lazebnik, S., Schmid, C., Ponce. J.: Beyond bags of features: spatial pyramid matching for recognizing natural scene categories. In: International Conference on Computer Vision and Pattern recognition IEEE/CVPR, pp. 2169–2178 (2006)
17. Li, L., Su, H., Xing, E., Fei-Fei, L.: Object bank: a high-level image representation for scene classification and semantic feature sparsification. In: Neural Information Processing Systems NIPS, pp. 1378–1386 (2010)
18. Maji, S., Bourdev, L., Malik, J.: Action recognition from a distributed representation of pose and appearance. In: International Conference on Computer Vision and Pattern recognition IEEE/CVPR, pp. 3177–3184 (2011)
19. Yao, B., Fei-Fei, L.: Grouplet: a structured image representation for recognizing human and object interactions. In: International Conference on Computer Vision and Pattern recognition IEEE/CVPR, pp. 9–16 (2010)

20. Yao, B., Fei-Fei, L.: Modeling mutual context of object and human pose in human-object interaction activities. In: International Conference on Computer Vision and Pattern recognition IEEE/CVPR, pp. 17–24 (2010)
21. Frontoni, E., Raspa, P., Mancini, A., Zingaretti, P., Placidi, V.: Customers' activity recognition in intelligent retail environments. Lecture Notes in Computer Science (including subseries Lecture Notes in Artificial Intelligence and Lecture Notes in Bioinformatics) LNCS, vol. 8158, pp. 509–516 (2013)
22. Frontoni, E., Raspa, P., Mancini, A., Zingaretti, P., Placidi, V.: Smart vision system for shelf analisys in intelligent retail environments. In: ASME/IEEE International Conference on Mechatronic and Embedded Systems and Applications (MESA2013), Portland, Oregon (2013)
23. Catani, L., Frontoni, E., Zingaretti, P., Di Pasquale, G.: Efficient traffic simulation using busses as active sensor network. In: Proceedings of the ASME Design Engineering Technical Conference (MESA 2011), pp. 889–894 (2011)
24. Catani, L., Frontoni, E., Zingaretti, P.: A framework based on vision sensors for the automatic management of exchange parking areas. In: IEEE/ASME International Conference on Mechatronic and Embedded Systems and Applications (MESA 2010)

MyCmon: Cloud-Based Smart Home Monitoring for Elderly People and People with Disabilities

Paolo Mongiovì, Ferdinando Grossi, Valentina Bianchi, Agostino Losard, Guido Matrella, Ilaria De Munari and Paolo Ciampolini

Abstract This paper describes the MyCmon Cloud-Based Smart Home Monitoring System including a home automation system permanently connected through the internet to a Cloud-based control system. MyCmon aims at supporting independent living of elderly people and people with disabilities, by making monitoring functionalities available to their loved ones or to caregivers, anywhere and at any time, through mobile devices. The project is under development, expecting to be ready for the market by March 2014.

Keywords Remote monitoring · Home automation · Cloud computing · Smart home · Assisted living · Intelligent system

---

P. Mongiovì
Sistema Casa, Milano, Italy
e-mail: mongiovi@sistemacasa.it

F. Grossi · A. Losardo
Dipartimento di Ingegneria dell'Informazione, Università degli Studi di Parma and I-cubo Srl, Parma, Italy
e-mail: ferdinando.grossi@i-cubo.it

A. Losardo
e-mail: agostino.losardo@i-cubo.it

V. Bianchi · G. Matrella · I. D. Munari · P. Ciampolini (✉)
Dipartimento di Ingegneria dell'Informazione, Università degli Studi di Parma, Parma, Italy
e-mail: valentina.bianchi@unipr.it

G. Matrella
e-mail: guido.matrella@unipr.it

I.D. Munari
e-mail: ilaria.demunari@unipr.it

P. Ciampolini
e-mail: paolo.ciampolini@unipr.it

S. Longhi et al. (eds.), *Ambient Assisted Living*, DOI: 10.1007/978-3-319-01119-6_14,

1 Introduction

The worldwide population ageing trends is especially pronounced in Italy, making the implementation of technologies and policies aimed at supporting ageing at home particularly appealing for bringing benefits to both the elderly people and the community. Although elderly people, in our country, usually prefer to continue living in their home as long as possible (much valuing their own independence), family members are often concerned about risks connected to possible unattended home accidents or health issues. Resorting to a senior facility, on the other hand, implies a severe economic burden for both the family and the community. Similarly, people with temporary or permanent disabilities of many kinds may attain a better quality of life through independence, providing more peace of mind to them and to their loved ones.

Independent living can be fostered by means of advanced systems for the control and supervision of health and well-being. Such systems, however, need to be practical and easy to use. Plenty of vital-sign monitoring systems, as well as of environmental control systems, are currently on the market. The MyCmon Remote Monitoring System integrates both functions in a common framework, allowing for simpler and more effective control of the person's needs and of the surrounding environment. Besides monitoring functionalities (enhancing personal and environmental safety) the system also manages automation functions (lights, blinds, temperature, ...) and allows for better awareness and control of the energy consumption, yielding energy savings.

2 MyCmon System Architecture

Basic feature of the Remote Monitoring System is its cloud-based implementation: this enables remote system control at any time and from any remote terminal in the network, especially from smartphones and tablets. All data coming from the household are available to the cloud (under strict data security control). Based on such data, cloud agents can be exploited to infer the health and wellness status of the user, and to detect if something noteworthy has happened.
The system architecture is illustrated in Fig. 1. Main components of the system include:

1. Home automation control unit and sensors
2. Data aggregation and cloud management
3. Remote access interfaces.

2.1 Home Automation Control Unit and Sensors

This layer includes all the devices installed at home. In many cases, MyCmon system will be implemented on top of pre-existing home automation systems, exploiting available sensors for monitoring purpose as well. Further devices will be added,

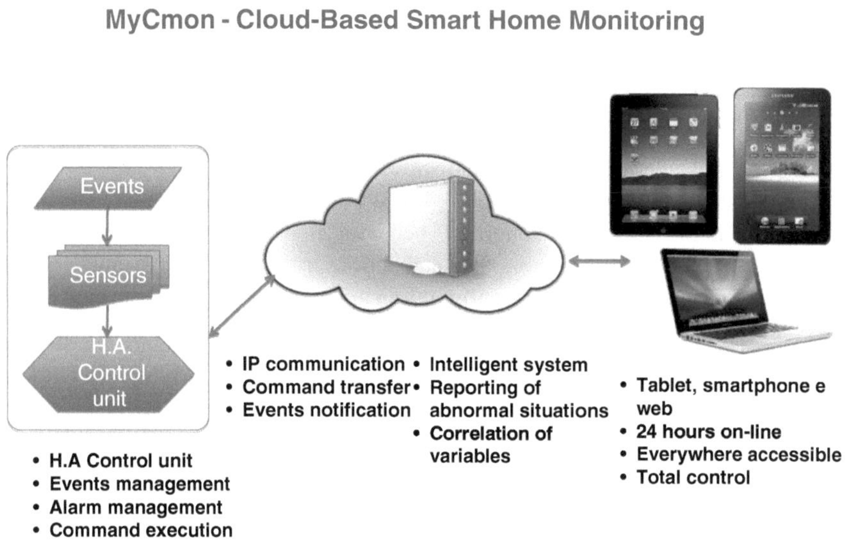

Fig. 1 System architecture view

dealing with personal monitoring: for instance, wearable sensors for the fall detection, or vital-sign monitors.

At the heart of the home automation system is the control unit: its function is to control all the sensors installed in the home and make decisions based on their state or change of. The control unit features a programmable decision logic: some decisions may imply local actions, whereas others will involve the cloud and result in remote notifications and management.

The home automation control unit connect sensors and actuators deployed into the home environment: such devices can feature either wired or wireless connections, depending on the specific case at hand: wired devices are usually expected to be used when the home is built or renovated; wireless devices, instead, are most appropriate for retrofitting existing homes with limited installation effort. The control system also acts as the gateway towards the internet and cloud applications.

2.2 Data Aggregation and Cloud Management

The second component is the cloud itself. The cloud is the key for making the system truly ubiquitous, accessible anywhere at any time, but also for optimizing service costs: the cloud approach, in fact, allows for the sharing computing resources, storage and network, thus resulting in a more scalable and cost-effective service implementation. A set of virtual servers, shared storage and network resources is made available to the home systems, managing the flow of information needed to monitoring functions. Web services are exploited to feed user interfaces applications running on different control devices (PCs, smartphones, tablets).

Cloud processing is exploited to observe "variables" describing the home control status, implementing the following functionalities:

- detection and calculation of time statistics for each variable
- showing status through dashboards
- detection and reporting of "abnormal" behavior
- evaluation of correlation among data and actions associated with events
- administration task, including configuration and management of user profiles.

Events can be combined, resulting in complex actions, for instance by defining a hierarchy of alarm reports, activating a given level if the underlying ones did not provide reactions, or implementing proactive home automation behaviors.

To check for "abnormal" conditions, the "distance" between the current situation and the reference, "ideal" one is evaluated; if such distance exceeds a given threshold, an attention flag is risen.

2.3 Remote Access Interfaces

The third component includes remote terminals. The easiest access implies accessing via a web browser from a PC. To account for ubiquity and mobility, however, applications are available for smartphones and tablets, providing caregivers and relatives with mobile monitoring tools, of course depending on the availability of an Internet access. A full range of function is available to the remote user, from basic monitoring to configuration and active home control. The application is available for devices running either Apple iOS or Android.

3 Smart Home

Home smartness comes from the integration of several technologies and systems into the home, aiming at creating comfortable, safe and easy-to-use home.

In fact, home automation:

- enables all users, including (to some extent) users with impairments or disabilities or elderly people, to access, use and interact with the environment;
- contributes to the redaction of accident risk, and at the same time accounts for emergency procedures in case of accidents;
- meets the needs of multiple categories of users, easily adapting to different needs by means of a flexible set of programming functions.

In addition, home automation systems have to feature open architecture, integrability, simplicity, reliability, service continuity and expandability. Expandability, in particular, is relevant to users with special needs, allowing for the inclusion of new technologies as they become available, following the evolution in time of the needs themselves and providing the user with simple and uniform interaction modes, regardless of the actual underlying specific technology and, as mentioned, suitable for different user interface devices (telephone, console, remote control, touch screen).

Home automation can therefore become a technology for all, adapting to different scenarios and to the different functions required by the situation at hand [1].

The Smart Home approach results in much more user friendly home environments, thanks to:

- the high degree of flexibility in programming the functions;
- the ability to empower users in dealing with daily living activities, providing support without depriving the user of full control;
- higher level of functionalities, allowing for combining or sequencing basic functions to cope with user-specific and articulated advanced tasks;
- user-friendly interfaces, enabling the user to effectively interact and programming the system according to one's own desires and needs.

It is also important to consider that home automation offers the ability to monitor energy consumption, supporting energy awareness and implementing energy-saving functions: in fact, the consumption of electrical power, water and gas are monitored and can be controlled, for instance providing for automatic shut-off.

4 Personal Sensors

Besides the environmental sensors and actuators, when it comes to wellness monitoring, a great deal of information may come from personal sensing devices. For instance, by means of integrated accelerometers, a simple device worn by the user gives information about the user's motion and posture, allows for detection of falls and for estimating the quantity and quality of the user's activity, as illustrated in the next paragraph. Another device useful to monitoring purposes is a bed occupancy sensor, which can be used for simple functions (for instance switching on the light when leaving the bed) as well as being exploited in fall prevention perspectives or in behavioral long term monitoring (wake-sleep cycle monitoring). To this purposes, MyCmon Remote Monitoring System includes personal wearable sensors such as MuSA (MUlti Sensor Assistant) [2, 3], which include on-board fall-detection features, as well as a call button, which can be exploited also to activate alternative system functions. MuSA is to be worn on the user's belt and is wirelessly connected to the main MyCmon control unit. Fall detection is based on a combined acceleration/posture sensing, to avoid false alarms. Similarly, the bed-occupancy monitor communicate with the system by means of wireless connection, making it more practical to install and less cumbersome. Special care has been devoted to the wireless devices energy consumption, accounting for longer battery lifetimes and thus reducing maintenance burden. Also, reliability was taken care of, ensuring early detection and prompt (remote) signaling of communication troubles, low battery status and other possible issues.

5 System Intelligence

MyCmon system continuously acquires data coming from environmental and personal sensors: their primary aim is the implementation of the control functions, according to the available programmable rules. Also, some sensors have explicit monitoring and alarm purposes. However, looking at data in a more general way, further monitoring features can be implemented, by tracking user's habits and behavioral changes which could be meaningful to early detection of health issues. For instance, the daily activity can be monitored by gathering data coming from all the sensors, and working out from such "big data" activity profiles, which provides insights on wellness status in an inexpensive and non-invasive way, as shown in [4]. The cloud environment onto which MyCmon system is based provides an ideal ground for big-data analysis: data coming from a manifold of households actually converge at a common data space, allowing for enough processing power to carry out accurate statistical analysis and inference of potential issues.

6 Implementation

MyCmon project exploits expertises of the three companies, joined in network funded by Regione Lombardia. Namely, the network includes: Sistema Casa, based in Milano and active, since more than 20 years, in the design, distribution and installation of home automation systems; Weblink, based in Varese, active in the field of software development and services for the Internet; Easycloud, based in Fino Mornasco (CO), active in the field of ICT services in the cloud

Also, I-Cubo, a technology spin-off participated by the University of Parma, cooperates into the project, bringing its expertise in the design and development of personal, assistive devices (such as fall and bed sensors) and in the evaluation of behavioral profiles for indirect monitoring.

Sistema Casa already has a large number (around 3.000) of home automation systems installed all over Italy and developed a network of skilled personnel dealing with system installation and service. Within this framework, the basic monitoring network, connecting all systems over the 24 h to the cloud is already available. More advanced monitoring apps are currently under testing and the user interface prototype is already available. Also, gateways toward I-Cubo wireless devices have been already developed.

Besides technical features, a key prerequisite for the project success is the availability of a suitable and effective technical support for the installation, operation and maintenance of the systems.

Such structure will be based on the aforementioned Sistema Casa network of partners, distributed on a territorial basis and under a central coordination. Training of personnel is of the utmost importance, when innovative functions are introduced: suitable training policies are being developed.

MyCmon remote monitoring system is offered for a monthly fee, covering the outlay for the setup of home automation control unit and sensors in the home.

7 Conclusions

In this paper, concepts, design and implementation of the MyCmon system are illustrated. The system is based on the availability of one among the largest networks of home-automation installation currently available in Italy, and aims at introducing into such a network innovative contents and functions. MyCmon system provides existing and new customers with monitoring functionalities supporting independent life of elderly people and people with disabilities. To this purpose, wireless, personal sensors are integrated in the system and a cloud-based strategy is devised for the management of data. This enables a flexible approach, allowing for tailoring each system to the specific needs and preferences of the user, providing her/him with a variety of local and remote interaction tools. It is worth emphasizing that the large application base onto which the project is based also required to pay attention at the technology scalability, as well as to the design and training of a suitable support network. The altogether makes MyCmon project a quite challenging, yet realistic and solid-grounded approach to ambient assisted living.

References

1. Quaranta G., Mongiovì P.: L'ABC della Domotica, Ed. Il Sole 24 Ore (2004)
2. Losardo, A., Grossi, F., Matrella, G., De Munari, I., Ciampolini, P.: Exploiting AAL environment for behavioral analysis. In: Assistive Technology: From Research to Practice, pp. 1121–1125. IOS Press, Amsterdam (2013)
3. Bianchi, V., Grossi, F., De Munari, I., Ciampolini, P.: Multi sensor assistant: a multisensor wearable device for ambient assisted living. J. Med. Imaging Health Inf. **2**(1), 70–75 (2012)
4. Montalto, F., Bianchi, V., De Munari, I., Ciampolini, P.: A wearable assistive device for AAL applications. In: Encarnação, P., Azevedo, L., Gelderblom, G.J., Newell, A., Mathiassen, N.-E. (eds.) Assistive Technology: From Research to Practice, pp. 101–108. IOS Press, Amsterdam (2013)

Part IV
Living with Chronic Conditions

Cognitive Home Rehabilitation in Alzheimer's Disease Patients by a Virtual Personal Trainer

Andrea Caroppo, Alessandro Leone, Pietro Siciliano, Daniele Sancarlo, Grazia D'Onofrio, Francesco Giuliani, Antonio Greco, Riccardo Valzania and Massimo Pistoia

Abstract Alzheimer's Disease is a problem of social and economic relevance as it involves both patients and their families/caregivers. Presently no cure is available. In the treatment of this disease, cognitive rehabilitation appears as an attractive choice for patients with a mild form. The aim of this work is the design and the development of a remote system, called "Virtual Personal Trainer", integrating advanced Natural User Interface technologies to support the patients during the rehabilitation process at home. The system allows both the autonomous execution of the required exercises and the data reporting and storing of the daily performance for every exercise. In this scenario, the patient avoids to physically move to a specialized center and the physician can use the platform in order to verify the response to the therapy and the compliance to the treatment.

A. Caroppo (✉) · A. Leone · P. Siciliano
CNR - Institute for Microelectronics and Microsystems, Via Monteroni,
73100 Lecce, Italy
e-mail: andrea.caroppo@le.imm.cnr.it

A. Leone
e-mail: alessandro.leone@le.imm.cnr.it

P. Siciliano
e-mail: pietro.siciliano@le.imm.cnr.it

D. Sancarlo · G. D'Onofrio · A. Greco
Geriatric Unit, IRCCS Casa Sollievo della Sofferenza, Viale Cappuccini, 71013 S.G.
Rotondo, Italy

D. Sancarlo
e-mail: d.sancarlo@operapadrepio.it

G. D'Onofrio
e-mail: g.donofrio@operapadrepio.it

A. Greco
e-mail: a.greco@operapadrepio.it

S. Longhi et al. (eds.), *Ambient Assisted Living*, DOI: 10.1007/978-3-319-01119-6_15,

1 Introduction

Alzheimer's Disease (AD) is a chronic degenerative disease involving cognitive areas leading progressively to a significant functional impairment. The evidence of possible pathogenetic mechanisms is poor. On this assumption, it is difficult to create rehabilitation models based on sufficiently solid theoretical foundations that go beyond simple theoretical assumptions. Overcoming the pessimistic view that has characterized the past, a large and lively debate is under way on the possible benefits of Cognitive Rehabilitation (CR), behavioral and psychosocial interventions in dementia. Several scientific evidences have been published [1–5], supporting the usefulness and relevance of CR in the management of patients with dementia.

Normally the CR process requires that patients go to the medical reference center once a week for several months, headed by a caregiver. Often, it is difficult for patients follow a CR cycle due to (a) motor deficits, (b) high comorbidity, (c) distance from the clinical center, (d) lack of private/public transport means to reach the clinical center and (e) presence of architectural barriers. In order to increase the chances of an effective care, the development of a cost-effective home-care service with CR functionalities could be very useful. Moreover, although education and ICT skills level among the elderly is often low, several pilot studies on small samples show that the introduction of ad-hoc ICT tools are accepted and used and could improve the quality of life [6] increasing the permanence at home. From the state of the art point of view, in the last 2 years, a community of researchers has investigated new solutions for cognitive assistance called serious games [7–11]. The research consists in exploiting videogames as a software platform allowing the support of new assistive tools, less expensive and more accessible, that could be used as a re-education tool helping to slow the cognitive decline of people suffering from AD. However, most of these serious game initiatives (including commercial products, such as Nintendo's Brain Age, Big Brain Academy or Vision Focus) provided only memory challenges or a series of random puzzles to be played few minutes per day with the aim of "improving brain performances".

F. Giuliani
ICT Unit, IRCCS Casa Sollievo della Sofferenza, Viale Cappuccini,
71013 S.G. Rotondo, Italy
e-mail: f.giuliani@operapadrepio.it

R. Valzania · M. Pistoia
eResult s.r.l., Via Ferruzzi Serafino, 47521 Cesena, Italy

R. Valzania
e-mail: riccardo.valzania@eresult.it

M. Pistoia
e-mail: massimo.pistoia@eresult.it

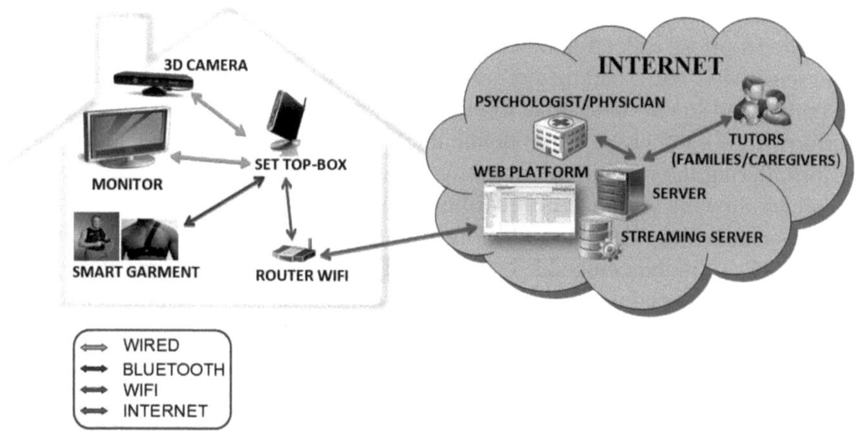

Fig. 1 Overview of AL.TR.U.I.S.M. platform

These initiatives suffer from multiple important limitations since they do not really suit perceptual and interaction needs of people suffering from AD.

2 AL.TR.U.I.S.M.: Platform Overview

The aim of AL.TR.U.I.S.M. platform (Alzheimer patient's home by a rehabilitation-based Virtual Personal Trainer Unique Information System Monitoring), is to provide a tele-rehabilitation tool through a "Virtual Personal Trainer" (VPT), by using advanced ICT technologies. The VPT allows to perform CR at home without the direct supervision of an health professional. Details about the platform are described in the following sub-sections.

2.1 Platform Architecture

In order to limit the costs of the proposed solution and allow its wide diffusion, the platform is made up by low-cost commercial components. The architecture of the platform is shown in Fig. 1. Each end-user (patient) will be equipped with a kit consisting of (a) a set-top-box connected to a TV monitor with Internet connection, (b) a 3D sensor camera and (c) an electronic device for biomedical parameters monitoring. According to the specific rehabilitation program provided by the physician, the set-top-box automatically downloads specific sequences of exercises from a remote server, taking into account the rehabilitation history and related performances. In order to make the system reliable, flexible and compliant with the international evaluation scales (ex. Mini Mental State Examination [12]), few input parameters need to be defined a-priori: execution time, maximum numbers of allowed errors and movement sensitivity. These parameters are set taking into account the residual abilities of the patient and so they determine a customized version of the system:. A web-based platform allows the physician to define the list of exercises that the patient should perform during the rehabilitation session. The platform integrates streaming functionalities allowing visual/audio recording for post-verification or online feedback to the physician which is able to follow the exercises execution by the patient through desktop PCs or mobile embedded solutions. The physician and the psychologist of the reference center could communicate to the patients through a remote connection, and monitor the progress or trouble in the execution of the different required tasks. At the end of the rehabilitation session, the central platform collects different kinds of data locally stored on the set-top-box. An ad-hoc multi-modal messaging procedure (e-mail, SMS, App, ...) is performed and relevant data are sent to the physician allowing instant verification of the performance through an easy-to-use Graphic User Interface. To increase the system acceptability at the begin of the experimentation phase the users will be tutored by a psychologist, could personalize the software background and could edit some specific characteristics. Hopefully this approach will permit to reduce the intrinsic and extrinsic factors that could limit the proper functionality of the tool.

2.2 *Active Vision System for Rehabilitative Practice*

The rehabilitation practice is achieved by a sequence of exercises of different categories, which have the aim to bring out specific cognitive activities, according to guidelines of the state-of-the-art international evaluation scales for Alzheimer patients. From the taxonomic point of view, the following categories have been implemented in the platform: temporal orientation, personnel guidance, topographical memory, visual memory, hearing attention, visual attention and categorization (see Fig. 2a for examples of User Interfaces). Each exercise is performed via a multi-modal contact-less Natural User Interface (NUI), as shown in Fig. 2b, available through the use of both the commercial low-cost Microsoft Kinect© sensor for gesture recognition [13] and the Microsoft Text-To-Speech engine (TTS) for human voice synthesis.

As other active vision systems, Microsoft Kinect©is designed for a 3D scene capture within 4 m range, providing people tracking and behavior/gesture recognition functionalities in a low-cost computational way. In order to interact with the VPT platform, the patient must move the required hand according to the specific rehabilitation exercise: hand tracking algorithms have been implemented providing a customized level of movement sensitivity which is manually tuned by the physician according to the residual abilities of the patient. Gesture recognition procedures have been developed in order to verify the correctness of the hand movement with respect to a previously recorded template, defining accurate score and performance metrics (amount of correct answers, amount of errors, amount of vocal/visual

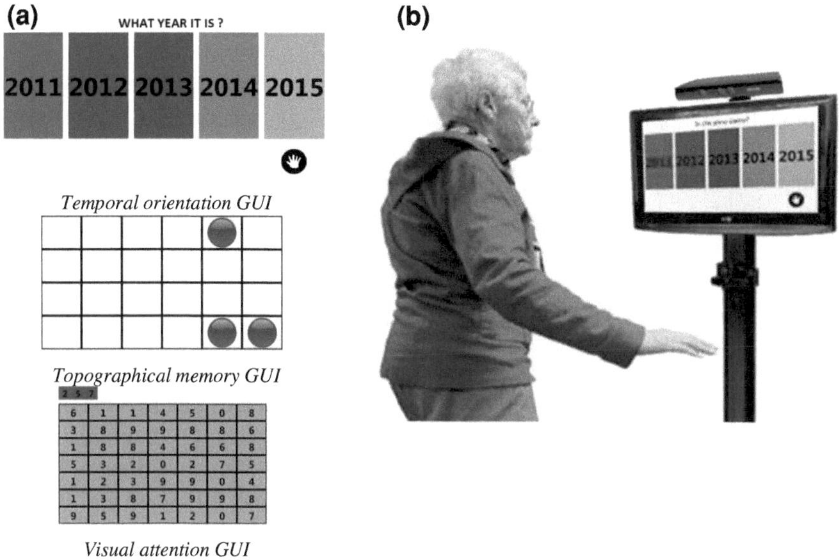

Fig. 2 **a** Graphic User Interfaces for rehabilitation practice. **b** Rehabilitation approach by Natural User Interface

aids, …) according to the quality of the carried out rehabilitation practice. In order to infer more knowledge during the rehabilitation activities, the main clinical parameters (heart rate, breathe rate, Electrocardiogram, …) are monitored by wearing the (optional) Smartex WWS e-shirt [14], integrating several sensing devices for biomedical applications. Through a Bluetooth radio link with the platform, each useful clinical parameter provided by the WWS device is stored on the set-top-box. Each performance metrics of the rehabilitation practice and the aforementioned clinical parameters are sent to the physician with a multi-modal paradigm for clinical evaluations. Moreover, the system allows the patients to establish a visual/audio link with the medical center, so the physician can interact with the patient during the CR, increasing the compliance and the efficacy of CR with the result of keeping the level of stimulation constant, making sure that the type and intensity of treatment is appropriate.

Since several sensible data are recorded, according to the national laws on the privacy, the platform keeps sensible data in the media storage for the shortest possible time, confidentially handled in accordance with the wishes of the patient avoiding data transfer without adequate protection (AES 128bit cryptography is used for data transmission to the remote service server).

3 Preliminary Results

At the beginning of the project a questionnaire to evaluate the expectations and reactions to the application of the proposed technology at home was administered to AD patients and their caregiver . Section 3.1 details the results obtained after the submission of the questionnaire, whereas in Sect. 3.2 a more technical contribution will be provided, clarifying setup parameters for the proper use of the VPT.

3.1 Questionnaire Results

Between January and April 2013 a survey was conducted at the Alzheimer Evaluation Unit of the "Casa Sollievo della Sofferenza" Research Hospital in San Giovanni Rotondo (FG). The inclusion criteria were: (1) age $\geq$65 years, (2) diagnosis of AD mild cognitive impairment according to the criteria of the National Institute of Neurological and Communicative Disorders and Stroke-Alzheimer's Disease and Related Disorders Association Work Group (NINCDS-ADRDA) [15], 3 and to the Diagnostic and Statistical Manual of Mental Disorders-Fourth Edition issues (DSM-IV) [16].

The presence of a signed informed consent from each patient or legal guardian if necessary, and the caregiver. The exclusion criterion was the inability to acquire data due to the poor clinical condition of the patient. For each patient the cognitive status, which considers the psychological aspects that drive the behavior of individuals on the basis of personal resources, both emotionally and mentally, was assessed. To explore this aspect Mini Mental State Examination (MMSE) was used. It allows an evaluation of attention, space-time orientation, short and long term memory, calculation, executive ability, writing, and the appropriate use of language. The score range is between 0 and 30 and includes a specific correction for age and education. In addition, we used the Clinical Dementia Rating scale (CDR) [17], which is a tool for the global assessment of the severity of dementia in elderly subjects with possible or probable cognitive impairment. The CDR was performed through semi-structured interview. In the CDR every domain score ranged between 1 and 5 (0 = absent, 0.5 = questionable, 1 = mild, 2 = moderate, 3 = severe).

The questionnaire included 17 items that explored different areas; the survey included 72 patients and their caregivers. The summary of the results are the following: 51.6 % of patients need care in self-hygiene, 15.6 % in moving at home, 12.5 % in moving out of home and 10.9 % in taking drugs. Among the patients with mild AD: 46.9 % had a slight deficit in personal care, 37.5 % had a mild motor disability, 6.3 % sever loco motor disability, 4.7 % present a serious lack in personal care, and 4.8 % significant sensory deficits. The 48.4 % of patients live with a spouse, 26.6 % with a child or sibling, 23.4 % alone and 1.6 % with a private caregiver. Patients are assisted by family members (92.2 %, with a mean age of 63.59 years $\pm$ 16.50) and private caregiver (7.8 %, with a mean age of 48.20 $\pm$ 2.78 years). Family members take care for patients for an average of 22.6 $\pm$ 2.1 days a week and 6.94 $\pm$ 5.44 h/day; private caregivers caring for patients for 6.00 $\pm$ 0.71 days a week and 12.0 $\pm$ 7.87 h/day.

In the sample 54.7 % of patients like to watch TV and 26.6 % watch preferably film. The 92.2 % of patients are capable to turn on and off the TV, 85.9 % is able to choose the programs in autonomy and use the remote control, the 67.2 % is able to maintain a level of attention from the beginning to the end of a television program; the 79.7 % of patients can use a PC only if helped and 50 % cannot use the PC keyboard. Family members/caregivers have reported in their preferences that the AL.TR.U.I.S.M. system could be useful to facilitate communication with the medical center of reference (34.4 %) and improve the quality of care through constant monitoring of the exercises performed (32.8 %). Moreover, the system could be very useful to improve the quality of life (73.4 %) and quality of care and assistance (73.4 %). Finally 62.5 % of patients could accept the use of interactive TV and devices to monitor vital signs and gesture recognition at a distance.

3.2 Virtual Personal Trainer Considerations

For proper interaction with the system, the patient must be visible (as frontally aligned as possible) from the 3D sensor in a distance range between 80 cm and 4 m: under these conditions, hands detection rate in almost equal to 100 % in every frame.

Each exercise is composed of a series of graphical user interfaces; the patient interacts with these objects (makes choice) if there is the overlap of an object, represented by a cursor which follows the movement of the hand closest to the monitor. In order to consider only significant responses, the system automatically determines whether a response is given by calculating the overlap time (2 s); the system was also tested in crowded environments, its operation has been successful because the procedure of detection and tracking of the patient is performed only on the skeleton closest to the sensor.

4 Conclusions

Great are the expectations from technology application in the support of caregivers and patients affected of AD. The home cognitive rehabilitation could represent an attractive option in this field. Physicians generally accept that cognitive stimulation can improve cognition and behavioural symptoms of patients with dementia and some studies reported a benefit in the integrated approach at home and in the hospital setting. Further development of CR paradigms holds promise in the promotion of ecologically-valid cognitive and functional gains and, potentially, in complementing the impact of new pharmacological agents under development. At the beginning of AL.TR.U.I.S.M. project we evaluated the expectations and reactions to the application of the NUI and ICT technology at home through the administration of a questionnaire to better define the target population. Response encourages the study since a great amount of patients friendly use a TV and about one half uses a PC without specific training. This study demonstrated that relatives/caregivers of older AD patients have a positive attitude towards innovative technology solutions that could help patients and their families to achieve a more independent lifestyle. Indeed, the VPT platform has been considered very useful by most of the interviewed people. The main idea of the proposed system is to promote home cognitive rehabilitation through the development of a "Virtual Personal Trainer" having several benefits from the patient and the caregiver point of views. One of the major task for the future will be the usability of the interface considering the comorbidity and peculiarity of these patients and their caregivers. As a future work, the implemented platform will be accessible to a wide class of patients suffering from AD in order to record the feedback during the usage of this technological tool. This will allow to tune up each component of the platform (Graphic User Interface, knowledge discovery logic, ...) in order to make it highly compliant with the needs and the requirements of the patients, according to the novel User Centered Design paradigm.

Acknowledgments The presented work has been carried out within AL.TR.U.I.S.M. project, driven by an aggregation of SME, and funded by Apulia Region under the health care call for the promotion of regional partnerships for innovation.

References

1. Clare, L., Woods, R.T.: Cognitive Rehabilitation in Dementia. Neuropsychological Rehabilitation (Special Issue). pp. 193–196. Psychology Press Ltd, Hove (2001)
2. Clare, L., Woods, R.T., Moniz Cook, E.D., Orrel, M., Spector, A.: Cognitive rehabilitation in Alzheimer's disease. Aging Clin. Exp. Res. **18**, 141–143 (2006)
3. Neal, M., Briggs, M.: Validation therapy for dementia (Cochrane Review). The Cochrane Library, Issue 2. Chichester, UK (2004)
4. Spector, A., Orrell, M., Davies, S., Woods, B.: Reality orientation for Dementia. Cochrane Database Syst. Rev. **4**:CD001119 (2000) Review
5. Woods, R.T., Bruce, E., Edwards, R.T., Hounsome, B., Keady, J., Moniz-Cook, E.D., Orrell, M., Russell, I.T.: Reminiscence groups for people with dementia and their family carers: pragmatic eight-center randomised trial of joint reminiscence and maintenance versus usual treatment: a protocol. Trials **10**, 64 (2009)
6. Pilotto, A., D'Onofrio, G., Benelli, E., Zanesco, A., Cabello, A., Margelí, M.C., Wanche-Politis, S., Seferis, K., Sancarlo, D., Kilias, D. and on behalf of the HOPE Investigators: Information and Communication Technology systems to improve quality of life and safety of Alzheimer's disease patients: a multicenter international survey. J. Alzheimers Dis. **23**, 131–141 (2011)
7. Hofmann, M., Rösler, A., Schwarz, W., Müller-Spahn, F., Kräuchi, K., Hock, C., Seifritz, E.: Interactive computer-training as a therapeutic tool in Alzheimer's disease. Compr. Psychiatry **44**(3), 213–9 (2003)
8. Imbeault, F., Bouchard, B., Bouzouane, A.: Serious games in cognitive training for Alzheimer's patients. In: IEEE International Conference on Serious Games and Applications for Health (IEEE-SeGAH), Braga, Portugal, November 16–18, pp. 122–129 (2011)
9. Jiang, C.-F., Chen, D.-K., Li, Y.-S., Kuo, J.-L.: Development of a computer-aided tool for evaluation and training in 3d spatial cognitive function. In: 19th IEEE Symposium on Computer-Based Medical Systems, pp. 241–244 (2006)

10. Nacke, L.E., Nacke, A., Lindley, C.A.: Brain training for silver aged gamers: effects of age and game form on effectiveness, self-assessment, and gameplay. Cyberpsychol. Behav. **12**(5), 493–499 (2009)
11. United Nations (UN): World Population Ageing 2009, Department of Economic and Social Affairs: Population Division, 129 p (2009)
12. Folstein, M., Folstein, S., McHugh, P.R.: MiniMentalState: a practical method for grading the cognitive state of patients for the clinician. J. Psychiatr. Res. **12**, 189–98 (1975)
13. http://www.microsoft.com/en-us/kinectforwindows/
14. http://www.smartex.it/index.php/it/prodotti/wearable-wellness-system
15. McKhann, G., Drachman, D., Folstein, M., Katzman, R., Price, D., Stadlan, E.M.: Clinical diagnosis of Alzheimer Disease: report of the NINCDS-ADRDA work group under the auspices of Department of Health and Human Service Task Force on Alzheimer's Disease. Neurology **34**, 939–44 (1984)
16. American Psychiatric Association: Diagnostic and Statistical Manual of Mental Disorders, 4th edn. American Psychiatric Association, Washington (1994)
17. Hughes, C.P., Berg, L., Danziger, W.L., Coben, L.A., Martin, R.L.: A new clinical scale for the staging of dementia. Br. J. Psychiatry **140**, 566–72 (1982).

Experimentation of an Integrated System of Services and AAL Solutions for Alzheimer's Disease Patients and Their Caregivers in Marche: The UP-TECH Project

F. Barbabella, C. Chiatti, F. Masera, F. Bonfranceschi, J.M. Rimland, K. Bartulewicz, L. Rossi and F. Lattanzio

Abstract Alzheimer's Disease (AD) is one of the most crucial challenges in public health today. In addition to the negative impact of AD on the patient, the effect of this disease on family caregivers also needs to be considered, as they are often deeply burdened psychologically and emotionally. Many projects have tried to exploit ambient assisted living (AAL) technologies to support AD patients and their families, but evidence is so far inconclusive. The UP-TECH project aims at investigating an integrated package of care services and AAL solutions to support family caregivers and to promote the security and the independence of the patient at home. The study is designed as a randomized controlled trial (RCT) and includes the recruitment of 438 patient-

On the behalf of the UP-TECH research team: The paper has been prepared on behalf of the UP-TECH research team: Social workers: Diletta Baldassarri, Luigina Bitti, Aurora Carosi, Maila Sabbatini, Elisabetta Paolasini, Tania Fiore, Cinzia Fronzi, Simona Giacchetta, Valeria Giacomini, Laura Giovagnoli, Giuli Lattanzi, Anna Maria Manca, Ida Marinelli, Valeria Pigini, Marina Pignotti, Maria Chiara Proietti, Antonia Quarticelli, Martina Rovedi, Valentina Valeri. Nurses: Andrea Antonioli, Mariella Barabucci, Mara Bassani, Paola Bollettini, M. Cristina Bruttapasta, Carla Buccolini, Rosa Carangella, Franca Carboni, Daniela Ceccolini, Patrizia D'Incecco, Mariangela Di Felice, Lara Dini, Giordano Gioia, Giuseppe Di Prima, Giancarlo Giusepponi, Luciano Farinelli, Cesarina Lanciotti, Alessia Loffreda, Oriana Luciani, Laura Mariani, Francesco Mastrorilli, Roberto Moroni, Izabela Piatowska, Roberta Rucoli, Giampaola Scoccia, Tiziana Tonelli. Coordinators of Social Districts and Management of Local Services: Daniela Alessandrini, Antonio De Santis, Brunetta Formica, Giuliano Tacchi, Silvia Tortorelli. Management of medical doctors: Rosa Anna Bratti, Giuseppe Bonafede, Severino Lorenzetti, Cristina Paci, Giovanna Picciotti, Donella Pezzola, Vincenzo Rea, Vincenzo Sciale. Marche Region: Gemma Marchegiani, Giovanni Santarelli. Scientific and Technical Committee: Giorgio Caraffa, Filippo Cavallo, Lucia Di Furia, Antonio Lacetera, Claudio M. Maffei, Lamberto Manzoli, Demetrio Postacchini. INRCA Research Team: Roberta Bevilacqua, Anna Rita Bonfigli, Silvia Bustacchini, Marina Capasso, Laura Cassetta, Patrizia Civerchia, Gianluca Furneri, Paolo Marinelli, Gabriella Melchiorre, Maria Elena Moraca, Andrea Principi, Cristina Rocchetti, Daniela Vincitorio

F. Barbabella (✉)
Centre for Socio-Economic Research on Ageing, Italian National Institute of Health and Science on Ageing (INRCA),Via Santa Margherita 5, 60124 Ancona, Italy
e-mail: f.barbabella@inrca.it

S. Longhi et al. (eds.), *Ambient Assisted Living*, DOI: 10.1007/978-3-319-01119-6_16,

caregiver dyads, assigned to one of the two experimental groups or to the control group. In particular, one experimental group will receive a package of case management services and assistive devices installed at home, including activity sensors and alarms. The project will assess the impact of the tested AAL solutions using validated scales at a follow-up of 12 months. A cost-effectiveness analysis will also be realized in order to assess the likelihood of implementing such services on a wider scale.

1 Background

In 2010, it has been estimated that worldwide there were 35.6 million people with dementia. It is thought that this number will almost double to 65.7 million by 2030 and reach 115.4 million by 2050 [1]. In 2005 in Italy, the Italian National Institute of Statistics (ISTAT) calculated that there were roughly 250,000 people affected by Alzheimer's disease and similar dementias [2].

Family caregivers provide the bulk of informal care and often are referred to as "hidden second patients" [3]. In fact, AD heavily affects patients' families, on whom the main burden of care falls, putting caregivers at high risk of stress, anxiety, mortality and lower quality of life [4]. As a result, patients living with distressed caregivers are at higher risk of behavioral disturbances, agitation, use of anti-psychotic drugs and institutionalization [5].

Despite controversies [6], studies have shown that specific interventions aimed at supporting caregivers of AD patients can engender considerable improvements in their physical and mental health, by reducing the burden of caregiving and stress [7]. In this field, ambient assisted living (AAL) technologies have great potential

C. Chiatti · F. Masera · F. Bonfranceschi · J.M. Rimland · K. Bartulewicz · F. Lattanzio
Scientific Direction, Italian National Institute of Health and Science on Ageing (INRCA), Via Santa Margherita 5, 60124 Ancona, Italy
e-mail: c.chiatti@inrca.it

F. Masera
e-mail: f.masera@inrca.it

F. Bonfranceschi
e-mail: f.bonfranceschi@inrca.it

J.M. Rimland
e-mail: j.rimland@inrca.it

K. Bartulewicz
e-mail: k.bartulewicz@inrca.it

F. Lattanzio
e-mail: f.lattanzio@inrca.it

L. Rossi
Laboratory of Bioinformatics, Bioengineering and Home Automation, Italian National Institute of Health and Science on Ageing (INRCA), Via Santa Margherita 5, 60124 Ancona, Italy
e-mail: l.rossi@inrca.it

to ameliorate the quality of life of AD patients and their caregivers. Notwithstanding the absence of definitive evidence [8], pilot studies have revealed that domestic adaptations designed to aid older people to carry-out their daily needs are especially valuable for users' quality of life and well-being [9]. Several studies assessing the effectiveness of assistive technology in AD have been undertaken; however, most of these interventions are in a prototype/testing phase and fail to be scaled-up and systematically implemented in daily practice. A few of these projects include: Talking lights [10], Home Assurance System [11], ROSETTA [12] and ENABLE [13]. The latter project involved the testing of an integrated system of technologies in the home, including automatic night lights, locators for lost objects, computerized calendars, programmable telephones, gas sensors, touch screen pads for music and a voice prompter when medicines should be taken. These technologies led to increased independence of the older person and reduced caregivers' burden.

On the basis of previous studies, the UP-TECH project has been designed as a randomized, controlled trial (RCT), in order to adapt the most promising evidence-based interventions in dementia care into 5 different areas of the Marche Region in Italy [14]. The main objectives of the project are: (1) to reduce the burden of family caregivers of AD patients; and (2) to maintain AD patients at home preventing premature institutionalization. This will be achieved by the use of case-management strategies, new technologies in the home of patients, preventive home visits by trained nurses and integration of existing services.

The UP-TECH project is financed by the Marche Region through the 2010 National Fund for Dependent People and coordinated by the Italian National Institute of Health and Science on Aging (INRCA), with the support of ASUR Marche and the local social services in Ancona, Fermo, Macerata, Pesaro, and San Benedetto del Tronto.

2 Materials and Methods

A brief overview of materials and methods used in the UP-TECH project is provided in the following paragraphs. For more details concerning the study protocol, please refer to the previous publication by Chiatti and colleagues [14].

2.1 Study Design

The UP-TECH project is a multi-component RCT, lasting 12 months, aimed at enrolling 450 AD patient-caregiver dyads, allocated to 3 different arms of 150 dyads each, which receive the following interventions:

1. experimental group 1 ("UP-group"): dyads are followed by a case manager and receive three nurse home visits in a 12-month period;
2. experimental group 2 ("UP-TECH group"): dyads are followed by a case manager as in the previous "Up-group", with additional AAL technologies installed at home with support from the case manager;
3. control group: dyads receive standard assistance, including three nurse home visits and informative brochures on the management of AD.

The case management intervention is carried out by appropriately trained social workers and includes counseling sessions, scheduled follow-up phone calls, training sessions for family caregivers on stress management and caregiving activities, social service administration and establishment of links between healthcare services and general practitioners. Nurse home visits include a multidimensional evaluation of patients, together with a training intervention for the caregivers addressing issues such as caregiving activities, nutrition, drug administration and ergonomics of the domestic environment. The AAL technologies (the UP-TECH kit) were defined using a bottom-up approach, through extensive consultations with patients, professionals and care staff (see below).

The RCT will assess different dimensions, including the following measures:

1. Caregiver burden: the Italian version of the Caregiver Burden Inventory (CBI) validated by Marvardi and colleagues [15];
2. The number of days spent at home by the AD patient during the observation period. This is a continuous variable calculated at each intermediate (after 6 months), final (after 12 months) and follow-up assessments (after 24 months) by subtracting, from the total days of observation, the number of days of: institutionalization in a residential care facility, care home and/or nursing home; hospitalization and brief stays in an observation unit of an emergency room;
3. Caregiver quality of life: the Italian validated version of the SF-12 questionnaire [16, 17];
4. Caregiver anxiety and depression: the Italian validated version of the Hospital Anxiety and Depression Scale (HADS) [18, 19].

2.2 Eligibility Criteria

Patients with AD were enrolled if they met specific inclusion criteria: they had a pre-existing diagnosis of AD, were in an intermediate stage—according to the 2011 criteria of the National Institute on Aging-Alzheimer's Association (NIA-AA) [20–22]—, with a mini-mental state examination (MMSE) score [23] between 10 and 20, and lived in the community with the assistance of at least one family caregiver.

Family caregivers were defined as those relatives who cared for and directly supported the AD patient with the activities of daily living (ADL) and the instrumental activities of daily living (IADL), for at least 1 hour per day within the last

6 months. Only the primary caregiver (i.e. the family member spending the most time in caregiving activities) was included in the study, if a patient was cared for by more than one family member.

2.3 Definition of the AAL Solutions

The selection of technologies installed in patients' homes was based on a user-centered design, in order to address specifically the needs of patients and their caregivers. One focus group with 9 family caregivers was organized, aimed at understanding what kind of assistive device solutions could address their needs, make them comfortable and respect privacy and ethical aspects (also in relation to the patient).

The focus group followed structured guidelines for moderators which included two stages: a discussion on users' needs and the ways in which technology can address problems in ADLs; a discussion focused on technology acceptance by patients and caregivers, as well as issues related to sustainability and integration with formal care services. In a second step, AAL solutions were discussed with professionals and care staff during the five preliminary site visits carried out by the UP-TECH management team.

3 Results

3.1 Recruitment Process and Description of the Study Participants

The recruitment timeframe of this study was less than 6 months (January 14th, 2013–June 7th, 2013). It led to the enrollment of 438 out of the 450 planned patient-caregiver dyads, which were interviewed by a nurse trained to collect data. At enrollment, patients had a mean MMSE score of 16 (±3) points (Table 1). A majority of the subjects were women (71.5 %) with a mean age (±±SD) of 81.5 ± 5.7 years. These patients had a significant impairment in IADLs (35.1 ± 13.4 points out of 48 points), while they were relatively independent performing ADLs (1.5 ± 1.6 points out of 6 points on the ADL hierarchy scale). 26.5 % of the patients displayed behavioral problems as reported by the caregivers. There were no statistically significant differences in patient characteristics among the 3 intervention groups. A higher percentage of caregivers were women (66.2 %) with an average age of 61.4 ± 13 years. 30.8 % were spouses, 55.7 % were children and 13.5 % were another relative. 30.1 % of the caregivers had no formal education, 24.9 % had a primary level of education (elementary/middle school), 36.7 % possessed an intermediate level of instruction (high school), while 8.2 % reached a high level of schooling (university). Also in this case, there were no statistically significant differences in caregiver characteristics among the 3 groups.

Table 1 Basic socio-demographic, health and psychological characteristics of the sample

	Up (N = 147)		UP-TECH (N = 146)		Usual care (N = 145)		Total (N = 438)		P*
Patients' characteristics									
Gender (female)	68.0		71.9		74.5		71.5		0.47
Age (years)	81.0	±5.2	82.0	±6.0	81.4	±5.8	81.5	±5.7	0.28
IADL impairment	34.5	±13.4	37.1	±11.6	33.8	±14.8	35.1	±13.4	0.08
ADL impairment	1.5	±1.6	1.6	±1.7	1.4	±1.5	1.5	±1.6	0.61
MMSE	15.9	±3.2	15.8	±3.0	16.3	±2.9	16.0	±3.0	0.29
Behavioral disturbances	27.2		27.4		24.8		26.5		0.85
Caregivers' characteristics									
Gender (female)	66.7		59.6		72.4		66.2		0.07
Age (years)	63.3	±13.6	61.2	±12.6	59.6	±13.0	61.4	±13.0	0.06
Relationship									0.26
Spouse	37.4		28.1		26.9		30.8		
Son/Daughter	49.0		59.6		58.6		55.7		
Other relative	13.6		12.3		14.5		13.5		
Education									0.11
No title	38.1		26.0		26.2		30.1		
Primary level	22.5		21.9		30.3		24.9		
Intermediate level	32.7		43.2		34.5		36.7		
High level	6.8		8.9		9.0		8.2		

Values are expressed in either "Percentage" or "Mean ± Standard Deviation" where appropriate.
* p-value of the Chi2 test for association of categorical variables and ANOVA test for continuous variables

3.2 Development of the AAL Solutions

The main functions, proposed by users in the focus group and integrated with care staff opinions and experiences, that could be fulfilled by AAL solutions, were the following:

- Controlling home access through camera systems with automatic recognition. A common problem was that some patients would open the door to anybody, even strangers, and let them in;
- Carrying out bureaucratic and administrative procedures for accessing formal care services through an integrated on-line system;
- Sharing patient health care data among professionals and units involved in the care process through information technology (IT) systems;
- Entertaining patients and providing companionship through multimedia and interactive/automated applications;
- Checking the activity and the health status of patients through monitoring systems;
- Sending an alarm, when an accident occurs, through emergency buttons and similar devices;

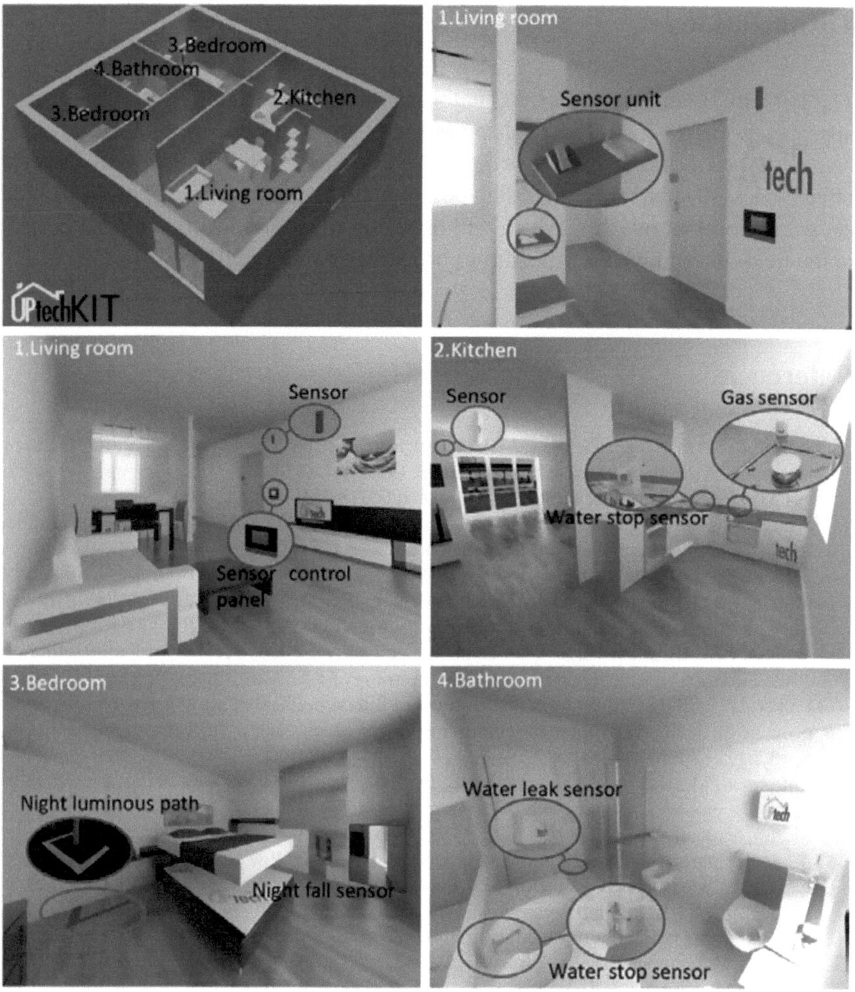

Fig. 1 The UP-TECH kit of AAL solutions

- Using sensors to send a passive alarm if the patient falls;
- Communicating in real time with the privately-employed care assistant and monitoring his/her activity.

Trying to address the expressed needs, the research team developed an AAL solution consisting of a kit that includes luminous paths, automatic lights and assorted detection sensors which activate when a person leaves the home, falls at night or when there is a gas or water leak. These devices are assembled by an external contractor and linked to a single-board microcontroller, which will transmit alarm messages, when triggered, through the cell phone system to the family caregivers. Assistive technologies will be installed by the case manager with the support of expert technicians who will also train caregivers on their use (Fig. 1).

4 Conclusions

The UP-TECH study is one of the major test-beds, worldwide, to investigate novel approaches to AD care. The forthcoming implementation of an AAL intervention on such a large scale will produce definitive evidence on the effectiveness and cost-effectiveness of such tools, providing policy-makers, industries and other stakeholders with important insights, which will be useful to tailor new policies and projects in the field. Interim results will start to be available during 2014.

References

1. Wimo, A., Prince, M.: The World Alzheimer Report 2010: The Global Economic Impact of Dementia. Alzheimer's Disease International, London (2010)
2. ISTAT: Health and Use of Health Services in 2005. ISTAT, Rome (2008)
3. Smith, G.C., Tobin, S.S., Robertson-Tchabo, E.A., Power, P.W.: Strengthening Aging Families: Diversity in Practice and Policy. Sage, Thousand Oaks, CA (1995)
4. Schulz, R., Burgio, L., Burns, R., Eisdorfer, C., Gallagher-Thompson, D., Gitlin, L.N., Mahoney, D.F.: Resources for enhancing Alzheimer's caregiver health (REACH): overview, site-specific outcomes, and future directions. Gerontologist **43**, 514–520 (2003)
5. Dunkin, J.J., Anderson-Hanley, C.: Dementia caregiver burden: a review of the literature and guidelines for assessment and intervention. Neurology **51**, S53–60, (discussion S65–57) (1998)
6. Parker, D., Mills, S., Abbey, J.: Effectiveness of interventions that assist caregivers to support people with dementia living in the community: a systematic review. Int. J. Evid. Based Healthc. **6**, 137–172 (2008)
7. Mittelman, M., Haley, W., Clay, O., Roth, D.: Improving caregiver well-being delays nursing home placement of patients with Alzheimer disease. Neurology **67**, 1592–1599 (2006)
8. Martin, S., Kelly, G., Kernohan, W.G., McCreight, B., Nugent, C.: Smart home technologies for health and social care support. Cochrane Database Syst. Rev. **8**(4), CD006412 (2008)
9. Fänge, A., Iwarsson, S.: Changes in ADL dependence and aspects of usability following housing adaptation—a longitudinal perspective. Am. J. Occup.Ther. **59**, 296–304 (2005)
10. Talking Lights project. www.talking-lights.com
11. Cuddihy, P., Hinman, R.T., Avestruz, A., Lupton, E.C., Livshin, G., Rodriguez, J.I., Leeb, S.B., Clark, C.M., Horvath, K.J., Volicer, L., Landfeldt, B., Kay, J.: Successful aging. IEEE Pervasive Comput. **3**, 48–50 (2004)
12. ENABLE project. www.enableproject.org
13. ROSETTA project. www.aal-rosetta.eu
14. Chiatti, C., Masera, F., Rimland, J., Cherubini, A., Scarpino, O., Spazzafumo, L., Lattanzio, F.: The UP-TECH project, an intervention to support caregivers of Alzheimer's disease patients in Italy: study protocol for a randomized controlled trial. Trials (2013). doi:10.1186/1745-6215-14-155
15. Marvardi, M., Mattioli, P., Spazzafumo, L., Mastriforti, R., Rinaldi, P., Polidori, M.C., Cherubini, A., Quartesan, R., Bartorelli, L.: The caregiver burden inventory in evaluating the burden of caregivers of elderly demented patients: results from a multicenter study. Aging Clin. Exp. Res. **17**, 46–53 (2005)
16. Gandek, B., Ware, J.E.: Methods for validating and norming translations of health status questionnaires: the IQOLA Project approach. International quality of life assessment. J. Clin. Epidemiol. **51**, 953–959 (1998)
17. Kodraliu, G., Mosconi, P., Groth, N., Carmosino, G., Perilli, A., Gianicolo, E.A., Rossi, C., Apolone, G.: Subjective health status assessment: evaluation of the Italian version of the SF-12 health survey. Results from the MiOS project. J. Epidemiol. Biostat. **6**, 305–316 (2001)

18. Zigmond, A.S., Snaith, R.P.: The hospital anxiety and depression scale. Acta Psychiatr. Scand. **67**, 361–370 (1983)
19. Costantini, M., Musso, M., Viterbori, P., Bonci, F., Del, M.L., Garrone, O., Venturini, M., Morasso, G.: Detecting psychological distress in cancer patients: validity of the Italian version of the Hospital Anxiety and Depression Scale. Support. Care Cancer **7**, 121–127 (1999)
20. Albert, M.S., Dekosky, S.T., Dickson, D., Dubois, B., Feldman, H.H., Fox, N.C., Gamst, A., Holtzman, D.M., Jagust, W.J., Petersen, R.C., Snyder, P.J., Carrillo, M.C., Thies, B., Phelps, C.H.: The diagnosis of mild cognitive impairment due to Alzheimer's disease: recommendations from the National Institute on Aging-Alzheimer's Association workgroups on diagnostic guidelines for Alzheimer's disease. Alzheimers Dement **7**, 270–279 (2011)
21. McKhann, G.M., Knopman, D.S., Chertkow, H., Hyman, B.T., Jack, C.R., Kawas, C.H., Klunk, W.E., Koroshetz, W.J., Manly, J.J., Mayeux, R., Mohs, R.C., Morris, J.C., Rossor, M.N., Scheltens, P., Carrillo, M.C., Thies, B., Weintraub, S., Phelps, C.H.: The diagnosis of dementia due to Alzheimer's disease: recommendations from the National Institute on Aging-Alzheimer's Association workgroups on diagnostic guidelines for Alzheimer's disease. Alzheimers Dement **7**, 263–269 (2011)
22. Sperling, R.A., Aisen, P.S., Beckett, L.A., Bennett, D.A., Craft, S., Fagan, A.M., Iwatsubo, T., Jack, C.R., Kaye, J., Montine, T.J., Park, D.C., Reiman, E.M., Rowe, C.C., Siemers, E., Stern, Y., Yaffe, K., Carrillo, M.C., Thies, B., Morrison-Bogorad, M., Wagster, M.V., Phelps, C.H.: Toward defining the preclinical stages of Alzheimer's disease: recommendations from the National Institute on Aging-Alzheimer's Association workgroups on diagnostic guidelines for Alzheimer's disease. Alzheimers Dement **7**, 280–292 (2011)
23. Folstein, M.F., Folstein, S.E., McHugh, P.R.: "Mini-mental state". A practical method for grading the cognitive state of patients for the clinician. J. Psychiatr. Res. **12**, 189–198 (1975)

Virtual Reality: A New Rehabilitative Approach in Neurological Disorders

Laura Di Biagio, Matteo Ferretti, Daniele Cingolani, Luca Buzzatti, Marianna Capecci and Maria Gabriella Ceravolo

Abstract Rehabilitation after brain damage is aimed to achieve functional recovery, through experience-based neuroplasticity. Virtual reality is an innovative rehabilitation approach that exploits the principles of modern "motor learning" theory. We assessed the effectiveness of rehabilitation approaches, integrating the use of an easily disposable device for virtual reality (Nintendo® Wii), in either upper limb recovery or balance recovery, in chronically disabled brain injured subjects. In hemiplegic stroke patients, a continuous training consisting of ten Nintendo Wii game sessions, 1 h a day, was useful to promote upper limb dexterity recovery; greater benefits were appreciated in subjects with mild cognitive impairment, with respect to those without cognitive troubles. In the second study, 20 sessions of Nintendo Wii Balance training, lasting 50 min each, 5 days in a week, were found effective at improving the balance and gait performance in patients suffering from Parkinson's disease. Both findings lead to believe that a continuous visual feedback, as provided by the Nintendo Wii game, may represent a facilitating stimulus, that helps brain injured patients to maintain a focused attention during task execution. Rehabilitation training through Nintendo Wii is a safe, easy and effective approach to be used by chronically disabled people, suffering from neurological disorders. It would represent a useful tool for continuous self-administered home training that stimulates motivation and enhances formal rehabilitation effects.

L. Di Biagio (✉) · M. Ferretti · D. Cingolani · L. Buzzatti · M. Capecci · M. G. Ceravolo
Department of Experimental and Clinical Medicine, Politecnica delle Marche University, Ancona, Italy
e-mail: lauradibiagio@gmail.com

M. Ferretti
e-mail: drmatteoferretti@gmail.com

M. G. Ceravolo
e-mail: m.g.ceravolo@univpm.it

S. Longhi et al. (eds.), *Ambient Assisted Living*, DOI: 10.1007/978-3-319-01119-6_17,

1 The Use of Virtual Reality in Neurological Rehabilitation

Task oriented techniques have acquired increasing importance in neurological rehabilitation. The use of a virtual reality (VR) environment, combined with task-oriented practice, allows neurologically disabled patients to perform repetitive, context and feedback-enhanced tasks, also exploiting motor imagery paradigm.

1.1 Bases of Motor Learning

Rehabilitation after brain damage is aimed to achieve functional recovery through experience-based neuroplasticity. It has been shown that synaptogenesis and brain connectivity may be enhanced through motor practice provided that "motor learning" principles are acknowledged. This means to comply with a few key elements such as: massive practice (both intensive and distributed over time), active subject participation, contextualization and generalization. Active participation of disabled individuals in the rehabilitation program is mandatory in order to stimulate motor learning. Therefore, it is crucial that the subject is aware of the exercise meaning, so to progressively refine movement strategies. Rehabilitative programs that make use of specific-goal approaches (task-oriented practice) prove to be more efficient as compared to those focusing on muscle strength increase, or muscle spasticity reduction. In this sense, it has been proposed that motor rehabilitation should be regarded by patients as a problem-solving process, rather than an exercise program made of different sequences of muscle movements to be memorized and repeated [1].

1.2 The Mirror Neuron Paradigm

Movement imagination and observation may represent an additional source of useful information for motor rehabilitation in stroke patients [2]. This is because brain areas normally involved in the planning and execution of a movement (i.e. the prefrontal cortex, the premotor cortex, the supplementary motor area, the cingulate cortex, the parietal cortex and the cerebellum) are also active when imagining that same movement. It has been suggested that a neuron population, called "mirror neurons", is involved in decoding the purpose of the action, and also in the understanding and predicting the actions of other subjects. Rehabilitation protocols exploiting action observation with the intention of movement imitation have been found more effective at promoting functional recovery after stroke, than conventional rehabilitation protocols [3]. A good practice point is to design exercises consisting of transitive actions, where the use of concrete daily objects is requested.

1.3 Virtual Reality

Virtual reality is an innovative rehabilitation approach that exploits "motor learning" theory. It is a computerized technology that artificially evokes sensory information and allows to produce experience and interaction with three-dimensional environments.
VR systems create a computer-simulated environment that can mimic physical presence in the real world or imaginary places. Most current virtual reality environments are visual experiences, displayed either on a computer screen or through special stereoscopic displays, but some simulations include additional sensory information, such as sound through speakers or headphones. Some advanced, haptic systems also include tactile information.
When applied to the rehabilitation of individuals with motor, cognitive and sensory disabilities, VR may allow for a high degree of emotional involvement, increased motivation and, in some cases, increased performances.
The "augmentative feedback" (through the visual, auditory, or somatosensory channels) is the key element designed to enhance the learning of the task. The results achieved through this technique can be easily generalized and applied to real life.
Expensive devices have been developed and successfully employed in clinical practice to provide patients with virtual environment illusions, that deliver intense visual feedbacks, during motor training. Cheaper tools, like Play Station equipped with Eye Sport software, and Nintendo Wii, are available in retail stores, thus offering many advantages in terms of cost and handiness.

1.4 Nintendo Wii

This technology uses a remote wireless "controller" (Wiimote), to grip with one or both hands, which detects movements of the patient or player. Through the infrared LEDs incorporated into the Wii Sensor Bar (placed above or below the television set) the controller perceives the pointing toward the screen, while the accelerometer and the gyroscope, integrated into the same controller, perceive the inclination and the rotation. This allows a representation of the patient on the screen, that represents a continuous visual feedback, thus providing the subject with a constant driving during task execution.
The graphics simplicity, the possibility to vary the game speed, the linearity of movement trajectories required by most games, make the device easy to use even by patients with cognitive impairment.
We investigated the effectiveness of rehabilitation approaches integrating the use of Nintendo Wii in either upper limb recovery or balance recovery, in chronically disabled brain injured subjects

2 Effectiveness of Virtual Reality Use Versus Task Oriented Practice in Post Stroke Upper Limb Recovery

It is estimated that 50–75 % of stroke survivors suffer from persistent upper limb (UL) impairment. There is a need to identify the best training strategies for retraining dexterity. Studies in computational neuroscience have demonstrated that VR technology, providing enhanced feedback of movement characteristics, improved motor performances in healthy subjects, as compared to traditional training. Following these observations, several authors have used VR-based rehabilitation to enhance motor relearning in post-stroke patients [4–6].
The purposes of this study are:

1. To assess the effectiveness of a rehabilitation treatment using virtual reality through "Nintendo Wii" technology, for upper limb recovery in post stroke hemiparesis, as compared to a task-oriented practice.
2. To verify the persistence of improvement at 3 months of treatment end.

2.1 Materials and Methods

Ten chronic stroke patients (seven men, three women) with mild to moderate upper limb paresis were enrolled and random assigned to two treatment groups (five subjects each). Rehabilitation consisted of 30-min sessions of either Nintendo Wii training (Cases) or upper limb task-oriented training (Controls), for two weeks, on five consecutive days per week. Each session was followed by 30 min treadmill training, in both groups.

2.1.1 Assessment Protocol

Clinical and functional assessment was realized before (T0) and after treatment (T1) as well as one (T2) and three months later (T3). The outcome measures were: upper limb Motricity Index to assess muscle strength, Fugl-Meyer (F-M) and Wolf Motor Function Test (WMFT) to evaluate dexterity.

2.1.2 Interventions

Cases: patients used the Nintendo-Wii technology to practice the following exercises: Shooting gallery, Hockey, Bowling, Jogging, Taurus, Tank, Fishing, Tennis, Baseball. Table 1 details upper limb movements required to complete each task.
Patients also performed balance and proprioceptive exercises, like Marbles, Penguin and Bubble Soap.

Table 1 Upper limb movements trained during the different Wii programs (balance games have been omitted)

Required upper limb movements(s)	Shoulder		Elbow		Wrist		Fingers
	Elevation	Abduction adduction	Flexion extension	Pronation supination	Flexion extension	Pronation supination	Flexion extension
Game							
Shooting gallery	X	X	X				X
Hockey		X	X	X	X		
Bowling	X		X	X			X
Jogging	X		X	X			
Taurus			X	X	X	X	
Tank			X	X	X		X
Fishing	X		X	X	X		
Tennis	X	X	X	X	X		X
Baseball		X	X	X	X		

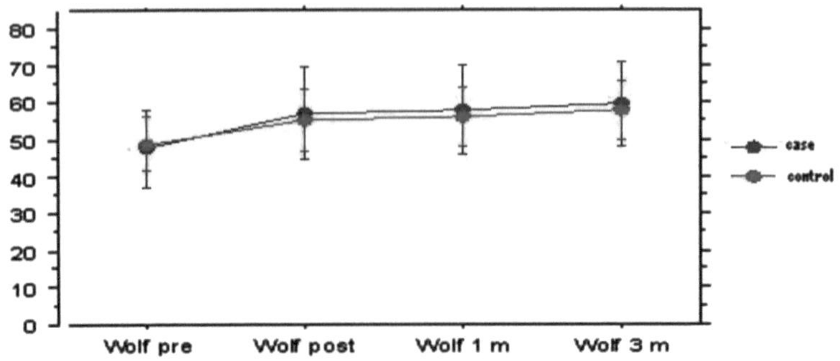

Fig. 1 Evolution of Wolf Motor Function Test score over time in the two treatment groups. No differences may be appreciated

Controls: the patients performed "task-oriented practice". They were sitting in front of a table opposite to the therapist, and practiced six different tasks that were relevant to activities of daily life (i.e.: combing hair, reaching, grasping and lifting a glass, tying a knot, using fork and knife).

2.2 Results

Statistical analysis used analysis of variance for repeated measures to document the trend in outcome measures over time (T0, T1, T2 and T3). At baseline, Cases and Controls did not differ by upper limb function as assessed through WMFT and F-M. Both measures showed a significant improvement after treatment in all subjects.
In particular: WMFT score improved by about 20 %, the benefit persisting unchanged throughout the study (Fig. 1)
When the performance of cognitively impaired patients was separately analyzed, it was possible to observe the following: (a) subjects with cognitive impairment (both in the Cases and Controls groups) had baseline scores lower than subjects with preserved cognition; (b) Cases suffering from cognitive impairment exhibited a greater relative improvement after treatment, as compared to all other subgroups (Fig. 2).
The same trend was appreciated when studying F-M score trend (Fig. 3)
The Analysis of variance for repeated measures confirmed the significant change in WMFT score after treatment in both groups ($p < 0.001$), and also documented a significant "time × treatment" interaction when comparing the trends in subgroups with and without cognitive impairment ($p < 0.05$).
The findings support the usefulness of a cheap, easily disposable technology to improve upper limb function in chronic stroke survivors with mild to moderate dexterity impairment. Furthermore, the investigation shows the feasibility and efficacy of an home-based VR-enhanced rehabilitation for subjects with mild cognitive

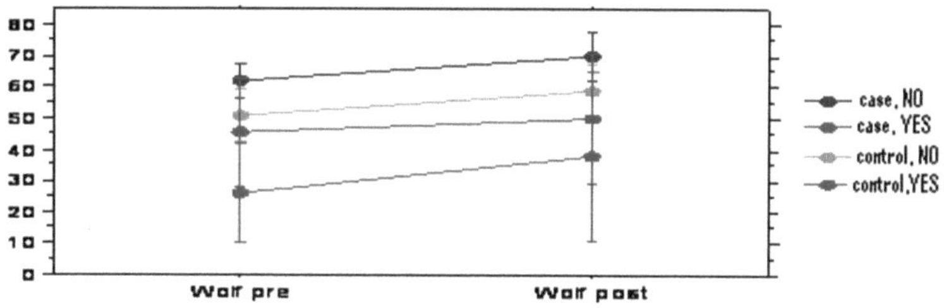

Fig. 2 Evolution of WMFT scores after treatment (Wolf pre:T0 Wolf post: T1) in the four subgroups of Cases and Controls with (Yes) and without (No) cognitive impairment

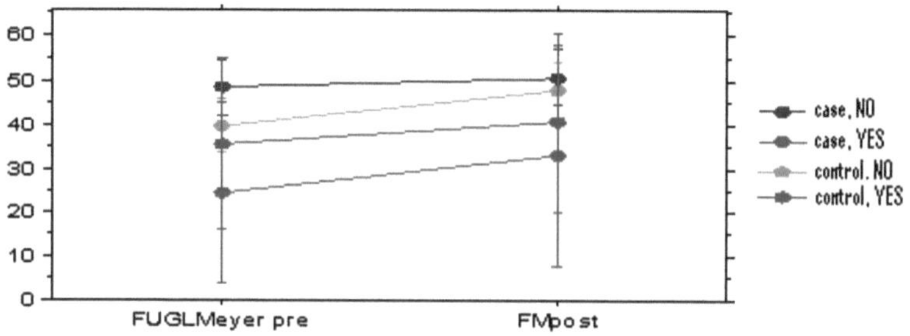

Fig. 3 Evolution of F-M scores after treatment (FUGLMeyer pre:T0 FMpost: T1) in the four subgroups of Cases and Controls with (Yes) and without (No) cognitive impairment

impairment. Such observation is in line with the hypothesis that VR approaches are especially useful in cases where learning abilities may be impaired. This approach, however, remains reserved for patients able to grasp and manipulate the joystick excluding subjects with moderate or severe paresis.

3 Additional Value of Virtual Reality in Balance Rehabilitation of Parkinson's Disease Patients

People with idiopathic Parkinson's disease (PD) commonly experience postural instability during daily activities. Impaired sensory and visual inputs, and a reduced base of support are reported to be related to postural instability in people with PD. Furthermore, deficits in the central integration of vision, somatosensation, and vestibular inputs as well as a cortically mediated decline in attention capacities, may contribute to postural instability in PD people. The effects of virtual reality-augmented balance training on sensory organization and attentional demand for postural control have been previously described in PD people [7]. The purposes of this study are: (a) to evaluate the effectiveness of 20 balance rehabilitation sessions exploiting

Nintendo-Wii-Balance device, as compared to training on a Stabilometric platform (Balance Platform Cosmogamma), at improving balance in PD patients, and (b) to follow-up balance performances for at least 3 months.

3.1 Study Design

Experimental Randomized Cross-Over Study with follow-up at the end of treatment and at 3 months.

3.2 Materials and Methods

Seven PD patients with balance impairment were randomized to receiving either 2-week Nintendo-Wii Balance rehabilitation followed by 2-week training on stabilometric platform (Group Wii-P) or the same treatments in the opposite sequence (Group P-Wii). Each week consisted of five daily one-hour sessions.
Training with Nintendo-Wii Balance was based upon the following exercises, shaped by the therapist according to patient's ability:

- Penguin: the task requests subjects to shift balance laterally from the right to the left side.
- Bubble: the task requests subjects to displace balance forward and sideways, growing more and more challenging over time and requiring substantial attentive effort.
- Marbles: the task requests subjects to move both forward and backward as well as sideways.

Outcome measures included: Timed 10-m Walk Test to assess walking speed, Timed Up and Go Test and Berg Balance Scale to assess balance and a Visual Analogue Scale—VAS of subjective perception of stability.
Assessment timing: at enrolment (T0), after the first 2-week treatment (T1), after the second 2-week treatment (T2), at one month (F1) and three months (F3) of first treatment end.

3.3 Results

Overall, a significant improvement was observed in all outcome measures, in all subjects. The benefit was maintained at 3 months.

No significant differences were appreciated when comparing changes obtained after each 2-week treatment period (Figs. 4, 5).

Greater improvement was achieved by PD patients with moderate balance disorder, as compared to those with mild impairment.

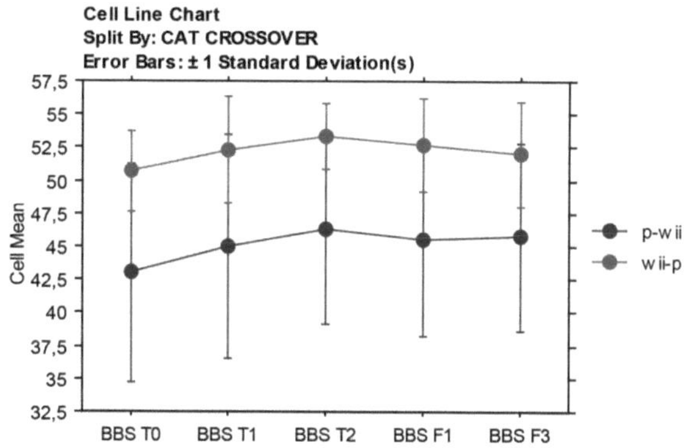

Fig. 4 Evolution of BBS score in the two groups

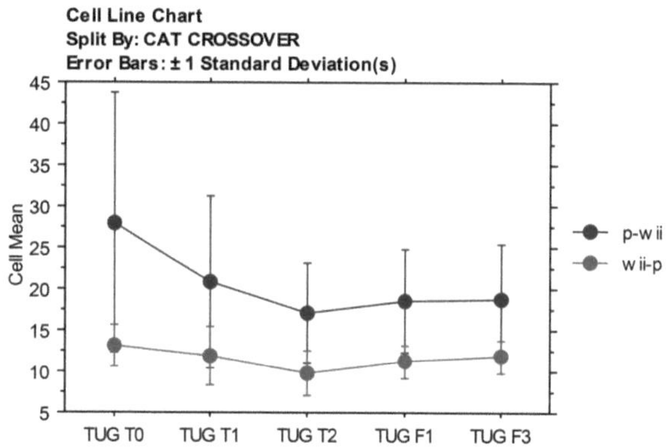

Fig. 5 Evolution of TUG score in the two groups

4 Conclusion

The findings from the two studies endorse the hypothesis that a rehabilitation training through Nintendo Wii is a safe, easy and effective approach to be used by chronically disabled people, suffering from neurological disorders. Nintendo Wii, characterized as a disposable device for home entertainment, is suitable to arouse patient's motivation. Given such premises, it may also be assumed that Nintendo-Wii would represent a useful tool for continuous self-administered home training, allowing for the enhancement of formal rehabilitation effects and the achievement of increasing levels of motor learning.

References

1. Ezekiel, H.J., Lehto, N.K., Marley, T.L., Wishart, L., Lee, T.D.: Application of motor learning principles: the physiotherapy client as a problem-solver. III. Augmented feedback. Physiother. Can. (Winter). **53**, 33–39 (2001)

2. Celnik, P., Webster, B., Glasser, D.M., Cohen, L.G.: Effects of action observation on physical training after stroke. Stroke. **39**(6), 1814–1820 (2008)
3. Buccino, G., Solodkin, A., Small, S.L.: Functions of the mirror neuron system: implications for neurorehabilitation. Cogn. Behav. Neurol. **19**, 55–63 (2006)
4. Henderson, A., Korner-Bitensky, N., Levin, M.: Virtual reality in stroke rehabilitation: a systematic review of its effectiveness for upper limb motor recovery. Stroke Rehabil. **14**(2), 52–61 (2007)
5. Merians, A.S., Poizner, H., Boian, R., Burdea, G., Adamovich, S.: Sensorimotor training in a virtual reality environment: does it improve functional recovery post stroke? Neurorehabil. Neural Repair **20**, 227–52 (2006)
6. Turolla, A., Dam, M., Ventura, L., Tonin, P., Agostini, M., Zucconi, C., Kiper, P., Cagnin, A., Piron, L.: Virtual reality for the rehabilitation of the upper limb motor function after stroke: a prospective controlled trial. J. NeuroEng. Rehabil. **10**, 85 (2013)
7. Yen, C.Y., Lin, K.H., Hu, M.H., Wu, R.M., Lu, T.W., Lin, C.H.: Effects of virtual reality-augmented balance training on sensory organization and attentional demand for postural control in people with Parkinson disease: a randomized controlled trial. Phys. Ther. **91**(6), 862–874 (2011)

Can the Current Mobile Technology Help for Medical Assistance? The Case of Freezing of Gait in Parkinson Disease

Lucia Pepa, Federica Verdini, Marianna Capecci, Maria Gabriella Ceravolo and Tommaso Leo

Abstract Parkinson's disease (PD) affects around 1.5 % people aged 65 years. Among PD features, freezing of gait (FOG) is frequent, involving almost 70 % PD people after 10 years of disease onset, and highly disabling. Effective management of FOG is a challenge for the limited responsiveness to both drug treatment and functional neurosurgery. As "cueing on demand" is the only strategy of proven efficacy on FOG, it would be crucial to develop a portable assistive device able to release suitable cues at every time the FOG occurs during the daily living (DL) of the patient, without interfering with his/her daily activities. The current smart mobile telephony devices are in principle apt to satisfy all the above mentioned requisites in terms of technological feasibility of ambulation monitoring devices and in terms of acceptability, because of their increasing widespread diffusion. In this paper we will outline a smart-phone based architecture able to detect FOG, to produce the proper cues, and to provide information for continuous monitoring of the events. The paper will specifically consider the clinical necessity, technical feasibility, economic sustainability of the solution proposed and its potential of application.

L. Pepa (✉) · F. Verdini · T. Leo
Dipartimento di Ingegneria dell'Informazione, UNIVPM, via Brecce Bianche, 60131 Ancona, AN, Italy
e-mail: l.pepa@univpm.it

F. Verdini
e-mail: f.verdini@univpm.it

T. Leo
e-mail: tommaso.leo@univpm.it

M. Capecci · M. G. Ceravolo
Dipartimento di Medicina Sperimentale e Clinica, UNIVPM, via Tronto, 60020 Ancona, AN, Italy
e-mail: m.capecci@univpm.it

M. G. Ceravolo
e-mail: m.g.ceravolo@univpm.it

S. Longhi et al. (eds.), *Ambient Assisted Living*, DOI: 10.1007/978-3-319-01119-6_18,

1 Introduction

Parkinson's disease (PD) is a progressive neurological disorder affecting the basal ganglia. It is caused by a progressive loss of dopaminergic and other sub-cortical neurons [5].

PD prevalence is high among elderly people ranging 1.5 % in people aged 65 years and beyond [26]. Although gait disorders are quite a distinctive feature across all disease stages, freezing of gait (FOG) may be regarded as the most frequent and disabling condition, involving up to 70 % PD people after 10 years of clinical onset. FOG typically manifests as a sudden and transitory inability to generate effective stepping [19] and particularly occurs when initiating gait, turning or negotiating an obstacle. This phenomenon is poorly known because of the difficulties of observing it in a clinical environment and of the different kinds of FOG observed by clinicians [14, 21]. FOG has been associated with increased risk of falls and poor quality of life [4]; it leads PD people to limit their community ambulation, thus reducing their activity levels and exercise capacity. Effective management of FOG is considered a challenge for clinicians given its episodic nature and limited responsiveness to both drug treatment and functional neurosurgery [12, 19].

There are various behavioral approaches developed by clinicians to overcome FOG and more generally to improve gait performance. These rehabilitation strategies are mainly characterized by the use of external rhythmical stimuli (cues). Cues can be of different nature: visual (e. g. colored lines on the floor), auditory (e. g. walking at the rhythm of music or metronome) or somatosensory (e. g. walking on a treadmill). In the literature there's good agreement that auditory cues are the most efficient in enhancing gait execution [13, 18].

The cues delivery during supervised walking exercises, has been shown effective in a laboratory setting, but it did not succeed at providing sustained functional benefits eventually [13, 18, 22, 23], probably owing to a disease-related impairment in motor learning mechanisms. Many researchers have therefore advocated a model of cue delivery "as needed".

The aim of the present study is to implement an architecture using every-day life devices for assisting PD patients during their daily life activities to overcome FOG: the portable assistive device would be able to release suitable cues at every time the FOG occurs during the DL of the patient.

2 Literature

A certain number of devices to be worn and able to detect the FOG event have been proposed into the literature.

Moore e al. [16] defined a freeze index (FI) using power spectrum analysis of vertical linear acceleration of the shank. This FI is defined as the ratio between the power in the "freeze" band (3–8 Hz) and the power in the "locomotor" band

(0.5–3 Hz). A threshold is applied to detect FOG. Recently the same group of study [17] used this parameter to detect FOG building an architecture composed of 7 inertial measurement units (IMUs) (3 for each leg and 1 on the lower back). The IMUs acquired triaxial linear acceleration and transmitted wirelessly to a computer for processing.

Bachlin e al. [2] proposed an on-line FOG detection algorithm based on the FI of Moore et al. [16]. Acceleration data from three sensors attached to the body (shank, thigh and belt) were transmitted to a wearable computer through wireless Bluetooth link. The wearable assistant is equipped with earphones in order to produce RAC whenever a FOG episode is detected. These authors improved the detection algorithm using a reduced window size to detect shorten FOG and a second threshold to avoid false positives during standing.

Tripoliti et al. [24] developed an algorithm of FOG detection based on signals received from wearable sensors (six accelerometers and two gyroscopes) placed on the patient's body. All sensors transmitted data to a portable PC using Bluetooth in order to process signals.

Mazilu et al. [15] propose a wearable assistant, composed of a smartphone as wearable computer and wearable accelerometers (hip, knee, ankle), for online detecting of FOG. When FOG is detected, the assistant provides rhythmic auditory cueing or vibrotactile feedback. The system is based on machine learning techniques. The classifier must be trained offline on a base station, then it is stored in a file and copied to the mobile-phone for its online use.

The class of systems discussed above presents some problems in term of acceptability: i.e. multiple sensors placed on patient's body, need to transmit data to a computing unit, which must be always nearby the patient. They are used in a rehabilitation facility context, but no solutions are apt to be used during the DL, except the one proposed at SFIT Zurich [15], asking for device training.

3 Acceptability

Acceptability is a crucial requisite for a device able to empower the patient during DL by means of a reliable and timely production of suitable cues and able to support remote monitoring of FOG events.

ICT Technology acceptance has been widely studied in a job related context and gave rise to the UTAUT (Unified Theory of Acceptance and Usage of Technology) [25]. On the contrary the medical technology seems to be driven almost exclusively by technical feasibility, medical necessity end economic issues so that number of devices are accepted for use only in a Hospital environment. Several studies [1, 27, 28] have shown that acceptability and usability barriers are the major obstacle to the approval and diffusion of medical assistive technology.

Furthermore acceptability studying of medical technology is far from the study of acceptability in a job related context. Such diversity is characterized mainly by two aspects [28]:

- medical monitoring generates the perception of breaking into one's own privacy;
- medical technology refers to "taboo related" areas, which are associated with disease and illness.

Interestingly ICT technology, particularly smartphones and other mobile technology, doesn't elicit these feelings because of their popularity. They also offer customizable user interface to increment perceived ease of use. The current smart mobile telephony devices are in principle apt to satisfy all the above mentioned requisites in terms of technological feasibility of ambulation monitoring devices and in terms of acceptability, because of their increasing widespread diffusion. At present according to the literature they appear to be mainly used as gateways to connect the Body Area Sensory Network with Internet or with another calculation unit except for the study of Mazilu et al. [15].

4 Architecture and Technical Aspects

In this paper we will outline a smart-phone based architecture able to detect FOG, to produce the proper cues, and to provide information for continuous monitoring of the events, and self-tuning on the individual patient characters.

The proposed gait assistant is composed of:

- a smartphone with an appropriate application installed;
- an elastic belt with an appropriate socket to hold the smartphone in the right position.

As can be noticed the architecture is extremely essential and doesn't need any sensor, wire, marker attached to the patient's body. The patient must hold only a smartphone in the socket.

The application collects acceleration data from the onboard accelerometer at 100 Hz, processes them to calculate a freeze index (FI), defined by Moore et al. [16], in combination with the signal energy (SE), defined by Bachlin et al. [2]. Differently from these previous works, our architecture works in real-time and performs sensing, processing and auditory feedback on the smartphone, without the need of an external calculation unit.

Acceleration data are passed into a time window of suitable length (Hamming window). On each window FI and SE are calculated and compared with their corresponding thresholds. When both the values of FI and SE exceed the threshold a FOG is detected and auditory feedback is provided. The window is applied every 0.4 s.

Thresholds are calculated from 20 s of standing posture. They are computed as the mean plus one standard deviation.

The features calculated by the application are stored in the smartphone as text files. During an every-day use the application is geared for sending periodically such files to an appropriate database of the clinical center.

5 Subjects, Experimental Protocol and Data Analysis

Subjects. The architecture was tested on 22 ambulant subjects (16 males and 6 females) suffering from Idiopathic PD (UKPDS Brain Bank criteria)[9] with a clinical history of FOG. Patients were $\leq$80 years old and not-severely demented (Mini-Mental State Examination—MMSE $\geq$20) [8]. The study was approved by local Ethics Committee and written informed consent was obtained.

Patients were assessed in the practically-defined 'off' state, after overnight withdrawal of the dopaminergic therapy. Patient characteristics were as follows: mean age 65.0 [SD 9.3], disease duration 15.3 years [SD 6.1], Hoehn and Yahr stage [10] 3.7 [SD 0.3], UPDRS Section I 2.9 [SD 2.5], UPDRS Section II 17.4 [SD 5.1], UPDRS Section III 18.5 [SD 9.1], UPDRS Section II item "Freezing of gait" 2.4 [SD .5], LEDD (Levodopa equivalent daily dose) 776.1 [SD 247].

Experimental protocol. Patients performed three kinds of timed up-and-go (TUG) tasks to provoke FOG on a standardized course: (i) the standard TUG test as described by Shumway-Cook et al. [20], (ii) the Cognitive Dual Task Timed Up and Go test as described by Hofheinz et al. [11] and, finally, (iii) the Manual Dual Task Timed Up and Go test, which was inspired to Baker's [3] experimental protocol. During the Manual Dual Task Timed Up and Go test, participants were seated in a chair, stood up, collected a tray with 2 cups of water placed on a table beside the chair, walked along the 3-m walkway carrying the tray, crossed a line marked on the floor, turned around, walked back, rested the tray on the table and sat down. The water levels in the cups and the positions of the cups on the tray were standardized. Each patient performed at least nine TUG tests, three for each kind of TUG. The overall aim of the investigators was to obtain a sequence of videos depicting a wide range of FOG episodes. Walking trials were recorded on a digital video camera from a consistent vantage point for later analysis, and each video showed a complete TUG trial starting and ending in the seated position. Simultaneous acceleration data was acquired from the trunk during the TUG trials described above. Synchronization of the video and accelerometer recordings was performed prior to data collection by alignment of the video camera and data acquisition computer clocks.

Data analysis. We calculated sensitivity, specificity and efficacy in order to evaluate the system performance. We defined true positives (*tp*) the windows correctly classified as FOG. We defined false negatives (*fn*) the windows classified as non-FOG despite the patient manifesting the symptom in that time instant. We defined true negative (*tn*) the windows correctly classified as non-FOG. We defined false positives (*fp*) the windows wrongly classified as FOG despite his absence. Based on these definitions we calculated sensitivity as $Se = \frac{tp}{tp+fn}$ and specificity as $Sp = \frac{tn}{tn+fp}$. In order to have a better overview of the system performance we can define another kind of sensitivity (Se_1) as the ratio between the number of FOG events identified by the applications and the total number of FOG events observed during all the trials. Since the objectives of such architecture are both symptom recording and treating, we are also interested in the ability of the architecture in improving motor

Table 1 Results

Paz.	Gender	Tremor	Se	Sp	n° FOG	FOG identified	FOG overcome	$Se_{overall}$
1	M	X	0.60	0.99	3	1	1	0.33
2	F		0.85	0.87	2	2	2	1
3	M		–	1	0	0	–	–
4	M		–	1	0	0	–	–
5	M		1	0.96	1	1	0	1
6	F	X	0.44	1	13	1	1	0.08
7	M		0.78	0.97	7	4	3	0.57
8	M		–	1	0	0	–	–
9	F		0.86	0.98	7	7	5	1
10	M		–	1	0	0	–	–
11	M		0.91	0.94	2	2	2	1
12	M		0.88	0.98	16	16	13	1
13	M		0.92	0.91	12	12	10	1
14	M		0.93	0.88	25	25	11	1
15	M		0.87	0.92	4	4	4	1
16	F		0.81	0.67	7	6	2	0.86
17	M		–	1	0	0	–	–
18	M		–	1	0	0	–	–
19	M		–	1	0	0	–	–
20	F		0.77	1	5	4	4	0.80
21	M		0.66	1	8	4	1	0.50
22	F		0.93	0.99	3	3	3	1
TOT			81.4 %	95.73 %	115	92	Eff = 68 %	80 %
TOT^{a}_{wt}			85.92 %	95.35 %	99	90	Eff = 67 %	91 %

[a] Results for subject without tremor

competences of the patient. For this reason we define efficacy (*Eff*) as the ratio between number of FOG overcome thanks to the cue supplied and number of FOG identified by the application. Among all the participants there were 2 patients affected by rest tremor: we can assess the effects of this disturbance on the algorithm performance. Studies show that frequency bands of FOG and tremor are quite entirely overlapped [6], so we can expect a great loss of sensibility in patients with rest tremor of distal arts. For this reason sensitivity results are both globally presented with respect to the whole sample and, separately, with reference to those with rest tremor.

6 Results

We observed 115 FOG events in total, as recognised by clinicians based on video recordings. The application correctly identified 92 of them. Results are presented in Table 1. Total results are displayed as the mean computed on all patients.

Our interest was also concerned with acceptability requirements and the only way for assessing this topic was asking the participants. With respect to the invasiveness and obtrusiveness, none of the patients felt embarrassment or discomfort in wearing the smartphone. Regarding the technical confidence, the major part of the patients taking part in our experiment didn't already possess the technical expertise needed to manage a smartphone, mainly due to their age. All of them were apt to learn how to launch the app and moreover their background always included a young relative much engaged with this kind of technology. Some patients expressed their strong desire in using the architecture at home due to the benefit found during the trials.

7 Discussion and Future Work

Results indicate that the proposed architecture is capable of identifying FOG episodes with suitable sensibility and specificity, compared to other studies [2, 15, 17]. Moore et al. [17] obtained 84.3 % of sensitivity and 78.4 % of specificity when using all the 7 sensors. Sensitivity and specificity become 86.8 and 82.4 % respectively for the hip sensor alone. Bachlin et al. [2] reported a mean sensitivity of 73.1 % and specificity of 81.6 %, these value incremented to 88.6 and 92.4 % respectively after an offline parameter optimization. Results in sensitivity and specificity obtained from Mazilu et al. [15] are divided for sensor placement and sensor axes. Therefore we can compare our results with their performance results for vertical axes of hip sensor which are 83.68 % of sensitivity and 98.26 % of specificity. This comparison is promising given the novelty of our architecture, which is, to our knowledge, the first attempt to manage FOG using only a smartphone-like device. We focus particularly on the aspect of acceptability of the technical device, thinking of a solution that neither labels the person as a patient, nor limits his movement, nor forces him to be equipped with obtrusive and unfamiliar technological hardware (e.g. sensors and wire attached to the body).

Algorithm performance manifests some user-dependent characteristics: for some patients it worked exceptionally, but there are some other patients that experienced a lower performance. This large variability was found also in other studies [2, 15] and is possibly related to different type of FOG (i.e. trembling vs. akinetic).

Results in efficacy are underestimated because some patients affirmed they didn't hear the cues. This problem can be easily overcome by using head-phones. In this stage of testing we couldn't adopt this solution because we need to evaluate the system and be aware of his functioning.

Both the high score of performance results and the unobtrusiveness demonstrate the potential use of the architecture in monitoring, gait assistance during daily living and rehabilitation therapy. In spite of these encouraging results, much work must be done in order to improve system performance. We are already working to include additional discriminative features to the algorithm since it is clear that multidimensional approaches [15, 24] give better results. As expected, rest tremor caused a loss of performance. It can be interpreted as a disturbance, since his frequency content is exactly between 3 and 8 Hz [6]. To overcome this problem we will integrate the algorithm with additional feature, apt to discriminate FOG from tremor of distal arts.

The proposed architecture can be easily applied to similar scenarios involving other diseases [7]. Clinicians can benefit from the information brought by aggregate and synthetic parameters clinically important, improving their knowledge and patient management.

References

1. Arning, K., Ziefle, M.: Different perspectives on technology acceptance: the role of technology type and age. USAB 2009. LNCS **5889**, 20–41 (2009)
2. Bächlin, M., Plotnik, M., Roggen, D., Maidan, I., Hausdorff, J.M., Giladi, N., Tröster, G.: Wearable assistant for Parkinson's disease patients with the freezing of gait symptom. IEEE Trans. Inf. Technol. Biomed. **14**(2), 436–446 (2010)
3. Baker, K., Rochester, L., Nieuwboer, A.: The immediate effect of attentional, auditory, and a combined cue strategy on gait during single and dual tasks in Parkinson's disease. Arch. Phys. Med. Rehab . **88**(12), 1593–1160
4. Bloem, B.R., Hausdorff, J.M., Visser, J.E., Giladi, N.: Falls and freezing of gait in Parkinson's disease: a review of two interconnected, episodic phenomena. Mov. Disord. **19**(8), 871–884 (2004)
5. Braak, H., Ghebremedhin, E., Rub, U., Bratzke, H., Del Tredici, K.: Stages in the development of Parkinson's disease-related pathology. Cell Tissue Res. **318**(1), 121–134 (2004)
6. Burne, J.A., Hayes, M.W., Fung, V.S., Yiannikas, C., Boljevac, D.: The contribution of tremor studies to diagnosis of parkinsonian and essential tremor: a statistical evaluation. J. Clin. Neurosci. **9**(3), 237–242 (2002)
7. Factor, S.A.: The clinical spectrum of freezing of gait in atypical parkinsonism. Mov. Disord. **23**(Suppl 2), S431–S438 (2008)
8. Folstein, M.F., Folstein, S.E., et al.: Mini-mental state. A practical method for grading the cognitive state of patients for the clinician. J. Psychiatr. Res. **12**, 189–198 (1975)
9. Gibb, W.R., Lees, A.J.: A comparison of clinical and pathological features of youngand old-onset Parkinson's disease. Neurology **38**(9): 1402e6 (1988)
10. Hoehn, M.M., Yahr, M.D.: Parkinsonism: onset, progression and mortality. Neurology **17**(5): 427e42 (1967)
11. Hofheinz, M., Schusterschitz, C.: Dual task interference in estimating the risk of falls and measuring change: a comparative, psychometric study of four measurements. Clin. Rehabil. **24**(9), 831–842 (2010)
12. Kim, S.D., Allen, N.E., Canning, C.G., Fung, V.S.: Postural instability in patients with Parkinson's disease. Epidemiology, pathophysiology and management. CNS Drugs **27**(2), 97–112 (2013)
13. Lim, I., Van Wegen, E., de Goede, C., Deutekom, M., Nieuwboer, A., Willems, A., Jones, D., Rochester, L., Kwakkel, G.: Effects of external rhythmical cueing in gait in patients with Parkinson's disease: a systematic review. Clin. Rehabil. **19**(7), 695–713 (2005)
14. Mancini, M., Priest, K.C., Nutt, J.G., Horak, F.B.: Quantifying freezing of gait in Parkinson's disease during the instrumented timed up and go test. Conf. Proc. IEEE Eng. Med. Biol. Soc. **2012**, 1198–1201 (2012)
15. Mazilu, S., Hardegger, M., Zhu, Z., Roggen, D., Troster, G., Plotnik, M., Hausdorff, J.M.: Online detection of freezing of gait with smartphones and machine learning techniques. 6th International Conference on Pervasive Computing Technologies for Healthcare (Pervasive-Health) and Workshops (2012)

16. Moore, S.T., MacDougall, H.G., Ondo, W.G.: Ambulatory monitoring of freezing of gait in Parkinson's disease. J. Neurosci. Methods **167**(2), 340–348 (2008)
17. Moore, S.T., Yungher, D.A., Morris, T.R., Dilda, V., MacDougall, H.G., Shine, J.M., Naismith, S.L., Lewis, S.J.: Autonomous identification of freezing of gait in Parkinson's disease from lower-body segmental accelerometry. J. Neuroeng. Rehabil. **10**, 19 (2013)
18. Nieuwboer, A., Baker, K., Willems, A.M., Jones, D., Spildooren, J., Lim, I., Kwakkel, G., Van Wegen, E., Rochester, L.: The short-term effects of different cueing modalities on turn speed in people with Parkinson's disease. Neurorehabil. Neural. Repair **23**(8), 831–836 (2009)
19. Nutt, J.G., Bloem, B.R., Giladi, N., Hallett, M., Horak, F.B., Nieuwboer, A.: Freezing of gait: moving forward on a mysterious clinical phenomenon. Lancet Neurol. **10**(8), 734–44 (2011)
20. Shumway-Cook, A., Brauer, S., et al.: Predicting the probability for falls in community-dwelling older adults using the Timed Up & Go Test. Phys. Ther. **80**(9), 896–903 (2000)
21. Snijders, A.H., Nijkrake, M.J., Bakker, M., Munneke, M., Wind, C., Bloem, B.R.: Clinimetrics of freezing of gait. Mov. Disord. **23**(Suppl 2), S468–S474 (2008)
22. Spaulding, S.J., Barber, B., Colby, M., Cormack, B., Mick, T., Jenkins, M.E.: Cueing and gait improvement among people with Parkinson's disease: a meta-analysis. Arch. Phys. Med. Rehab. **94**, 562–570 (2013)
23. Tomlinson, C.L., et al.: Physiotherapy versus placebo or no intervention in Parkinson's disease. Cochrane Database of Syst. Rev. Issue 9 (2013). Art. No.:CD002817
24. Tripoliti, E.E., Tzallas, A.T., Tsipouras, M.G., Rigas, G., Bougia, P., Leontiou, M., Konitsiotis, S., Chondrogiorgi, M., Tsouli, S., Fotiadis, D.I.: Automatic detection of freezing of gait events in patients with Parkinson's disease. Comput. Methods Programs Biomed. **110**(1), 12–26 (2012)
25. Venkatesh, V., Morris, M.G., David, F.D.: User acceptance of information technology: toward a unified view. MIS Q. **27**, 3 (2003)
26. Von Campenhausen, S., Bornschein, B., Wick, R., Bötzel, K., Sampaio, C., Poewe, W., Oertel, W., Siebert, U., Berger, K., Dodel, R.: Prevalence and incidence of Parkinson's disease in Europe. Eur. Neuropsychopharmacol. **15**(4), 473–490 (2005)
27. Ziefle, M., Röcker, C., Wilkowska, W., Kasugai, K., Klack, L., Möllering, C., Beul, S.: A multi-disciplinary approach to ambient assisted living. In: Röcker, C., Ziefle, M. (eds.) E-Health, Assistive Technologies and Applications for Assisted Living: Challenges and Solutions. IGI Global, Hershey (2010)
28. Ziefle, M., Wilkowska, W.: Technology acceptability for medical assistance. Pervasive Computing Technologies for Healthcare (PervasiveHealth), 2010 4th International Conference on-NO PERMISSIONS, pp. 1–9 (2010)

Electromagnetic Sensing of Obstacles for Visually Impaired Users

Lorenzo Scalise, Valentina Di Mattia, Paola Russo, Alfredo De Leo, Valter Mariani Primiani and Graziano Cerri

Abstract Vision impairment is a physical and sensory disability affecting a large number of subjects around the world. A large part of this subjects is aged 65 or older and their number grows faster than the overall population. In this work, a innovative sensing method, proposed by the research group of the Università Politecnica delle Marche based on EM pulses, is presented together with some experimental results achieved in obstacles detection. The proposed approach accomplishes most of the operative requirements of electronic travel aids for visually impaired subjects and can provide additional information (height form the ground, distance and position of the obstacle) on obstacles respect to the available assistive technologies currently used by subjects affected by visual impairments.

1 Visual Impairment and Assistive Technologies

Blindness and vision impairment (VI) is a physical and sensory disability affecting a large number of subjects around the world. The incidence of severe VI varies between industrialised (0.4 % of the population) and developing countries (>1 % of the population) [9]. Globally the World Health Organization (WHO) estimates 40–45 million of people to be totally blind and up to 314 million to have some kind of visual impairment; the 87 % of them live in the developing countries [10].

L. Scalise (✉)
Dipartimento di Ingegneria Industriale e Scienze Matematiche (DIISM), Università Politecnica delle Marche, via Brecce Bianche, 60131 Ancona, Italy
e-mail: l.scalise@univpm.it

V. Di Mattia · P. Russo · A. De Leo · V. Mariani Primiani · V. Petrini · G. Cerri
Dipartimento di Ingegneria dell'Informazione (DII), Università Politecnica delle Marche, via Brecce Bianche, 60131 Ancona, Italy

G. Cerri
e-mail: g.cerri@univpm.it

S. Longhi et al. (eds.), *Ambient Assisted Living*, DOI: 10.1007/978-3-319-01119-6_19,

The majority of blind and VI subjects is aged 65 or older; their number grows faster than the overall population and it is estimated that such figures will double by year 2020, making such issue of particular relevance for Europe strategic agenda.

From a social point of view, VI subjects live strong limitations on their social life, in fact the sight reduction or impairment limits the access to the information, the work opportunities and the interactions with other people and the surrounding environment. In particular autonomous mobility for VI subjects can be extremely difficult and in most of the case it requires the use of assistive devices.

The most common way to navigate for VI subjects is the white cane which is a pure mechanical device dedicated to the exploration of the ground in front of the user (about 1.5 m from his/her feet). Static obstacles and holes or steps can be detected by the user handing the cane simply by tactile-force feedback; floor characteristics (hard, soft, etc.) can also be inferred by the sound produced by the cane tip on the ground. This device is light, portable and cheap, but it suffers the limit to only provide information of static obstacles from the ground and a portion of less than 1.5 m in front of the user feet. The presence of trunk or head level obstacles is not detectable and consequently it can cause hits; this representing a major limitation on self-confidence in autonomous mobility for VI subjects, especially if aged.

The last decades have seen the proposal of many assistive technologies aiming to reduce the difficulties of VI subjects in mobility tasks. They typically involve the use of systems able to collect information of the space around the subject to provide orientation information and to warn for obstacle avoidance. Such devices are commonly known as electronic travel aids (ETAs) and are all based on exploring (with different sensing methods) the environment and to transfer the information gathered to the user by other sense (mainly hearing or touch). It is generally agreed that currently no available electronic travel aids (ETA) (commercial system or prototype) incorporates all the required features recommended by the Visually Impaired (VI) community and the International Guidelines to a satisfactory extent [4–6]. Moreover, it is interesting to note that in the literature there is a lack of studies which consider electromagnetic (EM) radiation in the radio frequency or microwave ranges as the physical quantity able to deliver information on obstacle presence for visually impaired users. Recently, the authors have demonstrated the possibility to sense the presence of obstacles in the volume in front of the user by EM pulse transmission [7].

In this work, an innovative sensing approach based on EM pulses is presented together with some tests conducted in real indoor/outdoor premises and achieved by the research group of the Università Politecnica delle Marche.

2 Operative Scenario and Scope

The ETA system proposed in this work has to give information on the presence and the distance of the obstacles eventually in front of the subject, exploring in elevation a region from ground to head level, and in azimuth an area corresponding to the subject's body.

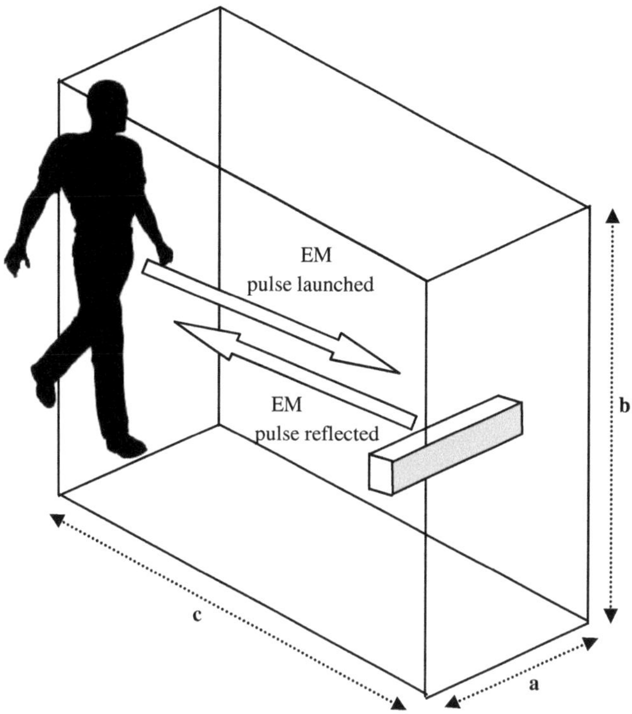

Fig. 1 Volume explored by the proposed system

The minimum distance or range over which this information is needed is a comfortable stopping distance at normal walking speed [4, 6]. The operative scenario explored in this study is represented in Fig. 1, where a = 1.5 m, b = 2 m, c = 3 m. The dimensions selected for the explored volume are a compromise between the necessity, in a real system, to give sufficient information to the user and the need to limit meaningless and multiple alarms thanks to continuous monitoring.

The present study aims to analyze the signal reflected from different types of obstacles placed at a different distance, level and position inside the above mentioned volume. In particular, it will be reported the case of two obstacle: the first obstacle is static (a person standing in front of the system) while the second is a moving obstacle (a person moving toward the system). The tests have been conducted in an outdoor space (parking pool) and in a corridor of the Faculty with standard metallic furniture.

3 The Proposed Solution

The experimental set-up realising the EM obstacle monitoring system is schematized in Fig. 2 and it is described in detail in [7]. The obstacle is illuminated using a helical antenna (matched from 4 to 6 GHz; half power beam of about 35.9° in the plane $\varphi = 0°$ and 36.9° in the plane $\varphi = 90°$).

Fig. 2 Photo of the electromagnetic obstacle detection sensor. The EM portable system (*left*); The system in use (*right*)

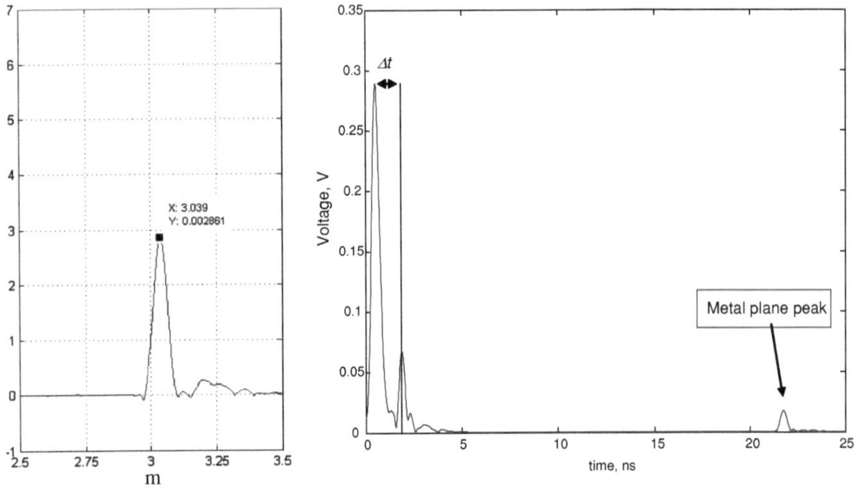

Fig. 3 Time response measured with a plastic plane (50 × 50 cm) placed at 3.04 m (*left*). Time response measured with a metal plane (200 × 200 cm) at 3 m (*right*)

A vectorial network analyzer (VNA) is used to measure the reflection coefficient at the antenna input. VNA provides the time domain response which has a pulse duration 0.4 ns (calculated at 50 % of the pulse amplitude), corresponding to 12 cm spatial resolution. The electromagnetic system is basically set up by a control unit, a Tx/Rx unit, and a radiating element. Each part has to be analysed and properly designed according to the requirements of the scenario. The radiated signal is a short pulse whose echo will be used to extract information of the object.

In Fig. 3, we report the time domain response measured in the case of a 50 × 50 cm plastic plane placed at 3.04 m (left) and of a metal plane (2 × 2 m) at a distance of 3 m (right). In Fig. 3-right, the x-axis is reporting time (ns) in order to appreciate the time scale of the whole phenomenon (the peak corresponding to the obstacle is detected after 21.8 ns for an obstacle at 3 m); the Δt (duration 1.32 ns) is the time interval between the coaxial-waveguide transition and the antenna aperture, during this interval the system is not able to detect the obstacle (it correspond to <28 cm).

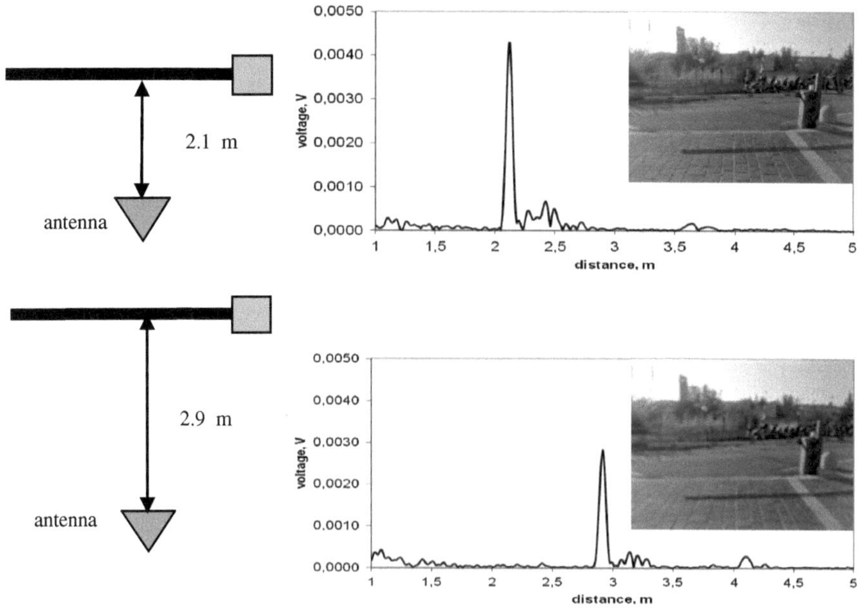

Fig. 4 Obstacle detection and distance measurement for the outdoor scenario

For the figure on the right, two peaks are well evident in the early time response corresponding to the coaxial-waveguide transition of the antenna and to the antenna aperture, they limit the minimum detectable distance to <28 cm.

4 Results

4.1 Outdoor Scenario: Static object

In Fig. 4, an example of an outdoor static obstacle (parking gate bar, height from the ground: 1.1 m)—not detectable by a cane—is reported for two different distances (2.1 and 2.9 m). The obstacle is clearly identified and its distance is determinable.

4.2 Indoor Scenario: Static and Dynamic Obstacles

In Fig. 5, it has been simulated a possible indoor scenario with static and dynamic obstacles: a corridor with one person standing on the right and one advancing the system). Signals measured by the proposed system are reported on the same figure on the right; the obstacles presence and their distances are clearly individuated by peaks. A sequence of measurements have been recorded while the subject on the right was moving toward the system.

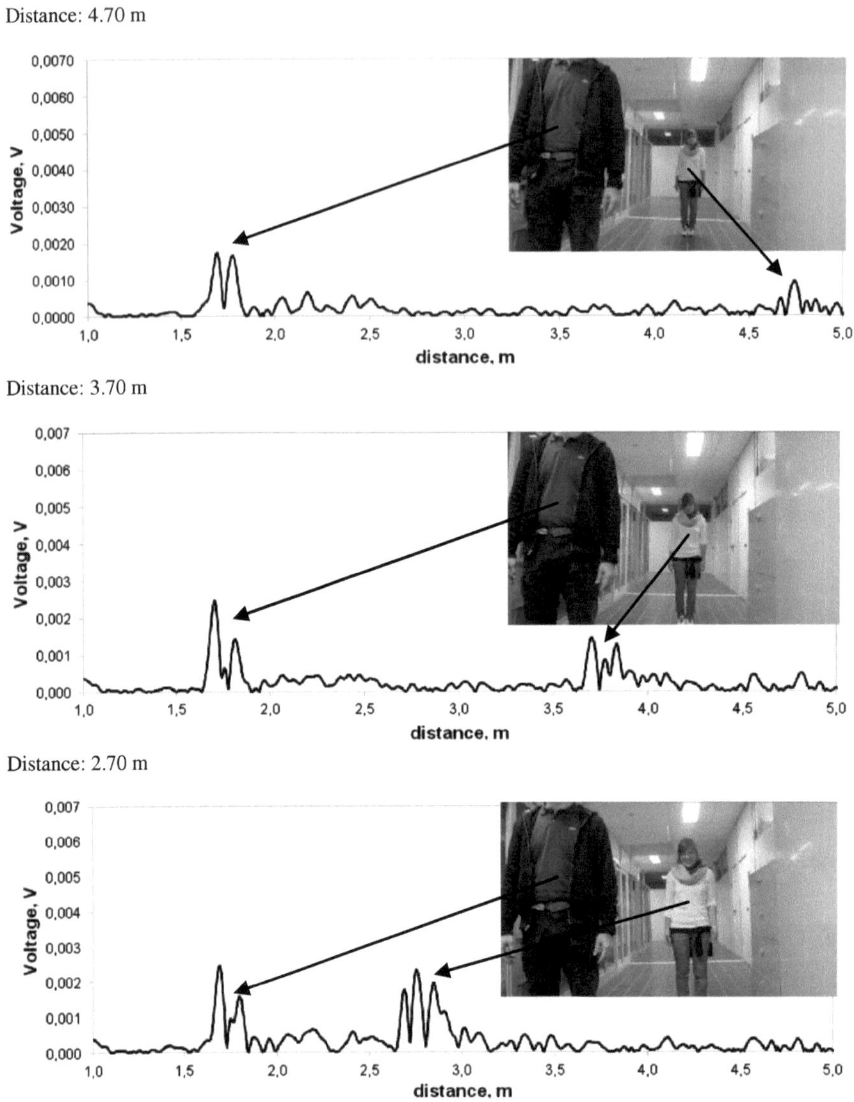

Fig. 5 Obstacle detection and distance measurement for the indoor scenario: test at different distances

Finally a test with a blind user has been operated with a portable version of the system and a specially portable antenna. The test has proved the ability of the user to avoid obstacles randomly placed, to identify in time obstacles at trunk height and to pass through a gate without hits (a video of the tests is available on line at [8]).

Distance: 1.80 m

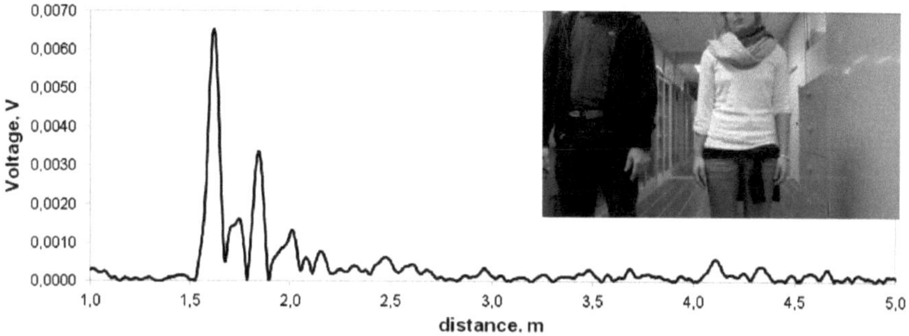

Distance: 1.20 m

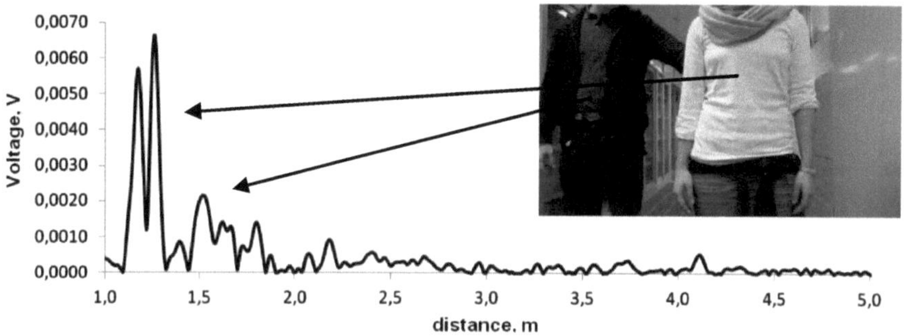

Fig. 5 (continued)

5 Conclusions

The use of EM pulses is suitable technique for obstacle detection in real mobility scenario for VI user. A first important characteristic of the proposed system is that, thanks to the high number of measurements operated per unit of time, the user can stand still or walk at an acceptable speed (<1 m/s). A second aspect is that the system is compatible with the cane and it can be also be mounted on the stick; if it is moved by the user in order to explore the volume of the space in front, (thereby intrinsically performing a space scanning like a real surveillance radar), mechanical motion is far slower than the speed of the EM pulse. Consequently the obstacle appears to be still with respect to the EM system. Therefore, the availability of a reflected pulse, providing the information on the distance of an obstacle, is the simplest and most effective way to inform the VI person.

The electromagnetic technology allows to achieve some advantages with respect to existing ETA systems (in particular the ultrasonic sensors):

- the frequency range allows the miniaturization of the antenna and of the device, more comfortable for the users.
- the possibility to integrate the antenna onto white cane. In particular more than one antenna could be used to detect obstacles at different height (ground, trunk, and head level). Finally a narrow beam in the horizontal plane would allow to determine the direction of the obstacle simply exploring the environment with the oscillatory movement of the cane.
- the penetrability of fabric from the electromagnetic waves, offering a full wearable ETA, improving acceptability and usability.
- the very high quantity of information collected in short time intervals, allowing the detection of moving obstacles whose velocity could be easily sensed.
- the dependence of the scattered field upon dimension and material of the radiated object could be usefully exploited to extract the obstacle characteristics.

It is important to note that future development of the proposed system will aim to improve system reliability and robustness, to design an advanced multi-sensory strategy allowing to optimise the system use and to study proper system-to-user-interface [1–3]. The use of EM sensing for VI subjects mobility is currently patent pending.

References

1. Andò, B., Graziani, S.: Multisensor strategies to assist blind people: a clear-path indicator. IEEE Trans. Instrum. Meas. **58**(8), 2488–2494 (2009)
2. Andò, B.: Electronic sensory systems for the visually impaired. IEEE Instrum. Meas. Mag. **6**(2), 62–67 (2003)
3. Andò, B., Baglio, S., La Malfa, S., Marletta, V.: Innovative smart sensing solutions for the visually impaired. In: Pereira, J. (ed.) Handbook of Research on Personal Autonomy Technologies and Disability Informatics. Medical information science reference, IGI Global, Hershey, New York (2011)
4. Blasch, B.B., Wiener, W.R., Welsh, R.L.: Foundations of Orientation and Mobility, 2nd edn. AFB Press, New York (1997)
5. Communications on Vision. Electronic Travel Aids: New Directions for Research. National Academy Press, Washington (1986)
6. Dakopoulos, D., Bourbakis, N.G.: Wearable obstacle avoidance electronic travel aids for blind: a survey. IEEE Trans. Syst. Man Cybern. Part C Appl. Rev. **40**(1), 25–35 (2010)
7. Scalise, L., Mariani Primiani, V., Russo, P., Shahu, D., Di Mattia, V., De Leo, A., Cerri, G.: Experimental investigation of electromagnetic obstacle detection for visually impaired users: a comparison with ultrasonic sensing. IEEE Trans. Instrum. Meas. **61**(11), 3047–3057 (2012)
8. Test demonstration of the EM system. http://www.youtube.com/watch?v=AyUngKPWtyE
9. Velázquez, R.: Wearable assistive devices for the blind. In: Lay-Ekuakille, A., Mukhopadhyay, S.C. (eds.) Wearable and Autonomous Biomedical Devices and Systems for Smart Environment, vol. 75, pp. 331–349. Springer, Berlin (2010)
10. World Health Organization. Visual impairment and blindness—Fact Sheet 282. http://www.who.int/mediacentre/factsheets/fs282/en/ (2012). Accessed 23 Sept 2013

Part V
Assistive Devices

Multi-Agent Simulation Model for Evacuation of Care Homes and Hospitals for Elderly and People with Disabilities in Motion

Joining Group Behaviors in Autonomous Elderly Motion Using a Social Force Model Approach

Marco D'Orazio, Luca Spalazzi, Enrico Quagliarini and Gabriele Bernardini

Abstract The progressive population aging provokes an increase of importance in safety aspects for the elderly and the people with disabilities that are housed in care homes and hospitals. Current regulations denote an inadequate approach to safety problems connected to possible evacuation due to events like earthquakes, floods, fires. The law approach implicates that patients are directly carried out by health workers assistants in evacuation. However, many patients can autonomously evacuate, helped by specific facilities for way finding and not assisted by the medical staff. Our research is intended to design "guidance" system for these categories, and to inquiry how these facilities interact with people and influence their motion. The understanding and the simulation of behaviors of this category become essential in order to reach these goal. This work proposes a multi-agent model for evacuation simulation, based on the Social Force motion approach and on experimental data. This paper focuses on joining group behaviors for autonomous elderly. The validation concerns various quantities describing group motion phenomena. The model will be integrated including aspects connected with eventual disabilities in motion for patients.

M. D'Orazio (✉) · E. Quagliarini · G. Bernardini
Department of Construction, Civil Engineering and Architecture (DICEA), Università Politecnica delle Marche, Via brecce bianche, 60131 Ancona, Italy
e-mail: m.dorazio@univpm.it

E. Quagliarini
e-mail: e.quagliarini@univpm.it

G. Bernardini
e-mail: g.bernardini@univpm.it

L. Spalazzi
Department of Information Engineering (DII), Università Politecnica delle Marche, Via brecce bianche, 60131 Ancona, Italy
e-mail: l.spalazzi@univpm.it

S. Longhi et al. (eds.), *Ambient Assisted Living*, DOI: 10.1007/978-3-319-01119-6_20,

1 Introduction

The population aging directly provokes an increasing request of accommodations for elderly and people with disabilities in motion: only in Italy, there are about 300,000 of these people housed in healthcare homes [1]. A particular attention must be posed on their safety, especially in evacuation due to earthquakes, floods or fires... Actually, regulations denote a schematic approach in building layout and in characteristics of healthcare assistance. It is considered that, in case of evacuation, patients are directly carried out by health workers. This approach implies a constant presence of a very large assistance staff (usually not adequate in the real cases), with consequent high evacuation times due to the transportation of each patient. Human behavioral aspects are omitted. However, many guests have a partial or complete motion autonomy: they could exit the building by themselves, even using a series of facilities in evacuation, such as "guidance" systems that doesn't need the intervention of an healthcare operator. Our research is intended to design "guidance" system for them. The two essential starting issues are: the understanding of behaviors and visual/motion difficulties for these categories in case of not-assisted evacuation; the design of punctual visual "guidance" systems (e.g: photoluminescent tiles) with experimental studies on their influence in space perceptions and motion in evacuation [2].

The behavioral inquiry in the first issue involves also the definition of a simulation model that is capable to analyze the influence of behavioral effects in evacuation. Many studies involves evacuation for different structures [3, 4], including hospitals and healthcare structures [5, 6], obtaining both numerical and behavioral data. Group motion phenomena, attraction between individuals and "Herd Behavior" [7, 8].

Actually, many models simulate human behavior and motion in evacuation conditions [9]; a limited number of them involves hospitals evacuation and the presence of disabled people [10, 11]. The social force model [8, 9] approaches analysis of real cases and describes motion in terms of attractive and repulsive social forces that lead individuals to achieve their goal by interacting with other people and the environment.

Our simulation model is based on the social force model, with modifications due to case study. The model is developed using an agent-based methodology, and a software is implemented on the same bases. The motion law is attributed to each single evacuating person; the interactions between them, and between them and the physical scenario, produce the global results retrieved from previous experimental analysis. In this paper, the definition of group attraction during motion is provided, in order to describe group motion phenomena. The model is also proposed as a tool for predictions of probable behaviors in different scenarios, in order to promote interventions for a good building design in relationship between the building layout and the "human" evacuation process.

2 Phases, Model Structure and Validation Criteria

2.1 Phases and Model Structure

This work is organized in the following steps; for each step, the relative paragraph is indicated in brackets, in italics:

- characterization of interactions in evacuation between the agents and joining group behaviors definition (*Intentional model and Criteria for joining group behavior*);
- model implementation in the simulation software (*Implementation*);
- validation of the criteria for joining group behaviors simulation (*Validation*).

The model uses a multi-agent architecture, with an evident "Lagrangian" approach: the motion law is applied to each single agent involved in the evacuation, and the results derive from the effective interactions between agents. Phenomena and quantitative values produced by the model have to be are similar to the experimental ones for the whole system. These interactions are developed adopting the i* language [12], founded on a graphical approach. The motion law is based on the Social Force Model one, with integrations to the case study due to experimental data.

The model is composed by two parts. The intentional model describes the general characteristics of the agents involved in the evacuation procedure and the relationships between them. The criteria for motion involves the effective motion law for a single agent. A software is implemented on these bases in order to test and validate the model.

2.2 Validation Criteria

The validation phase is affected by the aforementioned "Lagrangian" approach, in this paper, joining group behaviors are inquired. The same indoor scenario is tested by varying the number of involved people. Ten simulation are run for each test; average values and standard deviations are calculated.

The influence of attractive force on instantaneous accelerations is inquired in order to denote inadmissible values [9]. An experimental average value for distance between members of the same group is essentially retrieved by videotapes analysis (database at https://www.sugarsync.com/pf/D0452061_63556241_693560) of real evacuations of public spaces from all over the World, using the open source image analysis software "Tracker" [13]. Only distances lower than 3.0 m (maximum distance for interaction phenomena [9]) are considered, with an approximation of 0.1 m. Finally, relations between cohesion bound, group dimension and average quantities are offered.

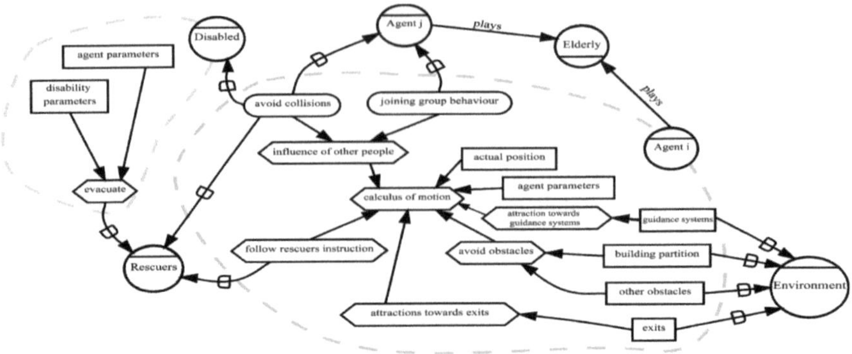

Fig. 1 The general intentional model expressed by i* language

3 Results

3.1 Intentional Model and Criteria for Joining Group Behaviors

Previous studies and experimental evidences suggest that, during his evacuation, a person refers to other pedestrians and to the physical scenario. Following an agent-based approach, Fig. 1 shows the global intentional model describing relationships between these actors, using the i* language [12].

Agents defined in order to consider the whole kinds of guests housed in healthcare structures. *Elderly* is an old person who is autonomous in motion, during the evacuation. *Disabled* is a guest who needs the *Rescuer* assistance in the evacuation; *disability parameters* allow to define different degree of disability. The healthcare assistants are represented by the *Rescuers* agent, that can interact also with *Elderly* giving them information during the procedure. In this paper, the attention is focused on the role *Elderly*, played by *Agent i* and *Agent j*.

The *agent parameters* describe his desired speed in motion, radius of pedestrian, mass of pedestrian [8, 9]. The *Elderly* makes decisions based on *Environment* (*avoid obstacles, attraction towards exit, attraction towards guidance systems*) and on other people's positions (*influence of other people*), in order to both avoid physical contact (*avoid collision*) and to maintain a certain distance from other people sharing a group bond (*joining group behaviors*). Equation 1 shows the general motion law, which terms are based on integration of the Social Force Model one [8].

$$m_i \cdot \frac{d\vec{v}\,(t)}{dt} = \vec{O}_g + \sum \vec{F}_{rep} + \sum \vec{F}_{attr} + \varepsilon\,(t) \tag{1}$$

The attractive term is modified as explained below in order to operatively describe joining group behaviors. *Elderly* who share a "group bound", including the sharing of the same evacuation target, are influenced in motion by attractive phenomena.

Equation 2 describes the definition of this attractive force, also according to previous studies [8].

$$\vec{F}_{attr}(t_n) = \sum_{gr,i} \vec{F}_{attr,gr,i}(t_n) = \sum_{gr,i} -\Theta\left(d_{ij}(t_n) - D_{\min,gr}\right) \cdot \left(C_{gr,i} \cdot \frac{\vec{p}_i(t_n) - \vec{p}_{gr,i}(t_n)}{\left\|\vec{p}_i(t_n) - \vec{p}_{gr,i}(t_n)\right\|}\right) \quad (2)$$

The target point is the geometric centre of the i's "reference group" (gr,i), composed by other *Elderly* sharing a "group bound" with i and surrounding i in a maximum radius $D_{min,gr} = 3\,\text{m}$ [9]. The attractive phenomenon is guaranteed by the minus sign: in fact, the force goes from i to i's reference group center. On the contrary, repulsive phenomena are described by vector pointing to i position (positive sign). The total attractive force depends on the overlapping of various attractions given by possible different "reference groups" (gr,i). The cohesion parameter $C_{gr,i}$ influences the modulus of this force.

3.2 Implementation

The model is implemented using the TROPOS methodology [14]; TAJ [15] is used as development environment, using Alan and Java languages. The implemented software uses a continuous 2D space representation. *Elderly* positions are updated solving the motion law (separately for the X and Y axes) at discrete time intervals of 0.1 s [16].

The *Environment* is described by a 2D DXF file, defining internal building partition (individuating the single rooms), doors and "guidance" systems. *Elderly* can be randomly generated in the different rooms, or their number can be directly imposed, room by room. An initial cohesion bound for group is given to the *Elderly* generated inside the same room. A straight corridor with 6 facing rooms is the tested *environment*.

For the validation, the Gaussian error $\varepsilon(t)$ in Eq. 1 is not considered, with the purpose to retrieve the "average" behavior. Both people sharing only the same evacuation target and also a group bond have the same $C_{gr,i}$ value. An *Elderly* has a characteristic radius equal to 0.35 m, a mass of 80 kg [9]; desired speeds are equal to 1.46 m/s or 1.22 m/s or 1.16 m/s [17], in order to considerate different attitude in motion. A maximum speed of 3 m/s is imposed according to previous studies [9].

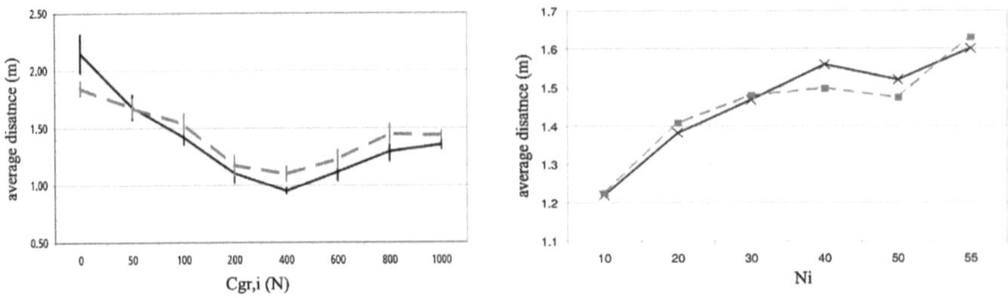

Fig. 2 Average distance between pedestrians in an evacuation group influenced by: *left* $C_{gr,i}$ variation; *right* number of people in the group *Ni* for $C_{gr,i} = 50\,N$

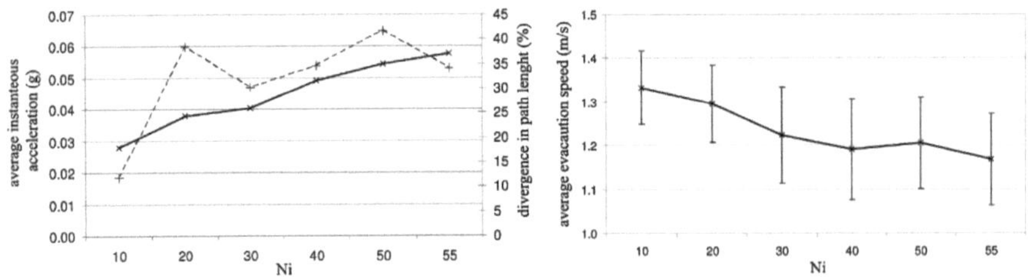

Fig. 3 Influence of the group dimension *Ni* ($C_{gr,i} = 50\,N$): *left* on average instantaneous acceleration and divergence in path length, *right* on average evacuation speed, including standard deviations

3.3 Validation

The validation phase follows the indications of paragraph 2.2. The average value of the instantaneous acceleration is lower than 0.2 g, and so acceptable [9], only when $C_{gr,i}$ is smaller than 500 N. Figure 2 shows a general view on average distance between members of a same evacuation group (dashed line) and people sharing also a group bound (continuous line).

On the right, the cohesion parameter influence is showed. Considering a $C_{gr,i} = 50\,N$, an average distance of about 1.7 m, that is near to the experimental one, is noticed. People sharing also a "group bound" stay closer in respect to people sharing only the same target: this fact is due to their initial positions (exiting from the same room) and to "herd behavior" influences [8]. On the right, the group dimension influence is showed. According to experimental valuation, people in smaller groups maintain a lower distance, as showed in. In wide groups, the average distance tends to the aforementioned value of 1.7 m. Data for high people density is influenced by the interposition of other pedestrians between members of a same groups.

Figure 3 inquires again motion quantities for different group dimension. People in a bigger group perceive highly the influence of other pedestrians, as demonstrated by both the instantaneous acceleration and the divergence in path length.

Moreover, an higher dimension of the group implicates a lower average speed in evacuation, for disturbs provoked by interactions expressed in the motion law.

4 Conclusions

A multi-agent simulation model for evacuation of care homes and hospitals for elderly and people with disabilities in motion based on the social force model approach is proposed; its software implementation is offered. In a particular way, group motion phenomena for autonomous elderly are investigated, by defining and testing of the formulation for attractive forces between members of an evacuation groups. For $C_{gr,i} = 50\,N$, an average distance between members of the same evacuation group of about 1.7 m, according to experimental data, is noticed. Results, in relation to the group dimension, evidence phenomena that are truly notable in real evacuations.

The model needs the operative integration of behavioral criteria for the other evidenced agents. Many tests with other environments are recommended. The definition of interaction forces between people and evacuation facilities, such as way finding elements or "guidance" systems, is required.

The complete model can be used as tool for evacuations previsions, in order to define strategies for the building design starting from the effective relations between environment and "human" behaviors in critical conditions, such as in evacuations. This work is proposed as one of the activities of our research group. Our goal is to define and design a series of evacuation facilities ("guidance" systems, interactive building components for evacuation) that allow people with in particular conditions (elderly, people with partial disabilities in motion) to properly evacuate in an autonomous way, with a reduction of the healthcare operators interventions to the really needed cases.

Appendix

- $C_{gr,i}$ cohesion parameter (N)
- F_{attr}, F_{rep} modulus of the attractive and repulsive forces (N)
- m_i pedestrian mass (kg)
- O_g modulus of the drive-to-target force (N)
- $\vec{p}_{gr_i}(t_n)$ position of the geometric centre of the i's group
- $\vec{p}_i(t_n)$ position of the pedestrian at instant t_n
- v(t) modulus of pedestrian velocity at the generic instant *t* (m/s)
- $\varepsilon(t)$ random variation of forces (N) at the instant t
- $\Theta(d_{ij}(t_n) - D_{min,gr})$ Heaviside step function related to the actual distance between *i* and *j* and the maximum value for group attraction; 0 if $d_{ij}(t_n) < D_{min,gr}$, 1 anywhere.

References

1. Istat: Annual report 2012 (Rapporto annuale 2012 - La situazione del Paese) (2012)
2. Tacconi, S.: Experimental evaluation of evacuation process from a theater using photoluminescent materials (Valutazione sperimentale dei processi di evacuazione all'interno della struttura

del teatro "Gentile da Fabriano" con l'ausilio di materiali foto luminescenti). Thesis, DICEA, Università Politecnica delle Marche, Ancona, Italy (2012)
3. Shi, L., Xie, Q., Cheng, X., Chen, L., Zhou, Y., Zhang, R.: Developing a database for emergency evacuation model. Build. Environ. **44**, 1724–1729 (2009)
4. Kobes, M., Helsloot, I., De Vries, B., Post, J.G.: Building safety and human behaviour in fire: a literature review. Fire Saf. J. **45**, 1–11 (2010)
5. De-Ching, H., Shen-Wen, C., Chien-Hung, L., Po-Ta, H., Yi-Ting, S., Huei-Ru, S.: A study for the evacuation of hospital on fire during construction. Procedia Eng. **11**, 139–146 (2011)
6. Wabo, N.C., Örtenwall, P., Khorram-Manesh, A.: Hospital evacuation; planning, assessment, performance and evaluation. J. Acute Dis. **1**, 58–64 (2012)
7. Mawson, A.R.: Mass Panic and Social Attachment: The Dynamics of Human Behavior. Ashgate, Brookfield (2007)
8. Helbing, D., Johansson, A.F.: Pedestrian, Crowd and Evacuation Dynamics. Encyclopedia of Complexity and Systems Science, pp. 6476–6495. Springer, New York (2010)
9. Lakoba, T.I., Finkelstein, N.M.: Modifications of the Helbing-Molnar-Farkas-Vicsek social force model for pedestrian evolution. Simulation **81**, 339–352 (2005)
10. Koo, J., Seog, Y., Kim, B., Christensen, K.M.: A comparative study of evacuation strategies for people with disabilities in high-rise building evacuation. Expert. Syst. Appl. **40**, 408–417 (2013)
11. Manley, M., Kim, Y.S.: Modeling emergency evacuation of individuals with disabilities (exitus): an agent-based public decision support system. Expert. Syst. Appl. **39**, 8300–8311 (2012)
12. Yu, E., Giorgini, P., Maiden, N., Mylopoulos, J.: Social Modeling for Requirements Engineering. The MIT Press, Cambridge (2011)
13. Brown, D., Christian, W.: Simulating what you see: combining computer modeling with video analysis. In: 8th International Conference on Hands on Science, Ljubljana, Slovenija (2011)
14. Bresciani, P., Perini, A., Giorgini, P., Giunchiglia, F., Mylopoulos, J.: Tropos: an agent-oriented software development methodology. Auton. Agent. Multi-Agent Syst. **8**, 203–236 (2004)
15. Paglierecci, F., Penserini, L., Spalazzi, L.: A goal-oriented framework tocope with requirements changes: tropos and alan. Int. Trans. Syst. Sci. Appl. **4**, 367–381 (2008)
16. Heliövaara, S., Korhonen, T., Hostikka, S., Ehtamo, H.: Counterflow model for agent-based simulation of crowd dynamics. Build. Environ. **48**, 89–100 (2012)
17. Gates, B.T.J., Noyce, D.A., Bill, A.R., Van Ee, N., Gates, T.J.: Recommended Walking Speeds for Pedestrian Clearance Timing Based on Pedestrian Characteristics. Transportation Research Record: Journal of the Transportation Research Board, Number 1982, pp. 38–47. Transportation Research Board, Washington (2006)

An Inertial and QR Code Landmarks-Based Navigation System for Impaired Wheelchair Users

Cimini Gionata, Ferracuti Francesco, Freddi Alessandro, Iarlori Sabrina and Monteriù Andrea

Abstract Personal mobility is a key factor in independent living for elderly people and people with motor disabilities, thus indoor navigation systems are of utmost concern in Ambient Assisted Living (AAL) applications. Driving an electric powered wheelchair in domestic environments becomes difficult for people with arms or hands impairments. Moreover, people affected by tetraplegia are completely unable to operate a joystick, and must rely on input interfaces, such as eye tracking and "sip and puff", which require tedious and repetitive tasks to be operated. Smart powered wheelchairs with autonomous navigation intelligence and their integration within AAL homes, may enhance independence and improve both the security and the perceived quality of life. Self-navigating systems combine different measurements provided by both absolute and relative sensors to improve localization accuracy. In this work, a low-cost localization system for autonomous wheelchairs, which takes advantage of Quick Response (QR) code landmarks information, is proposed. QR code is a low-cost pattern with fast readability and large storage capacity with respect to other landmarks solutions. The proposed wheelchair is equipped with an Inertial Measurement Unit (IMU) and a video camera: the inertial information, provided by the IMU, is fused with that provided by QR code recognition, thus reducing the error propagation caused by a Dead Reckoning (DR) approach. Autonomy and

C. Gionata · F. Francesco (✉) · F. Alessandro · I. Sabrina · M. Andrea
Dipartimento di Ingegneria dell'Informazione, Università Politecnica delle Marche, Via Brecce Bianche, 60131 Ancona, Italy
e-mail: f.ferracuti@univpm.it

C. Gionata
e-mail: g.cimini@univpm.it

F. Alessandro
e-mail: a.freddi@univpm.it

I. Sabrina
e-mail: s.iarlori@univpm.it

M. Andrea
e-mail: a.monteriu@univpm.it

S. Longhi et al. (eds.), *Ambient Assisted Living*, DOI: 10.1007/978-3-319-01119-6_21,

intelligence of the wheelchair is drastically increased by integrating within its navigation system both the knowledge about self localization and the environment (e.g. room identification). QR code landmarks are a suitable solution to store this information. This approach has been implemented and experimentally tested in an indoor scenario, demonstrating its feasibility and its good and reliable long-term performances.

1 Introduction

Assistive technology plays an important role in allowing vulnerable people (elderly or with motor disabilities) to lead and enjoy an independent lifestyle in their own homes, and thus improving their quality of life. For instance, wheeled mobility devices can significantly improve the quality of life of people who are unable to walk; however, in this case, attention must be paid also to the user's physical and mental condition since different symptoms combinations can benefit from different types of assistance and wheelchair form factors [1]. Several studies have shown that vulnerable people benefit substantially from access to independent mobility because it increases vocational and educational opportunities, reduces dependence on caregivers and family members, and promotes feelings of self-reliance. Conversely, a decrease of mobility can lead to emotional loss, isolation, stress, reduced self-esteem and fear of abandonment [2]. In order to assure a fully independent mobility, one has also to consider that many mobility impaired users are unable to operate a powered wheelchair safely, without causing harm to themselves, to others and to the environment. Thus, to fully support these users, several researchers have developed smart wheelchairs [3], employing technologies originally developed for mobile robots [4]. These devices typically consist of either a standard powered wheelchair, to which a computer and a collection of sensors have been added, or a mobile robot base to which a seat has been attached. Smart wheelchairs have been designed to help vulnerable users to navigate in a number of different ways, such as assuring collision-free movement, aiding the performance of specific tasks (e.g. passing through doorways) and autonomously transporting these users among different locations. Some smart wheelchairs operate similarly to autonomous robots: the user specifies a final destination and supervises as the smart wheelchair plans and executes a path to the target location, but they are usually unable to compensate for unplanned obstacles or travel in unknown areas. The technology developed for autonomous machines, which are capable of moving without human intervention, may offer certain multi-unable persons the needed assistance to achieve a fully freedom of mobility. A such mobility support device must be capable of working in unconstrained dynamic environments with complete regard for safety of its passenger, its surroundings and itself.

In this context, an important role concerns the indoor localization of the smart wheelchair. Moreover, a persistent navigation [5] is needed, and tags with information about absolute position, help to accomplish the indoor localization task [6]. In the literature different research works deal with this problem and different technologies are employed to solve it. Lin and Chen [7] have proposed special landmarks containing coordinates of the absolute position acquired by a camera and through a process of image segmentation. Nam [8] proposed a new approach to map-based indoor localization for walking people using Inertial Measurement Unit (IMU). Other technologies realize indoor positioning using Wireless Sensor Network (WSN) and Radio Frequency Identification (RFID), or an hybrid approach based on both, as proposed by Xiong et al. [9]. Instead of using sensors for building the navigation route map by the robotic wheelchair itself, various types of landmarks have been proposed. Courbon et al. [10], presented an indoor navigation system based on the use of a single camera and natural landmarks where the images of the environment are first sampled, stored, and than organized as a set of key images which the robot can follow as visual path to move [11].

The aim of this work is to develop a localization system which includes, at the same time, an IMU and a vision system based on Quick Response (QR) codes, used as landmarks. The proposed smart wheelchair is equipped with both an IMU and a camera: the inertial information provided by the IMU is fused with that one provided by the QR code recognition. This permits to drastically reduce the error propagation caused by the well-known dead reckoning approach. The idea is to replicate the same localization system most used in outdoor environments, i.e. the combination of an IMU and a Global Positioning System (GPS). IMUs are highly affected by both noise and disturbances, thus an exteroceptive localization sensor is required to correct the position estimation provided by the IMU. In indoor environments, however, the GPS cannot be used and an alternative exteroceptive sensor must be adopted for improving the position estimation. Localization methods based on features such as RFID (Radio Frequency IDentification) or visual systems, are often used. By using RFID technology, an RFID tag is recognized and so the location and the direction of the landmark, but in this case location accuracy is limited by the harsh propagation of the radio frequency signals caused, for instance, by the presence of obstacles among wireless nodes, multi-paths, interference, etc. These methods need dense deployments of RFID receivers, and are not able to track the mobile target [9]. An alternative is the use of visual sensors, however they can usually provide only a local estimation of the position unless the whole environment is known a priori. In this paper the IMU information is integrated with a method for indoor position and orientation based on two-dimensional barcode landmark: the QR code. QR codes are attached on the ceiling of the indoor environment and they store information about their position w.r.t. the absolute world coordinate system. By the use of image preprocessing and landmark recognition, it is possible to locate the landmark and get the information contained in the QR. The landmark shape is easily discriminated for its geometry and, by estimating the pose of it (and indirectly the pose of wheelchair w.r.t. the landmark), it is possible to calculate the absolute coordinates of the robot w.r.t. the environment. The contest in which the study has been realized is that of an autonomous wheelchair capable of driving itself in an indoor environment. This paper presents preliminary results where the considered trajectory has a “L” shape. The smart wheelchair follows the considered trajectory and, using the proposed navigation system, it is able to correct its position during the navigation.

The paper is organized as follows. Section 2 describes the system setup: the powered wheelchair, the sensors, the control unit and the adopted software. Section 3 deals with the description of the control strategy: localization, visual recognition and tracking. Section 4 presents preliminary experimental results of the proposed localization system tested on a "L" shape trajectory to be followed. Conclusive section summarizes the main results of the paper and illustrates possible improvements and future works.

2 System Setup

The proposed navigation algorithm has been developed in Robot Operating System (ROS), a framework for robotic applications which recently has been growing exponentially [12]. ROS is a thin, message-based, tool-based system designed for mobile manipulators. The system is composed of reusable libraries that are designed to work independently. The libraries are wrapped with a thin message passing layer which enables them to use and to be used by other ROS nodes. Messages are passed peer to peer and are not based on a specific programming language. ROS is based on a Unix-like philosophy of building many small tools that are designed to work together [13]. The hardware used for the experimental trials are a Microstrain 3DM-GX3-25 IMU [14] and a Logitech webcam HD C525 [15], which are connected via usb port to a laptop. An Arduino MEGA2560 board [16] has been used as an external joystick in order to set the directions and speeds motion of the wheelchair. Vision system algorithm and IMU data processing are implemented in ROS Groovy release [17]. In order to estimate the absolute pose of the considered wheelchair, two ROS nodes have been developed, one related to IMU and another one related to the camera. These nodes are capable to exchange messages between them, and thus provide the wheelchair absolute pose. The data sample rate of the IMU and the camera was set to 100–10 Hz, respectively. The camera resolution was set to 960×544 pixels. The complete system setup is presented in Fig. 1.

3 Control System

In this section the control system details are provided. In details, the algorithm for the pose estimation with IMU measurements, the QR code solution for information in the absolute reference coordinates and the visual system for recognition and tracking, are discussed.

3.1 Inertial Measurement Unit

Localization systems based on inertial measurements are capable to estimate the wheelchair position, from a known state and a previously determined location, by

Fig. 1 The smart electric-powered wheelchair with the IMU sensor

integration of internal measures. In the present paper an IMU is used to have internal measures such as angular velocity, linear acceleration and orientation. Dead reckoning system, based on a 2-D coordinate system, is described by the following mathematical model for the system at time $t_n = nT$, with sampling time T and $n \geq 0$ [18]

$$\begin{aligned} x_{t_n} &= x_{t_0} + \sum_{i=0}^{n-1} d_{t_i} \cos(\theta_{t_i}) \\ y_{t_n} &= y_{t_0} + \sum_{i=0}^{n-1} d_{t_i} \sin(\theta_{t_i}) \end{aligned} \tag{1}$$

where d_{t_i} and θ_{t_i} are the moving distance and the absolute angle at time t_i, respectively. The Cartesian coordinates at time t_n of the wheelchair are x_{t_n} and y_{t_n}. In detail, the IMU provides the linear acceleration and angular orientation, which are filtered by Kalman filter. The orientation is used to evaluate the θ parameter and the linear acceleration is doubly integrated to evaluate the d parameter. Unlike wheel encoders, an IMU is not affected by wheel slip, which is often encountered in mobile robot applications. However, there are several disadvantages on using an IMU: the errors caused by bias in the sensor readings accumulate with time, as show in Eq. (1), the misalignment of the unit's axes with respect to the local navigation frame can cause inaccurate readings. Moreover, the IMU measurements can be affected by electromagnetic interferences in the home environment such as mobile phone, laptop and wireless hotspots. An average of 100 samples acquired during system startup has been used to reduce the bias in the sensor readings.

3.2 QR Code

QR codes are barcodes that consist of black squares arranged in a square pattern on a white background. These images are capable of storing much information and handling all types of data (e.g. numeric and alphabetic characters, symbols and binary). Then any type of information can be associated and encoded in a QR code (e.g. website link, phone number, commercial content and messages). QR codes can be directly decoded by any camera and thus their information can be easily read. In the present paper, QR codes are used as landmarks, in which their absolute position is encoded. Various types of QR Code exist. For the experimental trials, the original version M2, which is able to encode up to 7,089 characters in one symbol, is chosen. The information encoded in the QR is a string of 21 characters formatted as

$$\#xx.xx\#yy.yy\#zz.zz\#rr \tag{2}$$

where $xx.xx$, $yy.yy$ and $zz.zz$ are the positions in meters of the QR code in the world coordinate system. The room, where the QR code is placed, is shown by the numeric character rr. This information is not used in the experiments but could be exploited to integrate AAL home automation technologies with the smart wheelchair. QR code detection algorithm is developed using the ZBar library [19]. This library allows to decode each QR code and thus to know the absolute position of each QR code in the environment.

3.3 Visual Recognition and Tracking

The vision system has been developed in ROS framework using different vision libraries such as ViSP, OpenCV and ZBar ([20, 21] and [19]). The algorithm is based mainly on six routines as shown in Fig. 2: camera calibration, QR code detection and decode, QR code tracking, relative pose estimation, absolute pose estimation and updating of the absolute pose. Camera calibration is a necessary step in 3D computer vision for extracting metric information from 2D images. The goal of the calibration is to estimate camera parameters that allow to make the relation between camera's natural units (pixel positions in the image) and real world units (normalized position in meters in the image plane). Furthermore, by calibration, it is possible also determinate the distortion parameters, which are intrinsic in cheap pinhole cameras, and using such parameters to reduce the distortion. The pinhole camera model is described as

$$s\begin{bmatrix} u \\ v \\ 1 \end{bmatrix} = \boldsymbol{A}\,[\boldsymbol{R}|\boldsymbol{t}]\begin{bmatrix} X \\ Y \\ Z \\ 1 \end{bmatrix} \tag{3}$$

with

$$A = \begin{bmatrix} f_x & 0 & c_x \\ 0 & f_y & c_y \\ 0 & 0 & 1 \end{bmatrix}, \; [\boldsymbol{R}|\boldsymbol{t}] = \begin{bmatrix} r_{11} & r_{12} & r_{13} & t_1 \\ r_{21} & r_{22} & r_{23} & t_2 \\ r_{31} & r_{32} & r_{33} & t_3 \end{bmatrix} . \tag{4}$$

In this model, a scene view is formed by projecting 3D points into the image plane using a perspective transformation. The matrix $\boldsymbol{A}$ is called intrinsic parameters matrix and the matrix $[\boldsymbol{R}|\boldsymbol{t}]$ is called the extrinsic parameters matrix which denotes the coordinate system transformations from 3D world coordinates to 3D coordinates fixed with respect to the camera. Camera calibration allows to estimate both intrinsic and extrinsic parameters. This step is carried out by the OpenCV software [20] that uses the algorithm described in [22]. In the present paper the extrinsic parameters are the position and the orientation of the origin of the QR code coordinate system expressed in coordinates of the camera coordinate system. This means that is possible to know the relative position $\boldsymbol{p_r}$ of the origin of the camera coordinate system respect to the origin of QR coordinate system as follows

$$\boldsymbol{p_r} = -\boldsymbol{R}^{-1} * \boldsymbol{t} = -\boldsymbol{R}' * \boldsymbol{t}. \tag{5}$$

After a QR code is detected, the vision system starts the tracking step using a combination of these two algorithms: the Kanade-Lucas-Tomasi (KLT) feature tracker described in [23] and the model-based edges (MBE) tracker which uses a 2D model of the QR code in which its edges are stored. These algorithms require some setting parameters that have been chosen in order to have the best tracking performance when maximum speed of wheelchair is under 1 km/s. During the tracking, the developed algorithm updates the $[\boldsymbol{R}|\boldsymbol{t}]$ matrix through two steps: in the first step the matrix is estimated using a linear approach which produces a first pose matrix that is used in the second step as initialization for the virtual visual serving non linear approach; this will converge to the solution with a lower residue than the linear approach [21]. Once the $[\boldsymbol{R}|\boldsymbol{t}]$ matrix is known, the relative pose between the QR code and the camera can be estimated using Eq. (5). Now the vision algorithm is able to estimate the absolute pose of the wheelchair by adding the relative pose between the QR code and the camera with the absolute pose decoded by QR code (Eq. 2). The absolute pose provided by the vision algorithm is fused with that carried out by the IMU in order to reduce error propagation. If the tracking routine loses the QR code (e.g. QR code is out of the camera field of view) the algorithm restarts from the detection step, otherwise it continues to perform the tracking step.

4 Experimental Results

The experiments have been performed in the hallway at the Dipartimento di Ingegneria dell'Informazione of Università Politecnica delle Marche. The wheelchair has been remoted controlled via ROS interface and the imposed trajectory reflects

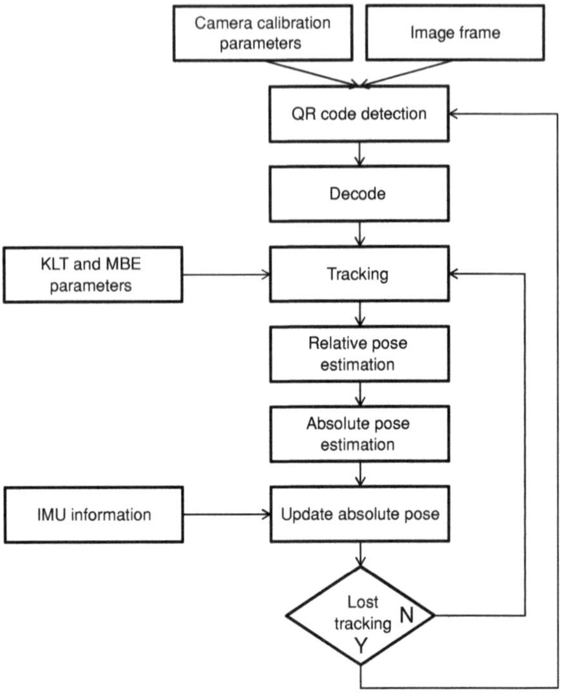

Fig. 2 Flow chart of the visual recognition and tracking system

a L shape of 8 m with 3.5 m route along x axis, a rotation of 90° at standstill and a 4.5 m route along y axis: the movements are expressed in accordance with the absolute reference system placed as in the laser scanned map in Fig. 3a. During the route the wheelchair is driven to move along the middle of the hallway which is 3 m wide. Three QR codes with $0.42 \times 0.42\ \text{m}^2$ dimension has been placed in the experimental setup and fixed to the roof of the hallway at an height of 3 m; thus the orthogonal vertical distance between the camera and the QR codes is about 2.2 m. Each QR code includes the information about its position relative to the absolute reference system, as in Eq. (2), nevertheless the room position has not been used for this experiment. The QR codes positions have been highlighted in Fig. 3b. The QR1, with coordinates (1.5; 4.5), is the first encountered during the path; the QR2, with coordinates (1.8; 1.5), has been placed near the point of 90° rotation and finally the QR3 has been placed in (4.5; 1.5) position. During the route the wheelchair has been forced to move at a constant speed of 1.16 m/s. The merge of both the IMU and the visual information gives a satisfactory pose estimation and the results can be easily evaluated in the same Fig. 3b: IMU measurements are useful in the absence of QR information, nevertheless during the periods without visual feedback, the pose error increases with time for the well known acceleration integration problem. In fact both the QR2 and the QR3 helps the algorithm to recognize the wheelchair in the absolute space as soon as they are detected.

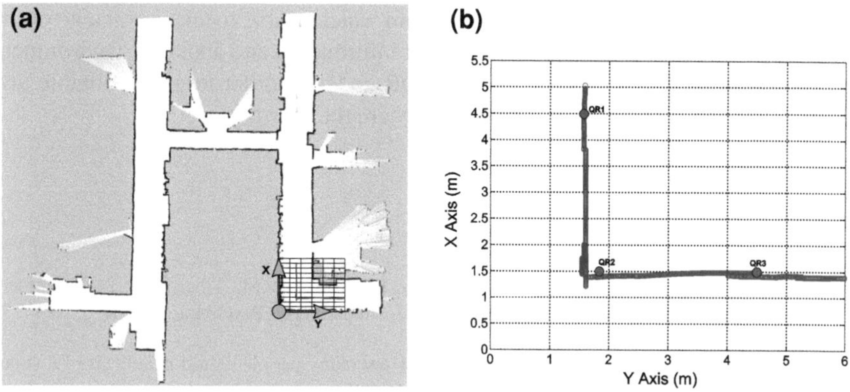

Fig. 3 Experimental Results for QR-code and IMU measurements localization algorithm. **a** Estimated pose on a setup scenario map. **b** Details for the wheelchair estimated pose

5 Conclusions

Smart powered wheelchairs with autonomous navigation intelligence and home integration ability are key points in AAL technologies as they could provide a possible sustainable solution to enhance independence, and improve quality of life for people unable to operate with classical powered wheelchairs. To ensure these aims a smart wheelchair has to be provided by a self-navigating system, which permits to localize it with good accuracy combining both absolute and relative sensors measurements. The authors main contribution is the development of an odometry and QR code landmarks-based indoor navigation system. The smart wheelchair is equipped with an IMU and a low-cost monocular camera. The vision system estimates the absolute pose of the wheelchair in the environment by adding the absolute pose decoded by the QR code with the relative pose between the wheelchair and the QR code using vision algorithms. The absolute pose provided by IMU sensor, is fused with the absolute position estimated by the vision system improving the localization accuracy. The navigation-system has been developed in ROS framework. The camera and the IMU sensor, with their relative software, have been developed as two ROS nodes that exchange messages between them in order to estimate the absolute pose of the wheelchair. According to the obtained results, the QR code localization has been proved to be a successful absolute localization method which helps to correct pose estimation algorithms based on dead-reckoning approach; as expected the improvements in the pose estimation deeply rely on the number of QR codes used. These preliminary results show that the present navigation system can be further developed and integrated with other navigation algorithms. In fact the authors are currently considering two possible future developments: the first is related to develop a voice navigation system in order to allow the users to move easily from a room to another. The system will be provided by a path planning algorithm based on the proposed QR code landmarks-based navigation system. The second is related to the integration of an home automation system which is able to take advantage of the knowledge about self localization structural information and about the environment (e.g. room identification) encoded in the QR codes in order to apply suitable and smart control actions in the environments where the user is placed.

References

1. Cesta, A., Cortellessa, G., Giuliani, V., Pecora, F., Scopelliti, M., Tiberio, L.: Psychological implications of domestic assistive technology for the elderly. PsychNology J. **5**(3), 229–252 (2007)
2. Simpson, R.C. Lo Presti, E.F., Cooper, R.A.: How many people would benefit from a smart wheelchair?. J. Rehabil. Res. Dev. **45**(1), 53–71 (2008)
3. Mikołajewska, E., Mikołajewski, D.: Wheelchair development from the perspective of physical therapists and biomedical engineers. J. Adv. Clin. Exp. Med. **19**(6), 771–776 (2010)
4. Siegwart, R., Nourbakhsh, I.R.: Introduction to Autonomous Mobile Robots. The MIT press, Cambridge (2004)
5. Milford, M., Wyeth, G.: Persistent navigation and mapping using a biologically inspired SLAM system. Int. J. Robot. Res. **29**(9), 1131–1153 (2010)
6. Becker, C., Salas, J., Tokusei, K., Latombe, J.-C.: Reliable navigation using landmarks. In: IEEE International Conference on Robotics and Automation, pp. 401–406. (1995)
7. Guoyu, L., Xu, C.: A robot indoor position and orientation method based on 2D barcode landmark. J. Comput. **6**(6), 1191–1197 (2011)
8. Nam, Y.: Map-based indoor people localization using an inertial measurement unit. J. Inf. Sci. Eng. **27**(4), 1233–1248 (2011)
9. Xiong, Z., Sottile, F., Spirito, M.A., Garello, R.: Hybrid indoor positioning approaches based on WSN and RFID. In: 4th IFIP International Conference on New Technologies, Mobility and Security (NTMS), pp. 1–5 (2011)
10. Courbon, J., Mezouar, Y., Eck, L., Martinet, P.: Efficient visual memory based navigation of indoor robot with a wide-field of view camera. In: 10th International Conference on Control, Automation, Robotics and Vision (ICARCV 2008), pp. 268–273 (2008)
11. Yanco, H.A.: Wheelesley: A robotic wheelchair system: indoor navigation and user interface. In: Assistive Technology and Artificial Intelligence, Lecture Notes in Computer Science, Springer, Berlin Heidelberg, vol. 1458, pp. 256–268 (1998)
12. Cousins, S.: Exponential growth of ROS [ROS Topics]. IEEE Robot. Autom. Mag. **18**(1), 19–20 (2011)
13. Cousins, S.: Welcome to ROS topics [ROS Topics]. IEEE Robot. Autom. Mag. **17**(1), 13–14 (2010)
14. LORD MicroStrain. http://www.microstrain.com/
15. Logitech. http://www.logitech.com/
16. ARDUINO. http://arduino.cc/
17. Robot Operating System. http://wiki.ros.org/
18. von der Hardt, H.J., Wolf, D., Husson, R.: The dead reckoning localization system of the wheeled mobile robot ROMANE. In: IEEE/SICE/RSJ International Conference on Multisensor Fusion and Integration for Intelligent Systems, pp. 603–610 (1996)
19. ZBar library. http://zbar.sourceforge.net
20. OpenCV 2.4.6. http://www.opencv.org
21. Visual servoing platform—Irisa. http://www.irisa.fr/lagadic/visp/visp.html
22. Zhang, Z.: A flexible new technique for camera calibration. IEEE Trans. Pattern Anal. Mach. Intell. **22**(11), 1330–1334 (2000)
23. Lucas, B.D., Kanade, T.: An iterative image registration technique with an application to stereo vision. In: Proceedings of the 7th International Joint Conference on Artificial Intelligence (IJCAI '81), pp. 674–679 (1981)

Chapter 22
Towards an Impact Assessment Framework for ICT-Based Systems Supporting Older People: Making Evaluation Comprehensive Through Appropriate Concepts and Metrics

Roberta Bevilacqua, M. Di Rosa, E. Felici, V. Stara, F. Barbabella and L. Rossi

Abstract Although it is internationally acknowledged that ICT-based systems have the potential to provide cost-effective services, their deployment is still limited. Through the User Centred Design, it is possible to draw up information on key concepts, such as attitude towards technology, acceptance and usability of any new products. The collected information will serve for the technological development on one side, and for the Impact Assessment Analysis on the other one, that will describe the future path of the devices. The paper describes in details the most important metrics—attitude towards technology, usability and accessibility—for conducting a prompt evaluation of a new device for the older people, suggesting common guidelines and critical issues to solve.

1 Introduction

Although more and more evidence suggests that ICT-based systems have the potential to provide efficient support services for the elderly at home—fostering their independence, quality of life and social participation—their deployment is still limited and far from being mainstreamed in most countries.

R. Bevilacqua (✉) · M. Di Rosa · E. Felici · V. Stara · F. Barbabella · L. Rossi
Polo Scientifico-Tecnologico, INRCA—Via Santa Margherita, 5-60124 Ancona, Italy
e-mail: r.bevilacqua@inrca.it

M. Di Rosa
e-mail: m.dirosa@inrca.it

E. Felic
e-mail: e.felici@inrca.it

V. Stara
e-mail: v.stara@inrca.it

F. Barbabella
e-mail: f.barbabella@inrca.it

S. Longhi et al. (eds.), *Ambient Assisted Living*, DOI: 10.1007/978-3-319-01119-6_22,

Many barriers contribute to this situation, such as, for example, the low digital skills of older people, the ethical concerns over privacy, intrusion and lifestyle monitoring, as well as a general resistance to change [1, 2].

Many studies have underlined that, when faced with new technologies, older people can find themselves in a relatively weak position, having prejudices about their usefulness and ease of use [3]. Often, such devices are too complex that older people prefer not to use them. Moreover, common limitations make new technologies difficult to use, and, even more often, the technological devices and services do not reflect the real needs of the end users.

Although some evidence concerning the effects of ICT-based systems for older people exists, much more effort would be necessary for improving the state-of-the-art knowledge in the field. In fact, despite the number of ongoing projects and interventions is rising, the quality of evaluations carried out and the comprehensiveness of results still seem to be limited, at least in the European context.

In this respect, it is obvious that impact assessment is an essential phase of the development and implementation of these initiatives and social innovations in general, especially concerning the characteristics of human-computer interaction (HCI) enabled by ICT systems, which require serious attention. For these reasons, our aim is to design and provide a basic framework for impact assessment that may be applied in every case ICTs are used for addressing needs of older people.

Through using the User Centred Design (UCD) as the main pillar of such a framework, it is possible to define a set of key criteria that should be assessed in a proper context of use, i.e. attitude towards technology, acceptance, usability and cost-benefit analysis. The collected information will serve, on one hand, for the technological development, and, on the other hand, for an adequate evaluation (ex ante, in itinere or ex post) of main direct effects of the user-system interaction.

2 The Elderly Profile

Even though solutions and services are conceived for offering help to the users, the application of the technological innovation has been limited until now, among the older people [4]. Some explanations can be found in:

1. *The general reluctance to use technology*. The main barriers in applying technologies to older people and caregivers originate from psychological factors, especially the perception of quality of life, prejudices, habits and education. Many elderly people strongly reject anything that could ask them to change their life or habits. Often, these people are not aware of the opportunity they could have to improve their quality of life.
2. *The unclear evidence of the real benefits of technologies*. The low level of technological literacy, especially in the Southern European countries as well as the Eastern ones, represents the most important issue to overcome, by spreading the knowledge on how to use the technology for achieving and supporting the well-being.

3. *The difficulty in using the technologies in an appropriate way.* In addition to the difficulty in learning the functioning of any new devices, also the ageing limitations can make a new technology too difficult to use for the elderly. The most common age-related impairments can be, for example, vision decline, hearing loss, motor skill diminishment, and the cognitive processes' deterioration in remembering names or maintaining the flow of the conversation, and the increased tendency to misplace things.
4. *The partial broadband coverage in many European geographical areas.* Broadband is a fundamental issue for the development of AAL technology, that allows the remote monitoring of the environments and elderly or non-autonomous people, as well as the connection of the users with the services. Unfortunately, broadband is not available throughout all the European areas: many rural areas and some Southeastern European countries are not covered by a broadband network. Older people who live in these specific areas often do not make a good use of all the public services, and, in particular, elderly people remain isolated from such applications.

3 The Impact Analysis Assessment

The final goal of every policy, intervention or project should be to produce positive effects on the target population. Since the range of ICTs for older people is tremendously wide, in this work we distinguish only a set of criteria that are applicable in every system delivering any kind of support service through technologies. The criteria described below can be of course integrated by practitioners and researchers with additional ones in relation to the real objective(s) and function(s) of the ICT-based system: dimensions of impact may include also users' quality of life, social participation, knowledge increase, empowerment in daily activities, cost-benefit analysis etc., strictly depending on the type of technology and services provided.

3.1 Attitude

With the term "attitude" is usually intended the individual predisposition directed toward some object, person or event, that can influence the behavior, for example, of technology usage [5]. Attitude towards technology depends on believes and values [6, 7], as well as the user's awareness of the technology and its purpose, the extent to which the features of the technology are consistent with the user's needs, the user's experience with the technology and the availability of support such as documentation and training [8]. Kai-ming and Enderwick [9] hypothesised that attitude is influenced by other internal believes such as the degree to which the perceived application of foreign technology is free of efforts (perceived difficulty), the history of adoption of new technology, the supplier's commitment and the perceived benefit.

According to the increasing literacy about ICT use by older adults, it is important to understand the nature of the attitude because it influences the willingness to accept and use technology, as well as it is modifiable during the life span, providing information and experience with ICT [10].

At this end, experimental researches have shown a rather positive attitude towards technological modification in the domestic environment of the elderly, suggesting a strong association between the elderly inclination to use a device and the problem they have to cope with [11, 12].

3.2 Usability

The definition of usability from ISO 9241-11 [13]—the extent to which a product can be used by specified users to achieve specified goals with effectiveness, efficiency and satisfaction in a specified context of use—is considered as the main reference of usability.

Effectiveness, efficiency and satisfaction are the first dimensions that have constituted the meaning of usability, followed by concepts like flexibility, learnability, safety, memorability and quality in use [14–17]. As it is shown, usability is not given an absolute definition, but is relative to the users, goals and contexts of use that are appropriate to the particular set of circumstances [18].

Designing any products or environments involves the consideration of many factors, but often such products and environments are designed for the average user, without taking into account the abilities and needs of specific target groups, like the older people. So, the conceptualization of usability, as it is currently formulated, needs to be extended to the concept of accessibility, essential prerequisite for the development of any assistive technologies.

The concept of accessibility is defined as "the usability of a product, service, environment or facility by people with the widest range of capabilities" [19]. For matching the accessibility of a given device, it is necessary to adopt the Inclusive Design approach, defined as "the design of products and environments to be usable by all people, to the greatest extent possible, without the need for adaptation or specialized design" [20].

Currently, there are many tools and techniques deeply described in literature, that can be applied for the analysis of the usability of any technologies, but they seem not fully appropriate in the case of older people, because they don't take into consideration the accessibility issue, on one side, and the particular characteristics of the elderly, on the other side. An example of this, can be offered by the application of the Nielsen's heuristics, for the evaluation of the human computer interaction [21, 22].

Typically, in fact, the heuristics evaluation involves evaluators examining the interface and judging its compliance with recognized usability principles, the heuristics. In this way, the risk of adopting an "expert perspective" seems to be higher than the adoption of the end-user perspective and, in addition, the accessibility remains unexplored.

Some researchers have made attempts in order to promote the development of guidelines on heuristics for the older people, that take into consideration key factors for evaluating their performance with a device.

In this case, the choice of the evaluation strategies represent a key issue to gain all the necessary information on the usability and accessibility of a technological artifact but also to keep the evaluators user-centered, whereas typical heuristic evaluations often become system focused or rule-focused [23].

Some researchers have shown that the method called "persona" can represent a good solution to understand if the artefact may be considered usable or it may not with the older users. More than a simple scenario, in fact, persona describes the user's goals and way of interaction, with the purpose of understanding the most important tasks in the User Interfaces and the user's motivations, allowing the observations of the artifact "through" the users-personas [24, 25]. In the case of persona, important characteristics of older users are addressed, such as: age (including chronological age and life experiences), ability (cognitive and physical) aptitude (expertise with the technology) and attitude (confidence levels and emotional state of mind) [23], even if issue of the great heterogeneity of the elderly population still remain to be solved.

3.3 Acceptance

The individual acceptance of ICT has been researched from multiple theoretical perspectives in order to better understand this complex construct. No theory is dominant, but there are some common themes across paradigms: acceptance behaviour is influenced by individual differences, social and situational influences, believes and attitudes [26].

On the research side, a step forward is necessary, in order to investigate which are the elements that can be considered as determinants of the technology use by the older people, not only in relation to impairments and/or unsatisfied needs.

In their model, Thinker and McCreadie suggested a linkage between the acceptance of technology and the felt need for assistance, based on individual characteristics and disability and attributes given to Assistive Technology (AT) [27]. For the authors, the AT offers the opportunity of narrowing the gap between individual capacity and the environment, depending on older people's willingness to use the artefact, which is strictly connected with its acceptability [27].

In this view, the acceptance of technology represents a more complex phenomenon respect to the analysis of older people needs per se and it could be defined as "the demonstrable willingness within a user group to employ technology for the task it is designed to support" [28].

To understand the kernel of the older people rejection of new technological artefacts means to understand deeply the personal believes that characterized the elderly and that can determine their closure to the innovation.

The positive or negative acceptance of a device is linked to intrinsic or extrinsic factors related to the technology. The overestimation or the underestimation of the devices' potentialities can lead to an overall erroneous vision of the technology, that can be modified and/or re-established through specific training, aimed at creating a technological literacy. As it was defined by ITEA, "technological literacy is the ability to use, manage, assess, and understand technology. It involves knowledge, abilities, and the application of both knowledge and abilities to real-world situations. Citizens of all ages benefit from technological literacy, whether it is obtained through formal or informal educational environments" [29].

Starting from the older person representation of the artefact, an adequate technological literacy training should provide skills and expertise to know how to use properly the artefact on one side, and to understand how technology can improve his/her daily life, on the other one [29].

4 Conclusion

The promotion of the technological literacy in the older people and the adoption of the Inclusive Design approach are the overall issues to follow for the development of new technologies for the elderly, as well as the progress in the usability content definition, through the use of effective evaluation strategies and the cooperation of a multisciplinary team, in order to provide guidelines for the specific target.

Even if the evaluation of the metrics described represents the kernel of impact assessment, some attempts should be made in order to find appropriate means for integrating evaluation with other dimensions of impact which depend on the peculiarities of the ICT-based system.

In this respect, the most useful and applicable dimension is quality of life of users. However, this construct is currently measured by instruments typically derived from social or health sciences. In many cases, they are not adapted for the evaluation of the improvement of the quality of life through the use of a new device, mainly due to their conceptualization derived from clinical contexts. For this reason, it is necessary to move further in the state of the art and develop adequate tools addressing the complexity of the range of technology-based systems developed in the field of gerontechnology.

References

1. Kubitschke, L., Cullen, K.: ICT and Ageing—European Study on Users, Markets and Technologies. Final report. Bonn: empirica Gesellschaft für Kommunikations- und Technologieforschung. Retrieved January 31, 2012, from www.ict-ageing.eu/ict-ageing-website/wp-content/uploads/2010/D18_final_report.pdf (2010)
2. Magnusson, L., Hanson, E.: Ethical issues arising from a research, technology and development project to support frail older people and their family carers at home. Health Soc. Care Community **11**(5), 431–9 (2003)

3. Empirica & WRC: Seniorwatch 2. Assessment of the Senior Market for ICT Progress and Developments. Final Report. Bonn-Brussels: empirica Gesellschaft für Kommunikations- und Technologieforschung. Retrieved January 31, 2012, from www.inst-informatica.pt/servicos/informacao-e-documentacao/biblioteca-digital/sociedade-da-informacao-1/2008/seniorwatch-2-assessment-of-the-senior-market-for (2008)
4. Clarke, A., Concejero, P.: The Digital divide–services for the elderly and disabled in 2010–the PRISMA project. Retrieved June 17 2013 from http://www.hft.org/HFT01/paper01/PSN/19_01.pdf (2010)
5. Regan, D.T., Fazio, R.: On the consistency between attitudes and behaviour: look to the method of attitude formation. J. Exp. Soc. Psychol. **13**, 28–45 (1977)
6. Pancer, S.M., George, M., Gebotys, R.J.: Understanding and predicting attitudes towards computers. Comput. Hum. Behav. **8**(2), 211–222 (1992)
7. Gardner, D.G., Dukes, R.L., Discenza, R.: Computer use, self-confidence and attitudes: a causal analysis. Comput. Hum. Behav. **9**(3), 427–440 (1993)
8. Mackie, R.R., Wylie, C.D.: Factors influencing acceptance of computer-based innovations. In: Helander, M. (ed.) Handbook of Human-Computer Interaction, pp. 1081–106. Elsevier Publishing Co, New York (1998)
9. Kai-ming, A.A., Enderwick, P.: A cognitive model on attitudes towards technology adoption. J. Manag. Psychol. **15**(4), 266–282 (2000)
10. Czaja, S.J., Sharit, J.: Age differences in attitudes toward computers. J. Gerontol. **53B**, 329–340 (1998)
11. Giuliani, M., Scopelliti, M., Fornara, F.: Coping strategies and technology in later life. In: Proceedings of AISB 2005–Symposium "Robot Companions: Hard Problems and Open Challenges in Robot-Human Interaction", Hatfield, UK (2005)
12. Roelands, M., Van Oost, P., Buysse, A., Depoorter, A.M.: Awareness among community-dwelling elderly of assistive devices for mobility and self-care and attitudes towars their use. Soc. Sci. Med. **54**, 1441–14451 (2002)
13. ISO/IEC: 9241–11 Ergonomic requirements for office work with visual display terminals (VDT)s—Part 11 Guidance on usability (1998) ISO/IEC 9241–11: 1998 (E)
14. Gould, J.D., Lewis, C.: Designing for usability: key principles and what designers think. Commun. ACM **28**(3), 300–11 (1985)
15. Shackel, B.: Usability–context, framework, definition, design and evaluation. In: Shackel, B., Richardson, S. (eds.) Human Factors for Informatics Usability, pp. 21–37. Cambridge University Press, Cambridge (1991)
16. Sharp, H., Rogers, Y., Preece, J.: Interaction Design: Beyond Human-Computer Interaction. Wiley, London (2007)
17. Stone, D., Jarrett, C., Woodroffe, M., Minocha, S.: User Interface Design and Evaluation. Morgan Kaufmann, CA (2005)
18. Petrie, H., Bevan, N.: The evaluation of usability, accessibility and user experience. In: Stepanidis, C. (ed.) The Universal Access Handbook. CRC Press (2009)
19. Iso: TS 16071 International Organization for Standardization: ISO 16071 Ergonomics of human-system interaction–guidance on software accessibility. Technical Specification. International Organization for Standardization, Switzerland (2000)
20. www.universaldesign.com
21. Nielsen, J., Molich, R.: Heuristic evaluation of user interfaces. In: Proceedings ACM CHI'90 Conference, Seattle, WA, 1–5 April, pp. 249–256 (1990)
22. Nielsen, J.: Heuristic evaluation. In: Nielsen, J., Mack, R.L. (eds.) Usability Inspection Methods. Wiley, NY (1994b)
23. Chisnell, D.E., Redish, J.C., Lee, A.: New heuristics for understanding older adults as Web users. Tech. Comm. **53**(1), 39–59 (2006)
24. Pruitt, J., Adlin, T.: The Persona Lifecycle: Keeping People in Mind Throughout Product Design. Morgan Kaufmann, San Francisco (2006). ISBN 0-12-566251-3
25. Chapman, C.N., Love, E., Milham, R.P., ElRif, P., Alford, J.L.: Quantitative evaluation of personas as information. In: Proceedings of the Human Factors and Ergonomics Society 52nd Annual Meeting, New York, NY, pp. 1107–1111 (2008)

26. Agarwal, R.: Individual acceptance of information technologies. In: Zmud, R.W. (ed.) Framing the Domains of it Management Research: Glimpsing the Future Through the Past, pp. 85–104. Cincinnati, Pinnaflex (2000)
27. McCreadie, C., Tinker, A.: The acceptability of assistive technology to older people. Ageing Soc. **25**(1), 91–110 (2005)
28. Mynatt, E.D., Essa, I., Rogers, W.: Increasing the opportunities for aging in place. In: Proceedings of the CUU 2000 Conference on Universal Usability. ACM, New York (2000)
29. ITEA: Standards for technological literacy: content for the study of technology. Reston, VA. http://www.iteaconnect.org/TAA/PDFs/xstnd.pdf (2007)

Brain.me: A Low-Cost Brain Computer Interface for AAL Applications

Niccolò Mora, V. Bianchi, I. De Munari and P. Ciampolini

Abstract A Brain-Computer Interface (BCI) is an alternative/augmentative communication device that can provide users (for example, individuals lacking voluntary muscle control) with an interaction path, based on the interpretation of his/her brain activity. In this paper, the design and implementation of a flexible, low-cost BCI development platform is presented; this platform could serve as a workbench to develop compact, standalone BCI embedded modules, specifically targeted to (even if not limited to) AAL control purposes. First, a low-cost, custom, bio-potential acquisition unit was realized; then, a Matlab-based environment was developed for EEG (ElectroEncephaloGram) signal analysis and processing. An application example involving a 4-class SSVEP-based BCI is presented, along with a novel classification algorithm which achieved 94.7 % classification accuracy.

N. Mora (✉) · V. Bianchi · I. De Munari · P. Ciampolini
Dipartimento di Ingegneria dell'Informazione, Università degli Studi di Parma, Pal. 3,
Parco Area delle Scienze 181/A, 43124 Parma, Italy
e-mail: niccolo.mora@nemo.unipr.it

V. Bianchi
e-mail: valentina.bianchi@unipr.it

I. De Munari
e-mail: ilaria.demunari@unipr.it

P. Ciampolini
e-mail: paolo.ciampolini@unipr.it

V. Bianchi · I. De Munari · P. Ciampolini
Dipartimento di Ingegneria dell'Informazione, i-cubo s.r.l., Pal. 3,
Viale G. P. Usberti 181/A, 43124 Parma, Italy

S. Longhi et al. (eds.), *Ambient Assisted Living*, DOI: 10.1007/978-3-319-01119-6_23,

1 Introduction

Ambient Assisted Living Technologies (AAL) aim at making the home environment more cooperative and intelligent, providing help to accomplish daily living tasks; AAL solutions have been successfully applied for supporting and promoting independent life of elderly people. However, individuals affected by severe impairments could be potential beneficiaries of AAL services too (such as those aimed at environmental safety and control): the main issue is to provide these users with an effective interaction path to the AAL system.

A possible approach to this problem involves Brain-Computer Interface (BCI) technologies. BCI are alternative/augmentative communication means [1] that aim at providing the user (for instance lacking voluntary muscle control) with an interaction path, based on the interpretation of her/his brain activity. In this paper, the development of an EEG (ElectroEncephaloGram)-based BCI communication unit is introduced, explicitly conceived for (even if not limited to) AAL control purposes. Therefore, with respect to most common BCI approaches, the proposed strategy focuses at relatively simpler tasks, leaving room for lowering costs, user's effort and invasiveness.

In our view, the BCI device must be seamlessly integrated into CARDEA [2], the flexible, LAN-based, AAL system developed by the University of Parma. CARDEA, like most common home automation systems, offers environmental control and safety services, but, in addition, it provides more advanced functionalities, related to Assistive Technologies (AT): examples are fall detection, vital sign monitoring [3] and indirect wellness monitoring [4]. Thus, CARDEA integrates both AT and AAL aspects under a unique, convergent vision.

In order to provide such services, CARDEA currently supports many user interaction paradigms, including button switches, touchscreen, vocal, remote internet control; in [5] the interaction scheme between CARDEA and a simple BCI is discussed. Other examples of AAL-focused BCI can be found in [6–9]. With respect to most literature works, the present approach aims at developing tools and methods for low-cost, standalone embedded BCI modules, making high-performance acquisition hardware or large computing powers unnecessary.

In this paper we describe the development of a platform conceived for prototyping of BCI embedded system; it includes three main units (Fig. 1): (i) an Analog Front End (AFE) for the acquisition EEG signal, (ii) a digital signal processing unit, implementing feature extraction and classification and, (iii), an output/feedback unit for display and implementation of active controls. We started from developing and testing a novel hardware AFE unit, aiming at a compact and inexpensive circuit. Signal processing and feedback units are currently implemented on a PC architecture, allowing for more flexibly testing and for better tuning performance. Nevertheless, the algorithms are specifically targeted for implementation on low-cost, portable devices, paying attention in devising computationally efficient methods.

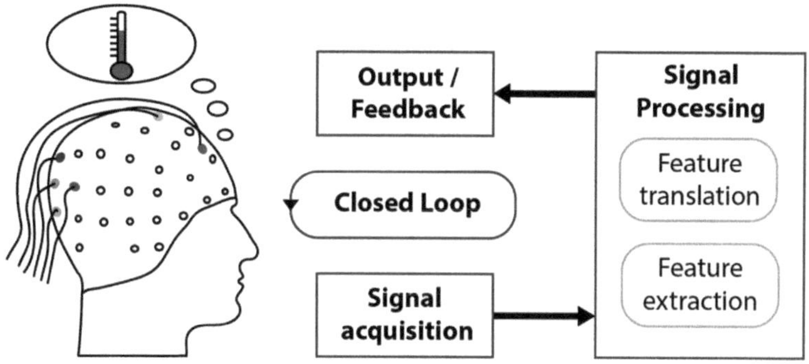

Fig. 1 Functional blocks of a BCI

Such environment enabled testing of different configurations, algorithms and methods: among them, a novel Steady State Visual Evoked Potentials (SSVEP) classification algorithm was devised, yielding good performance and requiring no initial user training. This allows for simplifying the overall design and also fosters its portability to different configurations. Eventually, all these functional units will be integrated in a compact, standalone embedded module, capable of local and autonomous signal processing, which will interface to the AAL system as a simple controller.

In Sect. 2 an overview of the proposed module is presented, demonstrating its capabilities and flexibility; Sect. 3 presents an application example of the BCI module using the Steady State Visual Evoked Potentials (SSVEP); finally, in Sect. 4 the results obtained in the previous section are presented and discussed, as well as the future work.

2 Overview of the Proposed BCI Module

Although many different methods can be used to extract information on the user's brain activity, such as the ones based on functional Magnetic Resonance Imaging (fMRI) [10], Near InfraRed Spectrography (NIRS) [11] or Positron Emission Technology (PET) [12], EEG is the most popular technique for practical BCI devices. In fact, EEG signals can be non-invasively acquired from the scalp, by sensing the electrical signal with (dry or wet) skin-surface electrodes. Moreover, given our purposes, EEG offers a satisfactory tradeoff between spatial and temporal resolution, as well as reasonable costs and compactness when compared to other methods.

As far as communication paradigms are involved, focusing on EEG-based BCI, many brain features can be exploited. Among the most popular ones we can cite: Slow Cortical Potentials (SCP) [13], Event Related De-synchronization (ERD) [14] and the related motor imagery paradigm, P300 [15, 16], Steady State Visual Evoked Potentials (SSVEP) [17]. The latter paradigm, in particular, exploits brainwave features elicited by the involuntary response to a continuous, repetitive stimulus, such as a blinking LED: within a given low-frequency range, the blinking frequency reflects on the onset of a isofrequency component in the brain power spectrum. With respect to other paradigms, the SSVEP paradigm makes synchronization issues less critical, and possibly lends itself for simpler acquisition architectures.

High-end, clinical EEG equipment with a large number of electrodes are scarcely suitable for the design of low-costs, small-size devices. In [5], an inexpensive EEG Analog Front End (AFE) module for extracting basic information on brain activity was presented; this module was realized using low-cost, standard electronic components to foster product interoperability.

In order to cope with such extremely small signals (whose amplitudes are as low as a few μV) low-noise design techniques were applied. In particular, signals had to be amplified before digitization, in order to prevent the ADC's (Analog to Digital Converter) noise floor from corrupting the informative content in the EEG waveforms. Obviously, the amplification chain should contribute as low noise as possible on the desired signal. This calls for AC-coupling and high gain applied to the input waveforms (intrinsic DC offset between the electrodes prevents from applying a high DC gain). Moreover, EEG signals generally exhibit poor Signal to Noise Ratio (SNR) due to, for example, electrical and power line interference. The use of active electrodes helps in mitigating those issues, having better performance than their passive counterparts, but this comes at the price of increased costs and complexity. Given the purpose of our work, focused more on cost effectiveness and versatility, we adopted passive electrode technology; this choice, however, does not preclude future deployment of different types of sensing elements.

Given the requirements discussed above, a multi-channel AFE for EEG signals was designed and realized on a standard 4 layer PCB. The module, which can be operated both in single supply or in split supply mode, features up to 6 differential/common reference EEG-specific channels, plus 2 spare, fully-differential channels. These additional channels can be used, for example, for simultaneous recording of other biopotentials, such as ElectroOculoGram (EOG), ElectroMyoGram (EMG), ElectroCardioGram (ECG); in this way, new types of analysis can be carried out, relating different biopotentials. Examples could be the correlation between the recorded EMG at the onset of a movement and the correspondent ERD in the sensorimotor cortex, or the detection of ocular artifacts by means of EOG signals.

Figure 2 shows the schematic of an EEG acquisition channel along with a Driven Right Leg circuit, introduced to improve common mode noise rejection. Each channel features a differential, passive AC-coupling stage, whose components do not affect the overall Common Mode Rejection Ratio (CMRR), nor their contributed noise impact significantly on the overall performance [18]. A differential, AC-coupled amplification stage applies a gain of 800 to the input EEG waveforms, and a second order, Bessel-response low-pass filter with a cutoff frequency of 250 Hz follows to limit the signal bandwidth: this value was chosen in order to allow easy recordings of other biopotentials, such as ECG. A high resolution 24 bit $\Sigma - \Delta$ ADC completes the signal chain.

Noise performance of the AFE was tested with input terminals shorted and the input referred noise was less than 1.8 μVpp, more than sufficient, in principle, to

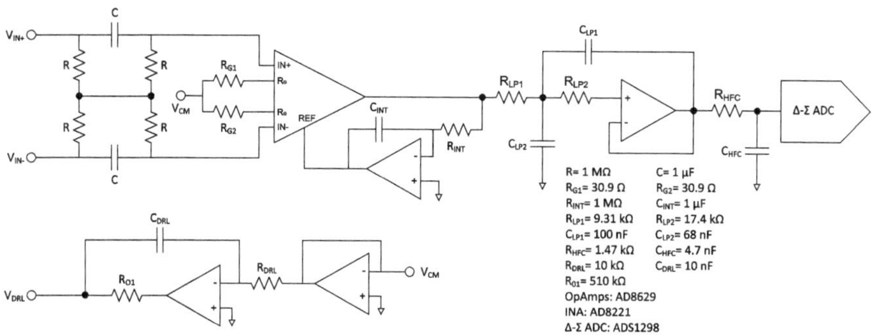

Fig. 2 Schematic of the proposed AFE

Fig. 3 Example of an 18 Hz SSVEP response. Electrode locations are reported in the legend

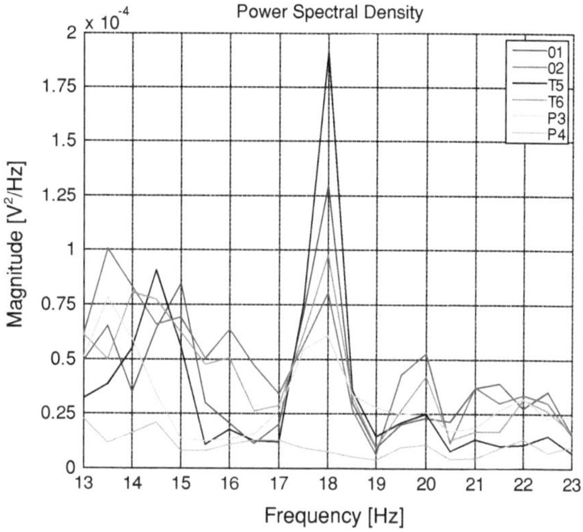

extract basic information of brain activity. As an example of functional validation, a Power Spectral Density (PSD) plot of an 18 Hz SSVEP response, acquired with the proposed AFE, is shown in Fig. 3; spectral peaks at the corresponding stimulation frequency are clearly visible.

At this stage of research, a Matlab-based platform for paradigm testing and algorithm prototyping was designed. From such environment, we can control several AFE parameters, such as ADC gain and data rate, as well as interact with the controllers of the stimulation units, such as flashing lights or arrays. Matlab-ADC interfacing is achieved by means of an Arduino Board, communicating via serial protocol. TCP/IP support is currently being developed.

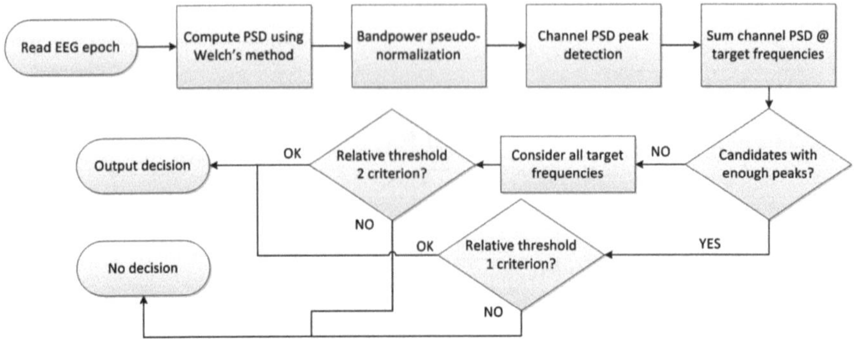

Fig. 4 The proposed lightweight, calibration-less SSVEP classification algorithm

3 Application Example: 4 Class SSVEP BCI

As an application example, a 4-class SSVEP BCI is presented, based on a novel, lightweight classification algorithm suitable for implementation on low-power embedded processors.

The experiment was devised as follows: participants were asked to stare at one of the four simultaneous flickering LED while resting on an armchair at approximately 1 m from the visual stimulus. Each trial lasted for 6 s, and each LED presented a different stimulation frequency (16, 18, 20, 22 Hz); EEG was acquired at 250 SPS from 6 scalp locations (namely O1, O2, T5, T6, P3, P4), using standard 10 mm Ag/AgCl disk electrodes with conductive paste.

In order to classify the SSVEP response, a novel algorithm was developed and tested, illustrated in Fig. 4.

First, the input EEG waveforms are low-pass filtered ($f_{cut} = 40$ Hz) for out-of-band noise reduction. Optionally, further pre-processing steps may include spatial filtering, such as a re-referencing of electrodes according to Common Average Reference (CAR) filter topology, or the creation of bipolar leads. Then, the Power Spectral Density (PSD) is estimated using Welch's method. At this point, given a pre-determined band of interest, the channel powers are equalized over this band.

The classification algorithm exploits the *a priori* knowledge of the actual set of stimulation frequencies, checking the conditions only on such set: the channel powers are summed at each target frequency. Candidate targets are selected whenever a given fraction (e.g., at least 50 %) of the channels exhibit a local maximum in the PSD at the target frequency. Then, candidates are compared; a two-step procedure has been devised: if a single candidate exists, the power of which exceeds all the remaining ones by a given threshold, the decision is made. Otherwise, a more selective comparison is made, considering all the target frequencies as candidates and raising the threshold.

Since the test involves only relative comparisons, the algorithm virtually requires no calibration at all. Fine tuning of the algorithm is obtained by adjusting theclassifier's

Table 1 Performance of the proposed algorithm as a function of the window length in terms of correct classification percent and, between brackets, ITR (bit/min)

	Window length			
Subject	3 s	4 s	5 s	6 s
1	90.0 (27.45)	83.3 (16.27)	86.7 (14.66)	90.0 (13.73)
2	91.7 (29.08)	91.7 (21.81)	100 (24.00)	100 (20.00)
3	87.5 (25.17)	95.0 (24.52)	92.5 (17.96)	97.5 (17.92)
4	94.1 (31.68)	94.1 (23.76)	94.1 (19.01)	91.2 (14.31)
Average	90.8 (28.35)	91.0 (21.59)	93.3 (18.91)	94.7 (16.49)

parameters, such as the fraction of channels required to pick a candidate frequency or the relative thresholds for comparing candidates as discussed above.

Moreover, signal processing involves operations particularly suitable for implementation on embedded devices, allowing to take full advantage of specialized digital signal processing hardware.

4 Results and Discussion

To evaluate performance of the aforementioned classification algorithm, experiments were conducted on four healthy volunteers (age 23–26, with normal or corrected to normal vision, five of them without any prior BCI experience); examples of acquired SSVEP responses are shown in Fig. 3. The algorithm was then tested on different window lengths in order to assess and optimize its performance. Two main indicators are usually considered in evaluating BCI setups:

- *Accuracy*, defined as the number of correctly classified trials over the total number of attempts
- *Information Transfer Rate* (ITR), defined in [1] as:

$$ITR = M\left[\log_2 N + \mathrm{P}\cdot\log_2 P + (1-P)\cdot\log_2\left(\frac{1-P}{N-1}\right)\right] \quad (1)$$

where M is the number of trials per minute, N the number of possible choices, and P the classification accuracy.

Table 1 summarizes the performance of the proposed algorithm in terms of accuracy and ITR, assuming a decision is made every *Window Length* seconds.

Even though the approach is intentionally simple, experiments showed that a maximum average accuracy of 94.7 % was achieved, making it practical for aimed AAL control purposes. Such accuracies are in line with state of the art multi-class SSVEP-based BCI. Higher classification speed (i.e. ITR) can be obtained at the price of more computationally intensive signal processing, such as in [19]. However, with respect to most works in literature, focuses on the development of compact, low-cost tools for researching and experimenting novel BCI algorithms, with a particular effort in producing computation-efficient methods, suitable for implementation on low-cost embedded systems. Furthermore, it is worth emphasizing how, focusing on AAL system control purpose, the accuracy of the selection process is actually more important than the selection speed.

The introduction of band-power normalization is shown to slightly improve the classification performance, especially in case of a strong inter-channel imbalance, due to, for example, different electrode impedance; the normalization improves the classification robustness by somehow self-adapting to variable scenarios.

In addition, the proposed algorithm is relatively simple and suitable for implementation on low-power, mobile digital processors, such as DSP or ARM microprocessors. Finally, it does not require synchronization between the stimulation and the acquisition unit, thus simplifying the overall design of an embedded BCI system. Further optimizations will lead to a compact, embedded system capable of processing the signals locally, and handling all the communication in a unique module.

Basic, cue-based online operation is supported, while online, self-paced operation is currently being investigated to provide the user with more flexibility and classification speed.

References

1. Wolpaw, J.R., Birbaumer, N., McFarland, D.J., Pfurtscheller, G., Vaughan, T.M.: Brain-computer interfaces for communication and control. Clin Neurophysiol. **113**(6), 767–791 (2002)
2. Grossi, F., Bianchi, V., Matrella, G., De Munari, I., Ciampolini, P.: An assistive home automation and monitoring system. In: ICCE 2008. Digest of Technical Papers, pp. 1–2 (2008)
3. Bianchi, V., Grossi, F., De Munari, I., Ciampolini, P.: MuSA: a multisensor wearable device for AAL. In: Federated Conference on Computer Science and Information Systems (FedCSIS), 2011, pp. 375–380 (2011)
4. Losardo, A., Matrella, G., Grossi, F., De Munari, I., Ciampolini, P.: Indirect wellness monitoring through AAL environments. Gerontechnology **11**(2), 330 (2012)
5. Mora, N., Bianchi, V., de Munari, I., Ciampolini, P.: Design for a low cost brain-computer interface for environmental and home control. Gerontechnology, **11**(2), 300 (2012)
6. Karmali, F., Polak, M., Kostov, A.: Environmental control by a brain-computer interface. In: Proceedings of the 22nd Annual International Conference of the IEEE Engineering in Medicine and Biology, vol. 4, pp. 2990–2992 (2000)
7. Edlinger, G., Holzner, C., Guger, C., Groenegress, C., Slater, M.: Brain-computer interfaces for goal orientated control of a virtual smart home environment. In: International IEEE/EMBS Conference on Neural Engineering, pp. 463–465, April 2009
8. Piccini, L., Parini, S., Maggi, L., Andreoni, G.: A wearable home BCI system: preliminary results with SSVEP protocol. In: 27th Annual International Conference Engineering in Medicine and Biology Society. IEEE-EMBS 2005, pp. 5384–5387, 17–18 Jan 2006 (2005)
9. Casale, P., Fernandez, J.M., Palou, X.R., Torrellas, S., Ratsgoo, M., Miralles, F.: Enhancing user experience with brain-computer-interfaces in smart home environments. In: 8th International Conference on Intelligent Environments (IE), 2012, pp. 307–310, 26–29 June 2012

10. Andersson, P., Viergever, M. A., Pluim, J., Ramsey, N. F., Siero, J.: fMRI based BCI control using spatial visual attention at 7T. In: 4th International IEEE/EMBS Conference on Neural Engineering, 2009, NER '09, 9 April–May 2 (2009)
11. Sitaram, R., Zhang, H., Guan, C., Thulasidas, M., Hoshi, Y., Ishikawa, A., Shimizu, K., Birbaumer, N.: Temporal classification of multichannel near-infrared spectroscopy signals of motor imagery for developing a brain-computer interface. NeuroImage **34**(4), 1416–27 (2007)
12. Grafton, S.T., Arbib, M.A., Fatiga, L., Rizzolatti, G.: Location of grasp representations in humans by positron emission tomography. Exp. Brain Res. **112**(1), 103–11 (1996)
13. Hinterberger, T., Schmidt, S., Neumann, N., Mellinger, J., Blankertz, B., Curio, G., Birbaumer, N.: Brain-computer communication and slow cortical potentials. IEEE Trans. Biomed. Eng. **51**(6), 1011–8 (2004)
14. Pfurtscheller, G., Brunner, C., Schlögl, A., Lopes da Silva, F.H.: Mu rhythm (de) synchronization and EEG single-trial classification of different motor imagery tasks. NeuroImage **31**(1), 153–9 (2006)
15. Nijboer, F., Sellers, E.W., Mellinger, J., Jordan, M.A., Matuz, T., Furdea, A., Halder, S., et al.: A P300-based brain-computer interface for people with amyotrophic lateral sclerosis. Clin. Neurophysiol. **119**(8), 1909–16 (2008)
16. Carabalona, R., Grossi, F., Tessadri, A., Castiglioni, P., Caracciolo, A., De Munari, I.: Light on! Real world evaluation of a P300-based brain-computer interface (BCI) for environment control in a smart home. Ergonomics **55**(5), 552–63 (2012)
17. Ortner, R., Allison, B.Z., Korisek, G., Gaggl, H., Pfurtscheller, G.: An SSVEP BCI to control a hand orthosis for persons with tetraplegia. IEEE Trans. Neural Syst. Rehabil. Eng. **19**(1), 1–5 (2011)
18. Spinelli, E.M., Pallas-Areny, R., Mayosky, M.A.: AC-coupled front-end for biopotential measurements. IEEE Trans. Biomed. Eng. **50**(3), 391–5 (2003)
19. Volosyak, I.: SSVEP-based Bremen-BCI interface—boosting information transfer rates. J. Neural. Eng. **8**(3), 036020 (2011)

A Speech Interaction System for an Ambient Assisted Living Scenario

Michele Alessandrini, Giorgio Biagetti, Alessandro Curzi and Claudio Turchetti

Abstract In this work we describe a speech recognition system aimed at controlling various apparatus of an intelligent home. The system is especially tailored, and ad-hoc optimizations and strategies have been implemented, to make it suitable to operate unobtrusively in the ambient, requiring that the user only installs small and cheap audio front-ends that will capture his spoken commands. A recognition back-end, running either as a network service reached over the Internet or on a PC in the user's home, performs the hard work of processing the data and turning it into commands, which are sent back to the desired actuator in the home. A case study involving the voice control of a DALI lighting system is presented, together with ideas and results on how to improve recognition accuracy and command spotting efficiency of a system which, by its very nature, might have to deal with audio captured from a distance and great amounts of background noise and unrelated sounds.

1 Introduction

Speech interfaces are commonly viewed as one of the most easy-to-use methods of controlling home automation systems [1]. Nevertheless, there are still many challenges to overcome in order to make them really suitable for ambient assisted living scenarios [2]. In our work we aim at making such an interface the least obtrusive

M. Alessandrini · G. Biagetti (✉) · A. Curzi · C. Turchetti
Dipartimento di Ingegneria dell'Informazione,
Università Politecnica delle Marche, Ancona, Italy
e-mail: g.biagetti@univpm.it

M. Alessandrini
e-mail: m.alessandrini@univpm.it

A. Curzi
e-mail: a.curzi@univpm.it

C. Turchetti
e-mail: c.turchetti@univpm.it

S. Longhi et al. (eds.), *Ambient Assisted Living*, DOI: 10.1007/978-3-319-01119-6_24,

possible, thus the scenario we envision is one in which the user only has to install small and cheap audio front-ends, equipped with a panoramic microphone, and somehow connected to the Internet. No other in-home installation is necessary (apart from the actuators and their network interface, of course), as the speech recognition software actually runs on a remote server following the distributed speech recognition (DSR) approach [3]. If, for some reason, Internet connectivity is not wanted or is lacking, or if privacy is a major concern,[1] the back-end running the speech recogniser can also be installed on a standard personal computer within the user's house.

This kind of setup is extremely demanding for the automatic speech recogniser (ASR), in fact, the microphone must be always on, and a system to detect the end-points of the possible commands and to tell them apart from ordinary noises, sounds, and unrelated speech going on must be developed.

In this work we focus our attention on two reasonably simple techniques that will help deal with this problem: an adaptive automatic voice activity detection (VAD) thresholding mechanism that enables the DSR system only when there is need to do so, and a mechanism that can easily be adapted to each particular application and which is able to effectively reject non-command utterances, background noise, and unrelated sounds. This technique is based on the generation of a *garbage model* [4], which includes suitably placed decoy words [5] that will help the ASR identify out-of-vocabulary (OOV) words, thus enabling it to discard non-command utterances.

2 Application Scenario

The application we focused on is schematically depicted in Fig. 1. It includes an audio front-end, an ASR running on a possibly remote server, and a command unit to control a DALI lighting system through a proprietary wireless interface.

The audio front-end we devised for our experiments includes either an inexpensive and power-efficient Raspberry Pi® equipped with a standard USB microphone, or a smartphone equipped with a custom app. In any case the front-end (FE) performs feature extraction and a first guess on where possible commands are located based on an adaptive sound energy thresholding, also known as VAD. When the VAD detects a possible utterance, the FE sends a compressed stream of features to the DSR back-end (BE), which replies with the recognised command, if any. The front-end can then give suitable feedbacks to the user, and talk to the local home automation system (in this case the DALI command unit) to actually perform the desired command, as shown in Fig. 2.

[1] Actually, it is not easy at all to rebuild the speech signal from the features that the DSR sends to the server, so some privacy is still preserved even if the Internet connection to the server, which should be encrypted anyhow, gets tapped.

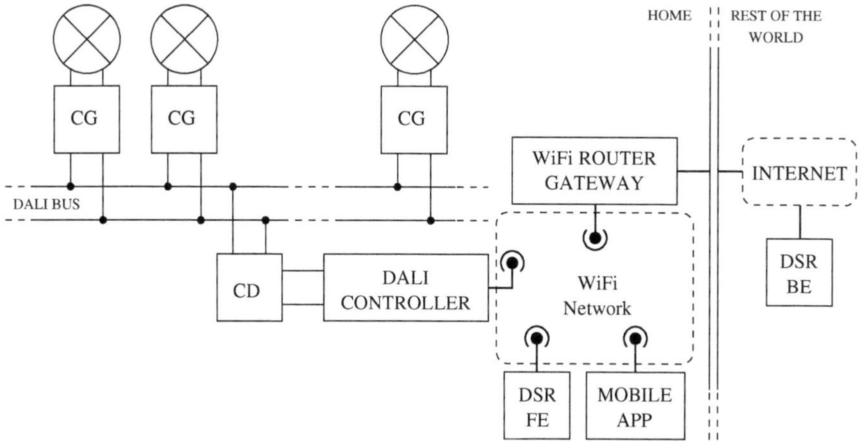

Fig. 1 System block diagram of the application used for the case study. CG and CD are, respectively, the DALI control gear (lamp drivers) and control device (bus controller). The DALI controller is a custom device that allows the control of the bus from a wireless network. The front-end, either a custom device or an app on a smartphone, sends commands via this wireless network, after having used the services of the back-end to help convert the audio stream into command words and having interpreted these according to an application-specific grammar

3 Experimental Setup

We performed our experiments with a setup as shown in Fig. 1, using a panoramic USB microphone connected to the front-end, and a hidden Markov model-based recogniser, Sphinx 4 [6], running on the back-end with a fine-tuned configuration optimized for the Italian language and command-spotting usage. The acoustic model was trained for a generic large-vocabulary continuous speech recognition task [7], the language model is grammar-based with a hand-crafted grammar suitable to control, besides other things, a lighting system, e.g., it included sentences like "accendi luce delle scale" (meaning "turn on stair lights"), "aumenta l'intensità della luce" (meaning "increase light level"), or "imposta luce del soggiorno colore giallo" (meaning "set living-room light color to yellow").

Two different garbage models have been implemented to test the system effectiveness in spotting commands during a typical usage session. A phone loop-based generic word model [8], essentially able to capture any out-of-grammar (OOG) sequence of phones, and a special decoy-based garbage model, also aimed at capturing OOG sequences.

The decoys are obtained through a technique we devised [9], which semi-automatically finds the erroneous words that the ASR most often mistakenly substitutes for the correct words. It works by iteratively trying the ASR engine over the desired spoken words, each time adjusting the grammar so that the words that can most likely be confused with the target word are identified and removed. As we will

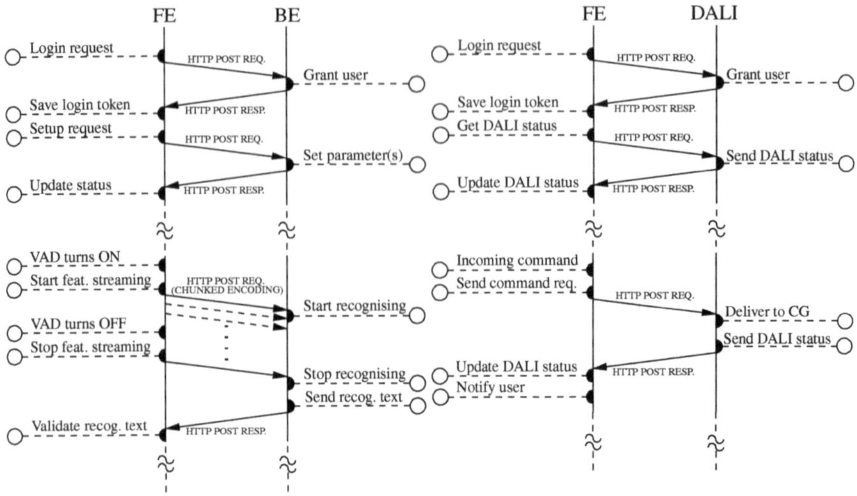

Fig. 2 Schematization of the dialogue between the various entities that comprise the system. To the *left*: data exchanges between the audio front-end and the recognition back-end. To the *right*: after a command is spotted and recognised, another exchange takes place between the front-end and the controller (in this case a DALI command unit). The *upper* portions of the exchanges deal with authentication and are only performed once at system startup

see, this technique alone proved to be very efficient in spotting commands within an unrelated stream of speech.

To test the system, we placed the microphone in an office room and recorded a session of about 40 min between two people doing their ordinary daily tasks, having asked them to speak a command from a given list from time to time. The number of actually spoken commands was also independently recorded, as well as the system response to them, so that the overall system performance could be statistically evaluated.

4 Results

Before testing the system with the final setup described above, we determined the receiver operating characteristic (ROC) curve, for the two garbage models used to help spot the commands (the standard phone-loop approach and our decoy-based approach). These are obtained by varying the garbage model insertion probability, i.e., the relative probability that a given utterance is not a command, used by the recogniser to decide how greedy it should be in discarding utterances. In order to speed up the process we performed a special recording session just for this experiment, having asked a few people to read a predefined sequence of utterances, half

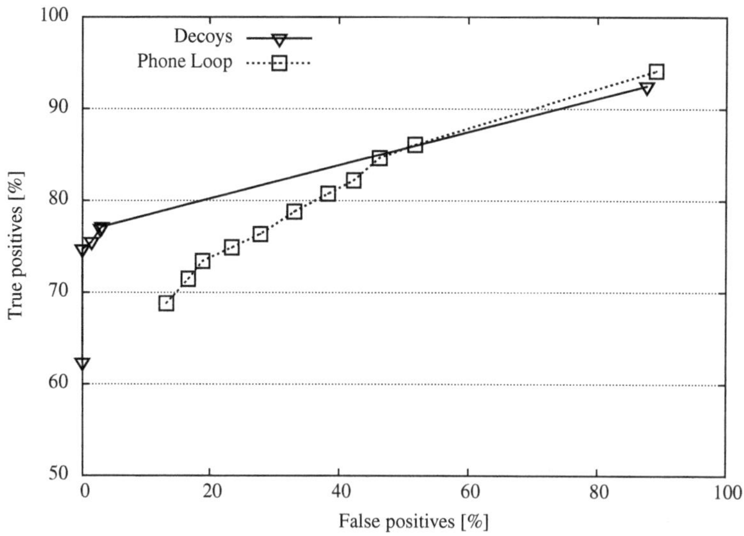

Fig. 3 ROC curves for the two command spotting methods

of which were valid commands. This way we were able to collect a statistically significant sample in a relatively small time.

The two curves can be seen in Fig. 3, which plots the true positive ratio (TPR), i.e., the fraction of correctly recognised commands over the total number of spoken commands, versus the false positive ratio (FPR), i.e., the fraction of commands that the recogniser erroneously deemed as valid when they were not, over the total number of non-command utterances spoken. It is clear that our decoy-based approach largely outperforms the standard phone loop-based approach, which is totally ineffective in this context.[2]

Having drawn the ROC curves, we tried the system with the more realistic setup discussed in Sect. 3. The insertion probability was set to the value that gave the best results in the ROC experiment (corresponding to the point closest to the upper left corner). This can be slightly biased because the command density in the actual speech is much lower, but experimental evidence suggested that there is little sensitivity near the optimal point.

The results are reported in Table 1. Here, TPR is the "hit rate", FPR the "false alarms", FNR is the "miss rate" (false negative ratio, FNR $= 1 -$ TPR), a is the system accuracy (i.e., fraction of correctly classified utterances over total number of utterances), p the precision (i.e., fraction of correctly interpreted commands over total number of recognised commands), and d, reported as a reference, the command

[2] An FPR significantly greater than zero will very soon irritate the user, as the system will start to behave erratically, performing actions when the user didn't mean to. We believe it is better to sacrifice some TPR to lower the FPR, but the phone-loop method simply didn't offer such an opportunity.

Table 1 Performance of the two command spotting methods compared

Method	TPR [%]	FPR [%]	FNR [%]	a [%]	p [%]	d [%]
Phone-loop	82.5	0.2	17.5	97.5	98.1	13.2
Decoys	88.6	0.5	11.4	98.7	93.9	7.3

density of the test speech (number of spoken commands over the total number of utterances). The two systems offer somewhat comparable performance in the real world-like test, with the decoy-based approach offering a higher hit-rate and a slightly better accuracy, though the need for larger-scale experiments is obvious to accurately compare the two methods (actually, the FPR figures given are caused by a single error for the phone-loop and just two errors for the decoy approach).

5 Conclusions

In this paper we described a speech-operated system suitable to be unobtrusively installed in a user home to control various apparatus. The effectiveness of the system in spotting and correctly interpreting the user commands, together with suitable solutions to improve its efficiency, has also been evaluated and discussed. Results show that the performance attainable by the proposed techniques is comparable or better than that of already published studies, and although further, extended experimentation is needed to optimally tune the parameters, our system appears to be already able to offer a viable way towards a complete home speech interface.

References

1. Anastasiou, D.: Survey on speech, machine translation and gestures in ambient assisted living. In: Tralogy, Session 4—Tools for Translators. Paris, France (2012). http://www.lodel.irevues.inist.fr/tralogy/index.php?id=125
2. Vacher, M., Portet, F., Fleury, A., Noury, N.: Challenges in the processing of audio channels for ambient assisted living. In: Proceedings of the 12th IEEE International Conference on e-Health Networking Applications and Services (Healthcom'10), pp. 330–337. Lyon, France (2010). doi:10.1109/HEALTH.2010.5556546
3. ETSI ES 202 050 V1.1.5: Speech Processing, Transmission and Quality Aspects (STQ); Distributed speech recognition; Advanced front-end feature extraction algorithm; Compression algorithms (2007)
4. Hirota, S., Hayasaka, N., Iiguni, Y.: Experimental evaluation of structure of garbage model generated from in-vocabulary words. In: Proceedings of the 2012 International Symposium on Communications and Information Technologies (ISCIT 2012), pp. 87–92. Surfers Paradise, Gold Coast, Australia (2012). doi:10.1109/ISCIT.2012.6381027
5. Levit, M., Chang, S., Buntschuh, B.: Garbage modeling with decoys for a sequential recognition scenario. In: Proceedings of the 11th IEEE Workshop on Automatic Speech Recognition & Understanding (ASRU 2009), pp. 468–473. Merano, Italy (2009). doi:10.1109/ASRU.2009.5372919

6. Walker, W., Lamere, P., Kwok, P., Raj, B., Singh, R., Gouvea, E., Wolf, P., Woelfel, J.: Sphinx-4: A flexible open source framework for speech recognition. Technical Report SMLI TR2004-0811, Sun Microsystems Inc. (2004)
7. Alessandrini, M., Biagetti, G., Curzi, A., Turchetti, C.: Semi-automatic acoustic model generation from large unsynchronized audio and text chunks. In: Proceedings of the 12th Annual Conference of the International Speech Communication Association (Interspeech 2011), pp. 1681–1684. Florence, Italy (2011)
8. Bazzi, I., Glass, J.R.: Modeling out-of-vocabulary words for robust speech recognition. In: Proceedings of the 6th International Conference on Spoken Language Processing (ICSLP 2000 / INTERSPEECH 2000), pp. 401–404. Beijing, China (2000)
9. Alessandrini, M., Biagetti, G., Curzi, A., Turchetti, C.: A garbage model generation technique for embedded speech recognisers. In: Proceedings of the 2013 Signal Processing: Algorithms, Architectures, Arrangements, and Applications (SPA 2013), pp. 318–322. Poznań, Poland (2013)

Part VI
ICT Tools in AAL

A Novel Tracking System for AAL Based on Smartphone Technology

Bruno Andò, Salvatore Baglio, Sebastiano Campisi, Cristian O. Lombardo, Vincenzo Marletta and Elisa Pergolizzi

Abstract This paper deals with a novel multi-sensor approach for the implementation of a tracking system exploiting features of smartphones. Actually, the widespread use of smartphones and performances provided by the variety of embedded sensors encourage their use in mobility oriented applications such as the exploitation of educational/job environments by technological skilled weak people. The proposed methodology exploits information provided by the multisensor features embedded in a standard smartphone and advanced paradigms to improve the efficiency of the system in performing user tracking tasks.

1 Introduction

Although several aids have been developed to assist weak people in the accomplishment of daily activities, indoor and outdoor navigation remains a critical task to be faced as a technological and social issue. Researchers are continually focusing on the

B. Andò (✉) · S. Baglio · S. Campisi · C. O. Lombardo · V. Marletta · E. Pergolizzi
DIEEI-University of Catania, V.le A. Doria, 6, Catania, Italy
e-mail: bruno.ando@dieei.unict.it

S. Baglio
e-mail: salvatore.baglio@dieei.unict.it

S. Campisi
e-mail: ianovir@gmail.com

C. O. Lombardo
e-mail: cristian.lombardo@dieei.unict.com

V. Marletta
e-mail: vincenzo.marletta@dieei.unict.it

E. Pergolizzi
e-mail: elisa.pergolizzi@dieei.unict.it

S. Longhi et al. (eds.), *Ambient Assisted Living*, DOI: 10.1007/978-3-319-01119-6_25,

development of new approaches (such as the use of smart multi-sensor architectures) allowing for the perception of the user-environment interaction in order to provide the user with an optimized form and degree of information for a safe and efficient exploitation of the environment.

Several methodological approaches and devices aimed to assist weak people during orientation and mobility tasks are available in the literature [1–13].

In particular, notable efforts have been dedicated to the development of systems for indoor positioning based on structured environment [1–3] to transmit some form of remote signal once the user gets into the range of the device. Anyway, such approaches suffer from the discontinuous information provided to the user as well as invasive and high installation costs.

Other solutions exploit some forms of indoor trilateration [4–6] which often require the installation of subsystems in the environment. Among such systems ultrasound based trilateration offers better performances than RF trilateration especially in terms of accuracy in tracking the user position.

An emerging approach to support weak people in mobility tasks exploits the use of inertial navigation systems [7–11], which often require the user to wear customized devices unsuitable for elderly or impaired people. The main problem with customized solutions is infact the use of non-standard user interface which often provide an arbitrary form of information. Moreover, such solutions require hard training efforts by weak users thus producing diffidence and compromising the possibility to actually improve their life quality.

Nowadays smartphones are largely diffused and provide a wonderful concentration of sensors, thus making them suitable candidates for new generation of monitoring and assistive systems for weak people [12–16].

Although the use of smartphones in AAL applications for elderly/disabled people is usually discouraged, due to difficulties in their use by not "technology-oriented" people and accessibility problems (e.g. small buttons, small displays and character dimensions, complicated navigation of functions), their use in applications oriented to support skilled weak people (e.g. young people, students) in navigation tasks for educational/job activities, can be very effective.

The main idea behind the platform presented in this contribution is the possibility to exploit embedded sensors in smartphone and advanced signal processing to perform suitable user tracking with particular regards to user walking distance estimation. Actually, the latter represents a fundamental task to implement user tracking functionality.

In particular, two different algorithms are adopted to implement standard and adaptive walking distance estimations. Such outputs are then processed by a dedicated data fusion paradigm exploiting data on user inertial behavior to extract an optimal information on the user stride.

Main advantages of the proposed architecture, as respect to State Of The Art, reside in the use of a full smartphone based architecture which does not involve customized sensing units and the exploitation of a novel data fusion approach for the walking distance estimation. Although, in its present form, the prototype developed uses algorithms running on a PC communicating with the smartphone, efforts have been dedicated to match computational requirements of algorithms with a standard smartphone performances.

The possibility to combine information on the user position with information on the environment allows to build awareness of the interaction between users and the environment itself such as the presence of obstacles or services (User Environment Interaction). This information delivered to the user allows for a safe and efficient exploitation of the environment and it can be strategic to support weak people while performing educational/job activities.

Moreover, it must be considered that interfacing capabilities of smart phones, such us wide touch screen and Bluetooth headset, are strategic to implement an effective interaction between the user and the AAL system.

As well depicted in Sect. 2, the idea behind the proposed user tracking methodology is the possibility to adapt the estimation strategy to the operating conditions.

Considering that the proposed platform could be exploited in both indoor and outdoor application, addressable application contexts could be museums, schools and public buildings.

2 The Proposed User Tracking Methodology

A flow diagram of the system developed is given in Fig. 1.

In the following a novel methodology implementing an adaptive walking distance estimation, to be exploited by the user tracking module, is presented. For the sake of completeness it must be observed that the system is intended to perform also Activity Daily Living (ADL) recognition tasks.

As well known, inertial user tracking concerns the measurement of both heading and walking distance. In the system proposed, heading is measured by the smartphone embedded gyroscope, while the walking distance is obtained by both steps counting and the exploitation of novel paradigms for an adaptive estimation of the step size.

The step number is identified by using signals provided by the tri-axial accelerometer embedded in the smart phone. To perform walking distance estimation, two different algorithms have been adopted: the first one exploits a fixed step size while the other one uses a procedure computing step sizes between each pair of steps.

The Static Step Algorithm, SSA, is based on the concept of "Static Step Size (ΔS_{stat})" and it uses a fixed dimension for the step size, which is empirically estimated during a preliminary tuning phase where the user is requested to walk along a well-defined path.

This technique produces critical estimation error when the user modifies his walking behavior (e.g. walking speed and step size). In this case, the Walking Distance (WD_{stat}) estimation is based on the following model:

$$WD_{stat} = N \cdot \Delta S_{stat} \tag{1}$$

where N counts for the step number.

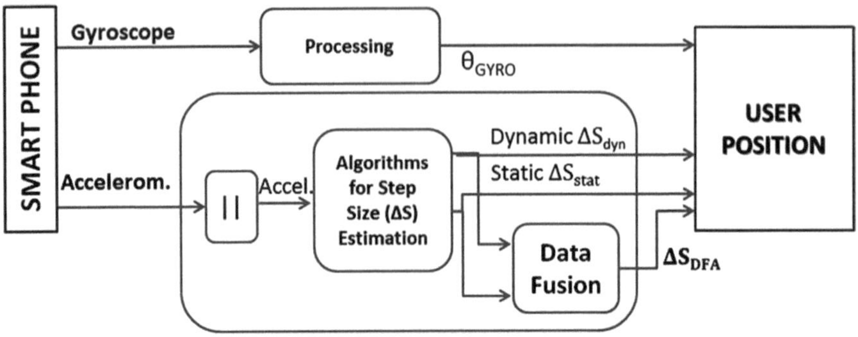

Fig. 1 A flow diagram of the user tracking methodology

The second paradigm, Dynamic Step Algorithm (DSA), provides an estimation of the step size taking into account the user inertial behavior. DSA exploits the step frequency, f_k, and the variance of the accelerometer signals between two steps, υ_k, for a real time estimation of the "Dynamic Step Size (ΔS_{dyn})" [16]. In this case the Walking Distance (WD_{dyn}), is computed by the following model:

$$WD_{dyn} = \sum_{k=1}^{N} \Delta S_{dyn}^{k} \quad (2)$$

with

$$\Delta S_{dyn}^{k} = \alpha \cdot f_k + \beta \cdot \upsilon_k + \gamma \quad (3)$$

$$\upsilon_k = \frac{\sum_{j=1}^{M} \left(a_j - \bar{a}\right)^2}{M}; \quad a_j = \sqrt{A_x^2 + A_z^2}; \quad f_k = \frac{1}{T_k - T_{k-1}} \quad (4)$$

where M is the number of samples between two consecutive steps; A_x and A_z are the longitudinal components of the acceleration, T_k is the time of the step k. The proposed algorithm requires one set of parameters for each user with the consequent need for a preliminary training.

As above mentioned, the main novelty introduced by this work is the use of a Data Fusion Algorithm, DFA, to properly estimate the user step size. The DFA algorithm weights outputs provided by SSA and DSA algorithms depending on the characteristics of inertial signals.

SSA and DSA algorithms require step identification, which is performed by comparing the module of the total acceleration with two thresholds (TL and TH). A step is identified when the accelerometer signal goes below the TL threshold after passing the TH threshold.

Performances of SSA, DSA and DFA algorithms in different operating conditions will be discussed in Sect. 3.

3 The Lab-Scale Prototype Developed and Experimental Results

In this section results obtained by a lab-scale prototype of the system above described are discussed. The prototype is based on the use of a commercial smartphone with embedded inertial sensors (accelerometer and gyroscope). A dedicated Android App has been developed which manages sensors' data acquisition and the WiFi communication with a PC. The latter run a GUI interface developed in the LabVIEW^{TM} environment which implements signal processing algorithms. The choice to use a PC based computational unit is motivated by the sake of convenience for debugging which is mandatory during the development phase. As already mentioned efforts have been dedicated to match computational requirements of developed algorithms to standard smartphone specifications.

The GUI allows for monitoring sensor signals (accelerometer and gyroscope) and it provides a graphical 2D representation of the estimated walking pattern.

Experiments have been performed in two phases: the learning phase aimed to the estimation of ΔS_{stat}, α, β, γ parameters, and the test phase to assess the system performances.

During the learning phase users are requested to walk along a predefined path. The total walking distance is 10 m and several patterns (acceleration components) have been generated by varying the walking speed (SLOW, NORMAL and FAST) and the walking step size (SMALL and MEDIUM).

The static step size, ΔS_{stat}, has been estimated by considering results obtained in case of a MEDIUM step size and a NORMAL walking speed.

Concerning the estimation of the Dynamic step size by model (3), step frequencies, f_k, and variances, $\upsilon_{k,}$ corresponding to the step k have been estimated for each learning pattern.

A Least Mean Square algorithm was used to obtain the best set of model parameters α, β, γ fitting the whole pattern set.

Test patterns have been created for each SLOW, NORMAL and FAST walking modes for the cases of SMALL and MEDIUM step size.

In order to assess model estimation capability in different operating conditions the following performance index has been defined:

$$J_{wd\%} = \frac{\left|WD^{Nom} - WD^{Estim}\right|}{WD^{Nom}} \cdot 100 \tag{5}$$

where WD_{Nom} is the expected walking distance while WD_{Estim} is the walking distance estimated by model (1) or model (2) (3).

Results obtained during the experimental survey are summarized in Table 1.

Rows report the performance indexes relative to WD_{dyn} and WD_{stat}. Columns represent the test patterns in case of LOW, NORMAL and FAST walking performed using SMALL and MEDIUM step size.

Underlined values show that SSA performs better than the DSA only in the case of MEDIUM step sizes for both SLOW and FAST walking speed. Experiments with

Table 1 Performances of SSA and DSA algorithms estimated during the test phase

J%	Low speed Step size		Normal speed Step size		Fast speed Step size	
	Small	Medium	Small	Medium	Small	Medium
WD_{dyn}	3,9	7,2	4,8	0,3	8,5	8,1
WD_{stat}	40,6	3,8	25,8	3,8	40,6	3,8

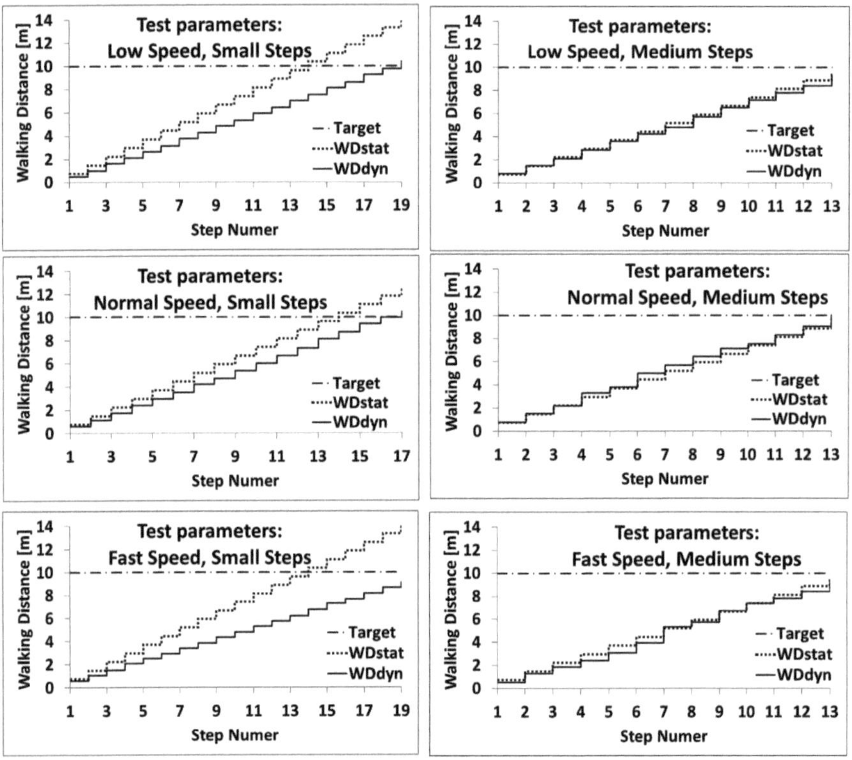

Fig. 2 WD_{stat} and WD_{dyn} performances in case of test patterns. Three different speeds and two different step sizes have been used

SMALL step size show the adaptability of the DSA which performs better than the SSA in any walking speed.

The complete walking distance evolution is represented in Fig. 2 for LOW, NORMAL and FAST speed test patterns. Performances of the developed algorithms are achievable comparing WD_{dyn} and WD_{stat} for each step detected.

In order to assess performances of the DFA algorithm a dedicated experiment was run. For this experiment users are requested to walk 3 times along a predefined trajectory of 15 m. The walking distance has been estimated by SSA, DSA and

Table 2 Performance index (5) obtained by SSA, DSA and DFA algorithms in case of the repeated trajectory experiment

	Gyro-SSA	Gyro-DSA	Gyro-DFA
$J_{wd\%}$	3.74	1.66	**1.29**

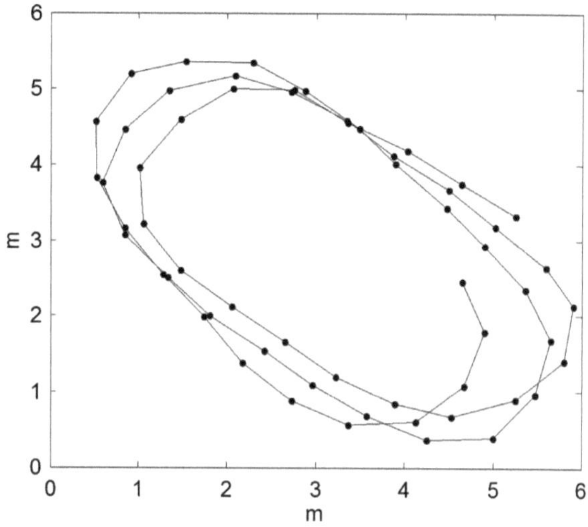

Fig. 3 Results obtained in case of an experiment aimed at the reconstruction of a repeated trajectory. Trajectories have been estimated by using the DFA algorithm and the gyroscope

DFA algorithms. Table 2 reports results obtained by this experiment aimed at the reconstruction of repeated trajectories.

For the sake of completeness Fig. 3 shows trajectories reconstruction performed by both DFA based walking distance and Gyroscope based heading estimations.

As it can be observed, despite Table 2 highlights interesting performance of the DFA paradigm in terms of total walking distance estimation, a drift is evincible in successive reconstructed trajectories which is quite usual in case of Gyroscope based heading estimation. Future efforts will be dedicated to the improvement of heading estimation task by using both gyroscope and compass sensors.

References

1. Wakuda, Y., Asano, S., Koshizuka, N., Sakamura, K.: Ubiquitous sensor-based pedestrian dead-reckoning for LBS applications. In: International Symposium on Micro-NanoMechatronics and Human Science (MHS), pp. 374–379. 4–7 Nov 2012
2. Leppakoski, H., Collin, J., Takala, J.: Pedestrian navigation based on inertial sensors, indoor map, and WLAN signals. In: Acoustics, Speech and Signal Pro-cessing (ICASSP), 2012 IEEE International Conference, pp. 1569–1572. 25–30 March 2012
3. Liu, Y., Wang, Q., Liu, J., Wark, T.: MCMC-based indoor localization with a smart phone and sparse WiFi access points. In: Pervasive Computing and Communications Workshops

(PERCOM Workshops), 2012 IEEE International Conference, pp. 247–252. 19–23 March 2012
4. Andò, B., Savalli, N.: CANBUS networked sensors use in orientation tools for the visually impaired wired versus wireless technology. IEEE Instrum. Measure. Mag. **11**(1), 49–52 (February 2008)
5. Andò, B., Baglio, S., La Malfa, S., Marletta, V., Savalli, N.: A distributed sensor network approach for orientation tasks. In: I2MTC 2009 - International In-strumentation and Measurement Technology Conference. Singapore, (2009)
6. Jung, J., Myung, H.: Range-based indoor user localization using reflected signal path model. In: Digital Ecosystems and Technologies Conference (DEST), 2011 Proceedings of the 5th IEEE International Conference, pp. 251–256. May 31–June 3 2011
7. Chen, W., Chen, R., Chen, Y., Kuusniemi, H., Wang, J.: An effec-tive pedestrian dead reckoning algorithm using a unified heading error model. In: Position Location and Navigation Symposium (PLANS), 2010 IEEE/ION, pp. 340–347. 4–6 May 2010
8. Elbes, M., Al-Fuqaha, A., Rayes, A.: Gyroscope drift correction based on TDoA technology in support of Pedestrian Dead Reckoning. In: Globecom Workshops (GC Workshops), 2012 IEEE, pp. 314–319. 3–7 Dec 2012
9. Afzal, M.H., Renaudin, V., Lachapelle, G.: Magnetic field based heading estimation for pedestrian navigation environments. In: Indoor Positioning and Indoor Navigation (IPIN), 2011 International Conference, pp. 1–10. 21–23 Sept 2011
10. Jimenez, A.R., Seco, F., Prieto, J.C., Guevara, J.: Indoor pedestrian navigation using an INS/EKF framework for yaw drift reduction and a foot-mounted IMU. In: Positioning Navigation and Communication (WPNC), 2010 7th, Workshop, pp. 135–143. 11–12 March 2010
11. Glanzer, G., Bernoulli, T., Wießflecker, T., Walder, U.: Semi-autonomous indoor positioning using MEMS-based inertial measurement units and building information. In: Proceedings of the 6th Workshop on Positioning, Navigation and, Communication 2009 (WPNC'09).
12. Tacconi, C., Mellone, S., Chiari, L.: Smartphone-based applications for investigating falls and mobility. In: 2011 5th International Conference on Pervasive Computing Technologies for Healthcare (PervasiveHealth) and Workshops.
13. Bujari, A., Licar, B., Palazzi, C.E.: Road crossing recognition through smartphone's accelerometer. In: Wireless Days (WD) IFIP, (2011)
14. Gadeke, T., Schmid, J., Zahnlecker, M., Stork, W., Muller-Glaser, K.D.: Smartphone pedestrian navigation by foot-IMU sensor fusion. In: Ubiquitous Positioning, Indoor Navigation, and Location Based Service (UPINLBS), 2012, Helsinki, pp. 1–8, 3–4 Oct 2012
15. Fahim, M., Fatima, I., Lee, S., Lee, Y-K: Daily life activity tracking application for smart homes using android smartphone. In: 14th International Conference on Advanced Communication Technology (ICACT), (2012)
16. Shin, S.H., Park, C.G., Kim, J.W., Hong, H.S., Lee, J.M.: Adaptive step length estimation algorithm using low-cost MEMS inertial sensors. In: IEEE Sensors Applications Symposium, (2007)

BAR LIS: A Web Tool for Italian Sign Language Synthesis and Interaction

Luca Nardi, Matteo Rubini, Stefano Squartini, Emanuele Principi and Francesco Piazza

Abstract This work presents the development and testing of BAR LIS (BAR in Italian Sign Language), a web application created in collaboration with the Ancona division of Ente Nazionale Sordi (ENS) and presented during X Masters Awards 2012 event in Senigallia, Italy. BAR LIS was structured as a small dictionary of words and related signs in LIS, Italian Sign Language (presented as 3D rendered animations) in order to ease communication between people attending the X Masters event and hearing impaired personnel of the ENS stall, which included a small coffee shop. A more extensive set of words was later created and tested with ENS members in order to study the impact of resolution of 3D models on comprehensibility and quality of the signs.

1 Introduction

Nowadays, the great development of WEB technologies and the increased capabilities of modern computers, smartphones and tablets, drive to the development of "open" language-to-signed-language systems, not linked to a particular

L. Nardi (✉) · M. Rubini · S. Squartini · E. Principi · F. Piazza
DII—Università Politecnica delle Marche, Ancona, Italy
e-mail: luca.nardi@live.it

M. Rubini
e-mail: rubini.matteo@hotmail.it

S. Squartini
e-mail: s.squartini@univpm.it

E. Principi
e-mail: e.principi@univpm.it

F. Piazza
e-mail: f.piazza@univpm.it

S. Longhi et al. (eds.), *Ambient Assisted Living*, DOI: 10.1007/978-3-319-01119-6_26,

hardware/software configuration or "machine" but accessible to any device capable of connecting to the internet and running a modern browser. In this context, the A3Lab research group of Università Politecnica delle Marche started in 2012 a collaboration with the Ancona division of ENS (Ente Nazionale Sordi). Objective of the agreement was the development of WEB interfaces and systems, as well as more traditional tools, to improve the quality of life of Italian hearing impaired people. In fact, while many international studies have been conducted on this field ([1–3]), only a few regard specifically the LIS. Moreover, excluding one notable exception ([4–6]), they are all quite dated, don't take into account the last technological developments and share the same "classical" approach, that is the development of a stand-alone, isolated system. The invitation of ENS people to X Masters Awards 2012 (a popular event taking place every year in Senigallia, with the aim to promote local associations operating in the field of social advancement) has given the opportunity to extend and field-test the tools developed during the collaboration between ENS and A3Lab in the context of the so-called A3LIS project. At the date, the A3LIS-147 Database had already been created (a video database of signs comprehending 147 words related to different situations of the common life [7]), and studies had been conducted in the field of automatic recognition of LIS signs and translation from LIS to Italian language [8]; what the A3LIS project still missed was an automatic synthesizer capable to translate Italian to LIS language. Since no prior work had been conducted in this particular field, a first phase of collection of the existing literature was carried out, followed by a study of the available technologies. Then, a new application was developed with these specifications:

1. The application should work "out of the box" on a standard notebook, without too demanding hw/sw prerequisites;
2. The application should work as a "video dictionary", allowing the selection of simple or composite words and showing the related LIS signs, reproduced by a 3D actor (avatar);
3. It should feature different variations of the same word (i.e. hot/cold milk) re-using the common part of the animation whenever possible;
4. The user should be given the possibility to choose between a male and a female avatar.
5. WebGL technology [9] should be tested together with more "traditional" solutions in order to study pros and cons of each technique.

2 Development of BAR LIS Web Application

2.1 3D Animation

3D animation is based on the same principles of traditional animation. Illusion of movement is produced through very fast reproduction of single frames, which in the case of 3D animation are obtained from digital models. Project A3LIS animations were created using the MakeHuman anthropomorphic model [10]. In total, 4 models were created (2 male and 2 female avatars), one with the highest possible resolution and skeleton complexity and the other with medium skeleton complexity and low resolution. All the models were then imported in Blender to make animations.

Blender is a very popular open source 3D rendering software. Among its features are the possibility of animating 3d models through different techniques, such as keyframes, animation curves or path-following animation. Animations for the A3LIS Project were created using keyframe animation. Each "keyframe" is obtained repositioning the models (moving the bones of their skeletons) using frames extracted by the original videos as a guide. After having created a sufficient number of keyframes, Blender is capable of obtaining through interpolation algorithms all the other frames needed to complete the animation.

2.2 WebGL and Javascript Picture Rotator

Since the aim of the project was to study the feasibility of WebGL, and not to develop a full-feature 3d model web application, a pre-existing WebGL engine was chosen, in order to minimize the development time. The chosen engine was the Levis WebGL Implementation, proposed by Marco Levis [11]. It was then enhanced by removing un-needed features and optimizing the 3d animation related code, which was too slow for the complex models needed to reproduce LIS signs and didnt take into account the time needed to load models. Then, a simple Javascript Picture Rotator was developped as an alternative to the WebGl Engine. In this case, the input is constituted by pictures created from the original 3d models. Both techniques were developed successfully and are worth considering in the development of a full-feature application. While WebGL is the most interesting one, because it allows to work directly with 3d models (which leads to the possibility to generate, edit and link animations in real time), but it requires much more computational power and the animations occupy more space. Therefore, in contexts with few words and little or no need to link them, such as a dictionary application, the Javascript Picture Rotator would be the better solution, while in contexts with many words which can change often or when linking words together to make whole phrases is important, such a translator application, the WebGL engine is preferable (Fig. 1).

2.3 Bar LIS Web App

BAR LIS was developed as a web application running on a simple USB Web Server. The Javascript Picture Rotator was preferred over the WebGL engine because its hardware requirements are much lower and, being the set of words related to the "coffee shop" context limited, the advantages of elaborating 3d models in real time were not needed. A simple animated web interface was created, using only CSS

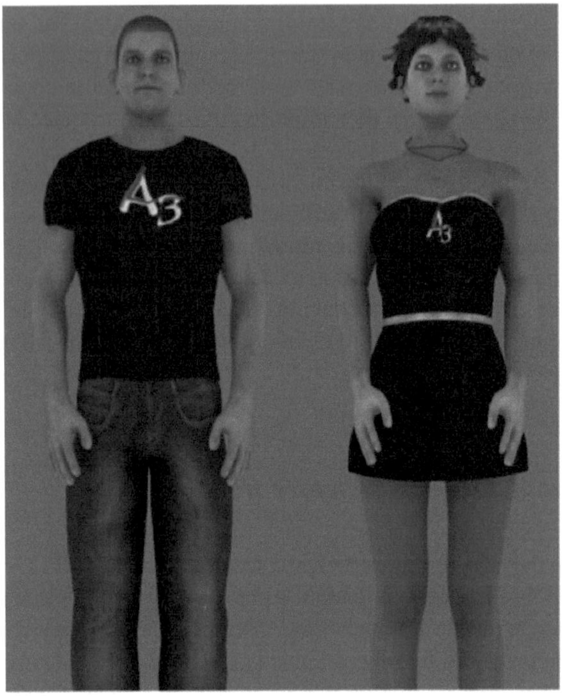

Fig. 1 Male and female avatars for the BAR LIS web app

Fig. 2 BAR LIS interface

and HTML to keep the application as compliant to web standards as possible. The structure of the application interface is shown in Fig. 2.

The words and expressions featured in the application are:

- Water (still or sparkling);
- Alcoholic/ alcohol-free drink;

- Bitter (tonic liquor);
- Beer;
- Good morning;
- Good evening;
- Coffee;
- Macchiato (coffee with a drop of milk);
- Milk (cold or hot);
- How much does it cost?;
- Euro;
- T-shirt;
- Please;
- Thank you;
- Offer;
- Redbull;
- Fruit-juice.

Since not all the words needed were part of the A3LIS-147 Database, a new campaign of video acquisition was conducted in collaboration with ENS members.

3 Validation Tests

After having developed a test application such as BAR LIS, it was possible to use the same structure to develop tests in order to evaluate the quality and comprehensibility of animations and the impact factors such as model/picture resolution have on the overall quality of the system.

3.1 System Evaluation

The quantitative evaluation of an automatic LIS signs video dictionary is not a trivial problem, because many factors concur to the overall quality of a "good" system, such as:

- Comprehensibility of signs;
- Complexity of the single words reproduced;
- Quality of sign linking/mixing algorithms and techniques;
- Realism of avatars/digital actors;
- Number of words/expressions featured;
- Ease of use of the system;
- Hardware/software requirements and software optimization;
- Extensibility of the system.

Moreover, the higher or lower importance of each of these factors changes depending on the context in which the system operates. For example, speed and sign linking/mixing would be a major requisite for a real time translator, while for a dictionary application aspects such as photorealism and precision of signs would be much more important. Another aspect to take into account is the not-complete standardisation of the Italian Sign Languages [12]. Each LIS signer expresses him/herself in his/her own way and the same word is never signed in exactly the same way by two different people and the inevitable differences have a similar effect on the quality of communication of speaking English with a person of another country [13].

Three different tests have been conducted to evaluate these different parameters:

1. Comprehensibility/quality of animations;
2. Impact of video resolution and model complexity on the overall quality of the system;
3. Quality of the Blender animation mixing algorithm in producing whole sentences.

All the tests were administered to the ENS members of Ancona. The tests were developed starting from the BAR LIS application and distributed both in the form of a USB stand-alone application and published online. The age and gender of each candidate is registered, and candidates are asked if the contributed to the creation of the A3LIS-147 Database.

3.2 *Test 1: Comprehensibility and Quality of Animations*

Objective of this test was to evaluate the comprehensibility of animations. The test is composed by 20 questions. In each question, the animation corresponding to a single word is shown; the candidate is asked to recognise the word represented and to express the confidence of his choice with a number from 1 (no confidence) to 5 (absolute confidence).

The test has been administered to a sample of 23 elements extracted from the members of the Ancona division of ENS. 13 of them are male (mean age 47.1), 10 are female (mean age 42.4). Of these, 3 people (2 men and 1 woman) had contributed to the creation of A3LIS-147 Database.

Two features were achieved in the development of this test:

1. The test is a free response test;
2. Each animation corresponds to a single word or concept, there is no context;
3. The test engine verifies that the candidate has played the animation before giving the answer.

Features 1 and 2 ensure that the candidate is given no clues from which deducing the meaning of the sign.

56.74 % of signs were recognised correctly, and the mean confidence is 61.96 %. In both cases, women had better results than men (58.50 % recognition and 62.80 % confidence vs 55.38 % and 61.31 %). Clustering results by age groups, it is possible to

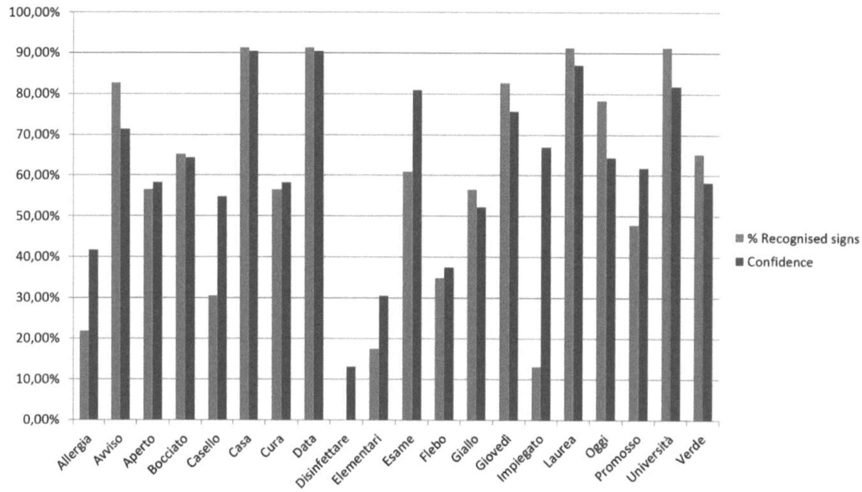

Fig. 3 Test 1 results: % of recognised signs and confidence of given answers

see that percentage of recognition and confidence decrease sensibly with increasing age while unanswered questions increase.

All the results were discussed with the help of an expert of LIS language from ENS. The overall results appear satisfactory, with much words recognised in more than 70 % of interviewed subjects, even if in LIS the meaning of a single sign depends on the sentence containing it much more than in the spoken language. Moreover, some signs differ from one another for the facial expression (i.e. "allergy" and "itch") [14], while all the animation created featured a neutral facial expression. In fact, in more than 80 % of wrong answers the word recognised has the same sign of the correct word (Fig. 3).

3.3 *Test 2: Impact of Video Resolution and Model Complexity on the Overall Quality of the System*

Objective of this test was to evaluate the impact of video resolution, quality of picture rendering and complexity of the 3D model (i.e. number of polygons) on the overall quality of the system. The test is composed by 20 questions. In each question, an original video taken from the A3LIS-147 Database is shown together with two animations created from that video using high quality and low quality settings. The candidate is asked to tell which animation is the most accurate reproduction of the original video; moreover, he/she is asked to evaluate the overall quality of each animation with a mark from 1 (not accurate at all) to 5 (extremely accurate).

The test has been administered to a sample of 14 elements extracted from the members of the Ancona division of ENS. 12 of them are male (mean age 45.3),

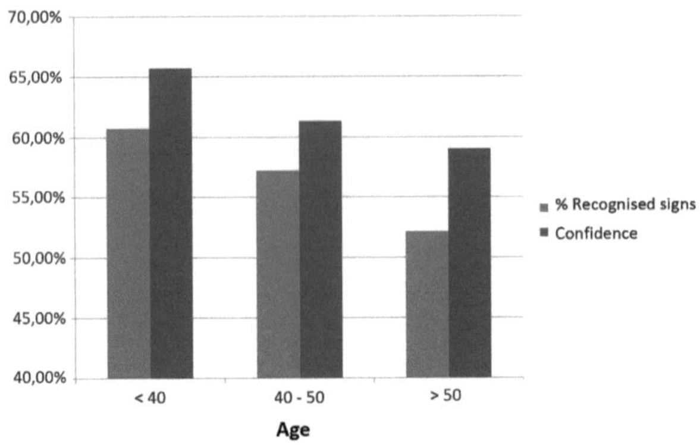

Fig. 4 Results by age

2 are female (mean age 47.5). Of these, 2 men had contributed to the creation of A3LIS-147 Database.

Globally, the high resolution animation is preferred in 50.64 % of cases, achieving an average quality mark of 2,5 out of 5, while the low resolution animation achieves an average mark of 2,3 (Fig. 4).

Even if the high quality animation is preferred over the low quality one, the difference in the evaluation of the two versions is not significant (5 %). This is particularly relevant if we consider that the high resolution model counts 7502 vertices against the 655 vertices of the low resolution model. The difference in number of polygons is not relevant in this particular scenario, in which the final application works with pre-rendered pictures, but it is decisive if the WebGL technique is adopted, because a lower number of vertices means .obj files 5 times smaller (600kB vs 3MB) and therefore an invaluable gain in terms of speed of data transfer and elaboration time. The overall judgement of both the high and low resolution models is not completely satisfactory. If test 1 shows that the comprehensibility of signs is good, this results suggest that photorealism and accurate reproduction of movements are factors to be improved in future development and that some features, such as facial expressions, are very important in the overall quality of signs.

3.4 Test 3: Quality of the Blender Animation Mixing Algorithm in Producing Whole Sentences

Objective of this test was to evaluate the possibility to use Blender animation mixing algorithms to obtain complex animations of whole sentences starting from animations of the single words. Contrary to the first two tests, Test 3 was conducted in a more informal way. The test is composed by 3 animations of these complete sentences:

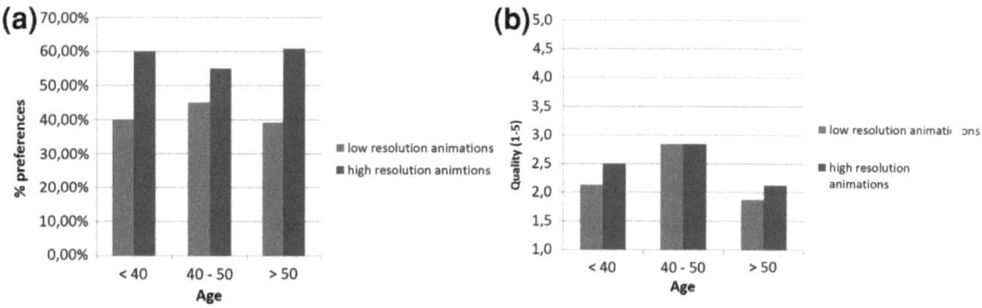

Fig. 5 Test 2 results: high resolution versus low resolution animations

Fig. 6 Blender animation mixing tool

- Where do I have to deliver the modules for the final examinations?
- The Ancona toll booth today is closed.
- How much is the train ticket to Rome?

The sentences were composed using only words already available in the A3LIS-147 Database and the related animations where connected with the Blender animation mixing tools in the correct order. The resulting complete animations where then administered to a selected group of expert LIS signers from the Ancona division to ENS, asking to evaluate the overall quality of the sentence (Fig. 5).

All the people consulted evaluated the overall quality of the sentences as very good to excellent.

The challenge of connecting two or more animations together was already dealt with while making animations of complex expressions for the BAR LIS application (i.e. expressions such as "cold milk" or "how much does it cost"). In that case, each animation was created starting and ending in the same neutral position, so that the animations could be reproduced one after the other without any particular expedient. The resulting complex animation, while perfectly comprehensible in most cases, is however perceived as strange or unnatural. The problem of the "neutral position" could be lessened without resorting to complex animation interpolation techniques by using an "intermediate position", with arms bent and hands at chest height, so that the distance traversed by the hands while going from the end of a sign to the intermediate position and from the intermediate position to the starting position of the next sign is shorter.

4 Conclusion

We developed BAR LIS, a simple "dictionary-like" web application and field-tested it with good results during the X Masters Awards 2012 event. During the development of such application, we were able to study the different techniques available to achieve reproduction of 3D animation inside an internet browser, WebGL and reproduction of pre-rendered frames. While WebGL is the most interesting technique for future development, because it allows to mix and edit the animations in real time, the reproduction of pre-rendered frames is the most advisable technique when data exchange and computational power are limited resources, such as in mobile applications. We also used BAR LIS as a base for the development of tests in order to evaluate the quality of animations. The results displayed a good overall comprehensibility of the signs, with many words recognised in more than 70 % of cases. The main difficulties were related to different words corresponding to the same sign, words not known at all by the people interviewed and facial expressions, which weren't taken into account in the making of the animations. Furthermore, use of a more compact, less complicated human model appears feasible, since the high resolution model marks were not significantly better than the low resolution model ones. This suggests that the making of a full WebGL application, at first discarded for the complexity of models and the amount of computation power required to deploy such an application in a real time context, is indeed possible using much simpler models without losing quality. Concerning the mixing of simple animations to obtain whole sentences, complex interpolation algorithms such as those used inside Blender achieve the best results, but it seems possible that comparable results could be obtained in the future by determining the best intermediate position of the single animations.

Acknowledgments The authors express their gratitude to all people at the Ancona Section of Ente Nazionale dei Sordi (ENS) for their support in the development of this work.

References

1. E.A.R. Group: EMBR character animation engine. http://embots.dfki.de/EMBR/ (2011)
2. Hanke, T.: HamNoSys—Representing Sing Language Data in Language Resources and Language Processing Contexts, University of Hamburg—Institute of German Sign Language and Communication of the Deaf
3. Moemedi, K.A.: Rendering an Avatar from signwriting notation for sign language animation. MSc Thesis, University of the Western Cape—Department of Computer Science (2010)
4. Bertoldi, N., Tiotto, G., Prinetto, P., Piccolo, E., Nunnari, F., Lombardo, V., Mazzei, A., Damiano, R., Lesmo, L., Del Principe, A.: On the creation and the annotation of a large-scale Italian-LIS parallel corpus, In: 4th Workshop on the Representation and Processing of Sign Languages: Corpora and Sign Language Technologies (2010)
5. Lombardo, V., Battaglino, C., Damiano, R., Nunnari, F.: An avatar based interface for the Italian sign language. In: 2011 International Conference on Complex, Intelligent, and Software Intensive Systems (2011)

6. Prinetto, P., Shoaib, U. Tiotto, G.: The Italian sign language sign Bank: using wordnet for sign language corpus creation. In: International Conference on Communications and Information Technology (ICCIT), Aqaba (2011)
7. Fagiani, M., Principi, E., Squartini, S., Piazza, F.: A New Italian Sign Language Database, Advances in Brain Inspired Cognitive Systems, vol. 7366. LNCS Springer, Heidelberg (2012)
8. Fagiani, M., Principi, E., Squartini, S., Piazza, F.: A new system for automatic recognition of Italian sign language. In: Proceedings of WIRN 2012, Springer LNAI (2012)
9. OpenGL—WebGL standard 2.0. http://www.khronos.org/webgl/
10. MakeHuman Project. http://www.makehuman.org/
11. Levis, A.P.: http://pralevis.com
12. Bertone, C.: La struttura del Sintagma Determinante nella Lingua dei Segni Italiana (LIS), Venezia: MSc Thesis—UniversitàCa' Foscari Venezia—Dipartimento di Scienze del Linguaggio (2006)
13. Sandler, W., Lillo-Martin, D.: Sign Language and Linguistic Universals. Cambridge University Press, Cambridge (2006)
14. Rossini, P., Reilly, J., Fabbretti, D., Volterra, V.: Aspetti non manuali nelle narrazioni LIS. In: C. Bagnara, Chiappini, P., M.P. Conte e M. Ott (a cura di), Viaggio nella citt invisibile. Atti del 2 Convegno nazionale sulla Lingua Italiana dei Segni. Genova, 25–27 settembre 1998. Pisa: Edizioni del Cerro

OMNIACARE: A Comprehensive Technological Platform for AAL

Paolo Casacci and Massimo Pistoia

Abstract Technological projects and products developed in the scope of Ambient Assisted Living provide a wide range of solutions for the elderly's and their caregivers' needs, but typically focused to single aspects of care and assistance. There are systems for monitoring of health signs, along with home automation or home assistance request systems. Multiple needs may therefore lead to adopt multiple systems at the user's premises, making it more difficult for the elderly and their circle of support to manage their usage and fully benefit from the advantages of technologies. The OMNIACARE platform brings along an innovative concept of integration and adaptation to users' needs. It comprises different modules, developed in ad-hoc projects, to cover different sides of home care and assistance with one interface in one single system.

1 Introduction

Quality of living is getting better worldwide. Due to improvement in nutrition and in therapies and, due to an increased attention to prevention of diseases, population is rapidly ageing. The number of older people, in good health conditions but still in need of constant, although non-obtrusive, monitoring will grow up in the next years. This may greatly impact on national health systems, for cost increases in assistance and hospitalisation. Technological solutions are a key factor to help people to actively live in their own homes despite aging.

P. Casacci (✉)
IT Manager eResult s.r.l., Ravenna, Italy
e-mail: paolo.casacci@eresult.it

M. Pistoia
Projects Department Manager eResult s.r.l., Cesena, Italy
e-mail: massimo.pistoia@eresult.it

S. Longhi et al. (eds.), *Ambient Assisted Living*, DOI: 10.1007/978-3-319-01119-6_27,

The OMNIACARE platform, developed by **eResult**, is built with the aim to realise an innovative and integrated system to help improve the quality of life of the elderly, fostering their autonomy to live in their homes, and to provide and manage for themselves in order to keep good health [1]. The system principally addresses the needs of self-sufficient older adults, who have fragilities or chronic conditions typical of ageing that require small or moderate assistance [2]. The proposed technology contributes to keep the elderly autonomous and confident but also prevent degradation of mental, physical and cognitive conditions, with the goal to also detect potential risk situations at their early stage.

2 OMNIACARE: System Overview

OMNIACARE is a software platform that provides modular and flexible functionalities. It supports several devices that can be employed singularly or in multiple combinations: smartphones, tablet PCs, TV set-top boxes, wireless sensors. The system can virtually be interfaced with any device which is capable to detect vital parameters or environment conditions and can continuously and automatically acquire data. Such data are processed by a home server to be later stored into a centralised back-end server. Through access to the back-end cloud server, caregivers can configure the behaviour of client devices, assess the status of local sensors, and view logs of past events or browse along the clinical history of the patient. In the event of critical situations, OMNIACARE can also send alerts to families or caregivers and activate IP cameras to remotely assess and control the situation.

2.1 System Architecture

The architecture of the system is shown in Fig. 1.

Each end-user (patients) is equipped with a device that can consist of:

- a set-top box connected to a TV monitor with Internet connection;
- a smartphone with 3G connection
- a tablet PC with 3G connection

Eventually, sensors and actuators can be located around the users' house to detect parameters or trigger motors or lights. 2D or 3D sensor cameras can also be installed to monitor sensitive areas and even user's behaviour. Wearable devices able of detecting biomedical parameters can be provided.

The set-top box, smartphone or tablet has the task to act as interface to the users for the services provided. It also provides live audio and video conferencing, along with the possibility to act as home server for first data processing and transfer to the cloud server, although a fixed PC is preferred for reliability due to power issues (portable devices are battery powered).

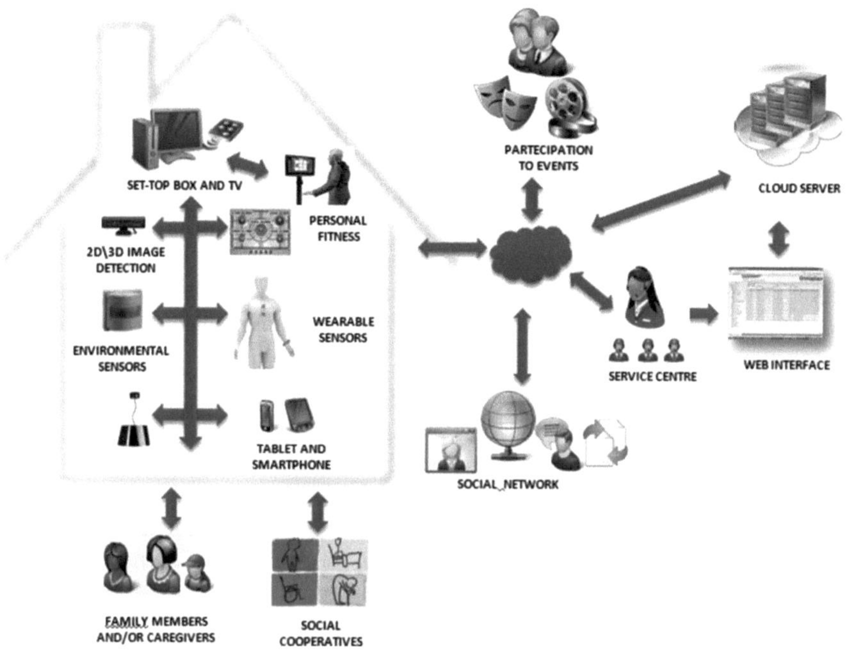

Fig. 1 OMNIACARE. Architecture of the system

The cloud server acts as a repository for data, vital parameters, video streams (if requested) and, through customized web interfaces, is the central point of access for families, caregivers and service centres. The cloud server can also provide services to the final users, which can range from home assistance and house maintenance to promotion and encouraging of leisure activities and access to dedicated social networks.

3 Modules

OMNIACARE's client-server architecture was developed in a project named *ADA*, in which the platform was devised and realised. A very special eye was kept on features of configurability, scalability and robustness. In the *Aid-Assistant* project, the platform's Monitoring Module was developed. The project aimed to bring along a product to help frail users to keep living in their own homes by supporting caregivers in their activities [3]. The platform was endowed and interfaced with devices able to:

- detect health status through wearable sensors for vital parameters (heart rate and breath)
- detect fall

- locate the user's position around the house (indoor localisation)
- monitor special events, like lack of response by the user to a reminder or wandering

As extension to the platform, attention has been paid not only to medical and care assistance, but also to social inclusion of the elderly. In the *HOST* project, the issue of providing customized and easy access to several e-Services for older people has been tackled [4]. The project objective is to provide an ICT architecture based on a friendly and well known technology such as the television, tablet PC or smartphone, enabling elderly independence and participation on a free choice basis to the self-serve society. The services are ensured by specialized and dedicated organizations such as social housing operators. The OMNIACARE platform has been consequently enriched with a Service Module to:

- provide sound or voice alerts to remind the user to perform a specific task (take drugs, call the physician, favourite TV programme...) and accept simple feedback in return, in order to record execution of the activities
- access services and better organize the way of life in social housing (request maintenance, browse events calendar, foster social and recreational activities)
- request home drugs delivery, accompaniment for medical appointments
- shop online

In the *HOST* project, a first version of an OMNIACARE's Social Module has also been created. The platform now allows to:

- send and receive messages to and from the people included in the elderly's own circle of support (manager of social housing association, tutors, caregivers or family members) that are registered in the system
- start Skype video calls by simply pushing a button.

OMNIACARE therefore provides the elderly with the opportunity to stay in touch with friends and relatives, close or distant (or even abroad) without call charges and without having to learn a new interface. In forthcoming projects, the creation of simplified social networks of elderly will be evaluated, to enrich older people's social life and offer socialization and human contact with the aim to foster their sense of usefulness, independence and, finally their well-being [5].

The platform is being enriched with extended capabilities for monitoring and assessment of health conditions. In the *AL.TR.U.I.S.M* project, OMNIACARE is being employed to develop a remote system for cognitive home rehabilitation [6] through a customized "Virtual Personal Trainer" that can allow the patients to perform some cognitive rehabilitation process at home [7]. The system architecture is realized by a set-top-box, with Internet connection, a TV monitor for Graphic User Interface visualization, a commercial 3D sensor (Microsoft Kinect) and a t-shirt for biomedical parameters detection.

Along with indoor monitoring and service offer, the platform is being trialled even for outdoor mobility for senior citizens and favour elderly's tourist activities. The *Sweet Mobility* project, now in course, exploits new specifically designed OMNI-ACARE Modules in order to advise and support the senior to choose and successfully reach a predefined touristic destination. Based on the information on the senior's

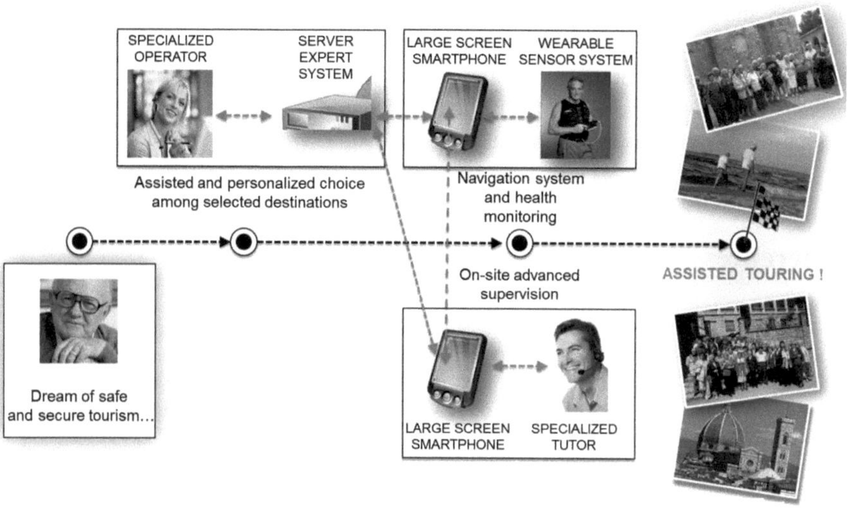

Fig. 2 Sweet Mobility: Architecture of the system

physical and mental history, the system estimates the adequacy of the chosen destination and may propose a more feasible alternative. During the touristic visit, the system continuously and adaptively monitors the senior's physical and mental state, by means of wearable sensors, and based on the knowledge of the track characteristics, it provides orientation and navigation during the walk, gives information on the available services along the path (toilettes, resting points, bars, etc.) and, when needed, may suggest to take action (have a pause, rest, have refreshments, etc.) in order to keep the older user's physical state within health parameter limits. The information is provided by means of easy visual communication on large screen mobile devices and acoustic communication via headsets or loudspeaker, thus minimizing the user cognitive workload. The localization and monitoring features provided by the system are also useful tools for caregivers and operators who may accompany or supervise the elderly's walk. All the necessary information is available to them in real-time via web access to the platform's cloud server, thus allowing to rapidly and effectively intervene if the person is in need. The service architecture of the system is shown in Fig. 2.

4 Conclusions

The OMNIACARE platform enables new application scenarios of ICT to improve the quality of life of people with special needs such as the elderly and frail persons, by including a wide range of services in modules that are easily integrated in the overall architecture. The work done so far has confirmed that the context of AAL is definitely a scope and impact of particular relevance where access to effective and reliable service packages selected directly by users and a better quality of communication adhering to the expectations and the ability of older people are able to give them greater autonomy and a stronger social inclusion, preventing improper forms of sanitation or institutionalization.

References

1. Wolf, P., Schmidt, A., Klein, M.: SOPRANO: an extensible, open AAL platform for elderly people based on semantical contracts. In: Proceeding of the 3rd Workshop on Artificial Intelligence Techniques for Ambient Intelligence (AITAm IŠ08), collocated with the 18th European Conference on Artificial Intelligence 17 (ECAI 08), Patras, Greece, 21–24 July 2008
2. Bagüés, M.I., Bermúdez, J., Illarramendi, A., Tablado, A., Goñi, A.: Using ontologies in the development of an innovating system for elderly people tele-assistance. In: Proceeding of the 2003 Intenational l Conference on Ontologies, Databases and Applications of Semantics (ODBASE'03), Catania, Italy, 3–7 Nov 2003
3. Chen, H., Finin T., Joshi, A.: An ontology for context-aware pervasive computing environments. In: Procedings of 18th International l Joint Conference on Artificial Intelligence (IJCAI-03), Acapulco, Mexico, 9–15 Aug 2003
4. Biocca, L., Morini, A., Padula, M., Paraciani, N.: HOST: Tecnologie intelligenti per servizi personalizzati rivolti agli anziani residenti in alloggi sociali, L'Edilizia, Speciale ITC-CNR 2011, pp. 123–126, estratto dalla rivista Structural, n. 171 N/D (2011)
5. Umberson, D., Montez, J.K.: Social relationships and health: a flashpoint for health policy. J. Health Soc. Behav. **51**, S54–S66 (2010)
6. Clare, L., Woods, R.T., Moniz Cook, E.D., Orrel, M., Spector, A.: Cognitive rehabilitation in alzheimer's disease. Aging Clin. Exp. Res. **18**, 141–143 (2006)
7. Folstein, M., Folstein, S., McHugh, P.R.: Mini mental state: a practical method for grading the cognitive state of patients for the clinician. J. Psychiatr. Res. **12**, 189–198 (1975)

From Sensor Data to User Services in GIRAFFPLUS

Amedeo Cesta, Luca Coraci, Gabriella Cortellessa, Riccardo De Benedictis, Francesco Furfari, Andrea Orlandini, Filippo Palumbo and Aleš Štimec

Abstract This paper presents an ongoing effort to create added value complete services for end users on top of an AAL continuous data gathering environment. It first presents the general architecture proposed by the GIRAFFPLUS project for long term monitoring of older users at home, then describes the main choices concerning the middleware (UNIVERSAAL compliant), the long-term data storage, and a user-oriented component, called DVPIS, dedicated to interaction with different users of the AAL environment. The current status of a deployed version of the system is then shortly described.

A. Cesta (✉) · L. Coraci · G. Cortellessa · R. De Benedictis · A. Orlandini
CNR-National Research Council of Italy, ISTC, Rome, Italy
e-mail: amedeo.cesta@istc.cnr.it

L. Coraci
e-mail: luca.coraci@istc.cnr.it

G. Cortellessa
e-mail: gabriella.cortellessa@istc.cnr.it

R. De Benedictis
e-mail: riccardo.de.benedictis@istc.cnr.it

A. Orlandini
e-mail: andrea.orlandini@istc.cnr.it

F. Furfari · F. Palumbo
CNR-National Research Council of Italy, ISTI, Pisa, Italy
e-mail: francesco.furfari@isti.cnr.it

F. Palumbo
e-mail: filippo.palumbo@isti.cnr.it

A. Štimec
XLAB, Ljubljana, Slovenja
e-mail: ales.stimec@xlab.si

S. Longhi et al. (eds.), *Ambient Assisted Living*, DOI: 10.1007/978-3-319-01119-6_28,

1 The GiraffPlus General Idea

Ambient Assisted Living (AAL) research and development has massively invested on issues like gathering continuous information at home, standardizing formats in order to create environments more easily, extracting further information from raw data using different techniques to reconstruct context. An aspect relatively less developed but very important to actually arrive to the market is the design of complete, and hopefully personalized, end-to-end services for the users of such technology being them either primary (older people) or secondary (medical doctors, caregiver, relatives).

These authors are working in the GiraffPlus project (http://www.giraffplus.eu/) whose general goals are the following: (1) to develop and thoroughly evaluate a complete system able to collect elderly people's daily behavior and physiological measures from distributed sensors, to perform context recognition and long-term trend analysis; (2) to organize the gathered information so as to provide customizable visualization and monitoring services for caregivers; (3) to foster social interaction between primary users (elderly) and secondary users (formal and informal caregivers). In broad terms, secondary users can virtually enter the home for a visit or to respond to an event generated by the system, while primary users can request a virtual visit at any time.

A detailed description of the GiraffPlus project is given in [1] while a short overview of its architecture is presented in the next section. A key motivation for our work is the development of effective services for end users (both primary and secondary). In this respect GiraffPlus has the ambitious goals of fielding a complete systems in 15 homes distributed over three European countries and performing long term experimentations in place. During the first year of the project an intensive activity of user requirements elicitation has taken place, then converted in a functional architectural specification. New technical development is being performed, and preliminary deployment in Living Labs place, while the start of a first set of test cases is expected during the second half of project year two. The preliminary deployment in Living Labs has been useful to sketch a number of user services in a safe environment that maintains the difficulty of a real house without involving the criticality of involving frail old people.

The fielding in real environments of an AAL system is relevant for a number of reasons:

- it forces technology developers to focus on the complete system rather than to single technologies;
- it underscores the need for robustness in AAL solutions, and this is even more critical since such complete systems integrate heterogeneous technologies;
- it points out the problem of hiding technological complexity in favor of easiness of use for a wide spectrum of users;
- it forces technologists to face the different perspective that usually guide real people using a system hence creating preconditions for designing more mature systems when compared to "laboratory tailored designed" ones.

This paper shows the path from sensor data to end user services as implemented in GIRAFFPLUS. It discusses some of the technical work done to create meaningful services dedicated to different users. In the final section it shortly describes the first deployment experience in an Italian Living Lab.

2 System Architecture

Figure 1 depicts an abstract component diagram of the GIRAFFPLUS system. In particular, three main components can be identified: (a) the Physical Environment and Software Infrastructure, (b) the Middleware Infrastructure and (c) the Service Layer.

The *Physical Environment and Software Infrastructure* coupled with the *Middleware Infrastructure* (see Fig. 1) offers the basic level of functionalities of the GIRAFFPLUS system. The Middleware Infrastructure is in charge of providing a common and interoperable communication service, constituting a *gateway* shared among all the system components whose aim is to support all the data services.

Then, the *Sensor Network*, composed by both *Physiological and Environmental sensors*, gathers the information generated in the home environment as well as provides the (possibly pre-processed) collected data. Finally, the *Telepresence Robot* provides the GIRAFFPLUS social interaction functionalities enabling remote access in the environment through a Pilot software embedded in the visualization and interaction services (see Sect. 5). The *Long-Term Data Storage* component (see again Fig. 1) is responsible for providing a general database service for all the data generated by parts of the system and providing data access functionalities. Specifically, the main role of this component is to manage a database containing all the data collected through the Middleware Infrastructure and generated by the Sensors Network (see Sect. 4). The *Context Recognition and Configuration Planning* (bottom-right in Fig. 1) is the component responsible for context/activity recognition and system configuration planning, i.e., two high-level reasoning systems in charge of (respectively) implementing the monitoring activities by means of context/activity recognition [2] and providing suitable configuration settings for the Sensors Network according to the requested monitoring activities [3].[1] Finally, the *Data Visualization, Personalization and Interaction Service* (bottom-left in Fig. 1) is the part of the system responsible for creating user-oriented service. A broad way to summarize the module is to provide different end-users with suitable interaction modalities for the available services. Sect. 5 provides a more thorough presentation of this module.

[1] The description of the Context Recognition and Configuration Planning components is out of the scope of the present paper. The reader may refer to [1] for further details.

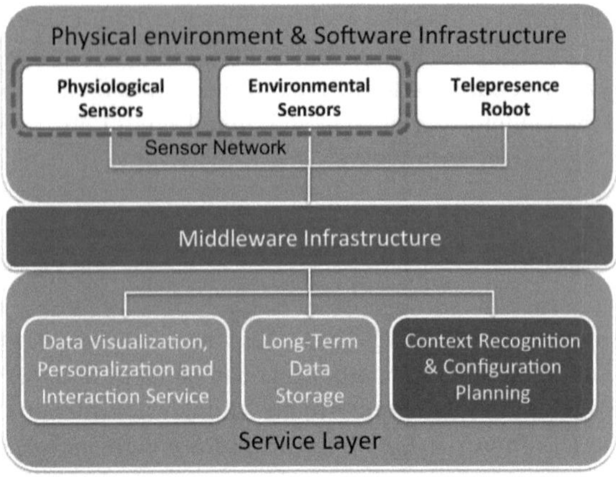

Fig. 1 The GIRAFFPLUS components

3 AAL Middleware for GIRAFFPLUS

In the GIRAFFPLUS system, a crucial role is played by the Middleware Infrastructure as it provides the central connection point that is shared by all the components according to the needed information exchanges. In fact, given the inner context-aware nature of GIRAFFPLUS system, the presence of a pervasive solution that provides any kind of information about the interaction between the user and the surrounding environment become a key aspect for its effectiveness. The sensors, the services, and the components integrated in the GIRAFFPLUS system need a software infrastructure, which is based on a middleware that hides heterogeneity and distribution of the computational resources in the environment. Moreover, the integration of such components is demanding, especially if we consider that the system is composed of different services written in different languages and it may need to be accessible by a number of remote healthcare centers which may use different protocols. To this aim, an AAL middleware solution well suited to the GIRAFFPLUS context is proposed.

The fragmentation in this sector is still high, but there are initiatives working to build converging solutions. Service interoperability is a key point to build an ecosystem of applications that helps the growth of an AAL consumer market. In this regards, several European projects have intensely worked to the definition and standardization of a common platform for AAL, on top of which to develop intelligent software applications for the end users. In this regard, the final objective of the GIRAFFPLUS project is to design and develop a system compliant to the results of the most promising research projects in this field, i.e., UNIVERSAAL [4].

It is intention of this work to be aligned to the UNIVERSAAL results, to reuse the Open Source software released by UNIVERSAAL as much as possible, and to enrich the UNIVERSAAL platform with the technological requirements deriving from the GIRAFFPLUS project. The concrete architecture currently selected by UNIVERSAAL is based on the Java/OSGi platform and it is aligned with the reference platform selected by GIRAFFPLUS.

Middleware Architecture. Within the proposed AAL ecosystem, hardware as well as software components can be able to share their capabilities. In the GIRAFFPLUS space, the proposed platform facilitates the sharing of two types of capabilities: Service (description, discovery and control of components) and Context (data based on shared models). Therefore, connecting components to the platform is equivalent to using the brokerage mechanism of the middleware in these two areas for interacting with other components in the system. The concrete middleware architecture is made up of two layers: a core middleware API layer and a communication layer that includes a publish/subscribe connector and a RESTful connector.

A generic service built upon the middleware can discover which sensors are present in the environment and other services together with their functionalities using methods from the middleware API layer. The underlying layer fulfills these requests exploiting the connectors available. In the communication layer, an MQTT [5] and a RESTful [6] connector are present. By mean of these connectors, the middleware realizes a publish/subscribe as well as a methods description/invocation mechanism in a transparent way.

Two buses form the heart of the proposed middleware: a context bus and a service bus. All communications between applications (i.e., the GIRAFFPLUS services) can happen in a round-about way via one of them, even if physically, the applications are located on the same hardware node. Each of the buses handles a specific type of message/request and is realized by different kinds of topics. The aim of the middleware is to provide a publish/subscribe mechanism for accessing the context information about the physical environment and physiological data. This information will be exposed as different topics: topics for discovery and description of devices and services that form the service bus and topics for publishing and retrieving data from devices and services that form the context bus. The middleware is in charge of presenting the available sensors and services in the system implementing an announce mechanism on the service bus. These resources are presented with a message on the relative topic in the service bus. The message is a descriptor file containing an id, a description, a type (i.e exporter or service), a set of resources (i.e. sensors or components), and a set of methods. Once a resource has been announced on the service bus a generic service can search for it filtering on the descriptor fields and use it. The topics used for announce and discovery of devices and service, the so called service bus, has this format

```
«location»\serviceBus\«serviceID»
```

where `location` identifies, e.g., the room in the assisted persons apartment, `serviceBus` is the keyword to identify the topic as a service bus topic and `serviceID` is the unique identifier of the service. The message of this topic is a JSON descriptor file. The middleware takes care of dispatching information about the state of the resources among services by means of a context bus. Any service that wants to make available his data (sensors readings and events or data analysis results) can use the middleware API to publish it. Any service interested of monitoring these data can subscribe to the relative context bus topics indicated in the descriptor using the middleware API. The topics used for gathering data from devices and service, the so called context bus, has this format:

```
«location»\contextBus\«serviceID»\«subtreefield»
```

where `location` identifies the room, `contextBus` is the keyword to identify the topic as a context bus topic, `serviceID` is the unique identifier of the service and the `subtreefield` identifies all the resources of that service that can be monitored. For each resources there will be a dedicated context bus sub-topic. The message of these topics is a string value.

4 Long-Term Data Storage Service

The Long-Term Data Storage module aims to provide a central storage for the GIRAFFPLUS system, responsible for storing all the data collected by the Physical Environment, as well as the data produced by other components (High-Level Services). Additionally, it also stores configuration data (sensor data, security restrictions, etc.) and logging data from all the software components in the system (all logs are centrally available, and therefore it is easier to maintain and debug the system). Such service consists of three connected components: (i) the database, which stores the actual data; (ii) the GIRAFFPLUS Middleware Listener component, which is instantiated in the middleware and forwards all relevant data to the database in a secure fashion; (iii) the RESTful web service, which enables secure access to the database and (iv) the GIRAFFPLUS Middleware Storage component, which enables other GIRAFFPLUS components to directly access the data via the middleware.

The guiding principles in the design of the GIRAFFPLUS Long-Term Data Storage system have been **security**, **flexibility**, **reliability**, **efficiency** and **scalability**. Security is paramount as we are dealing with potentially sensitive information. Flexibility is important as GIRAFFPLUS is a research project and then it is impossible to foresee all the possible data structures that might be stored in the system, as well as all possible connections between them. The storage system needs to be able to reliably store vast amounts of data and enable users and other GIRAFFPLUS components to efficiently access the stored data. Last but not least, the system needs to be scalable as we expect the amount of data to vastly increase during the course of the project and hopefully, during the commercialization phase of the project.

The GIRAFFPLUS **database.** After a careful examination of available database solutions, we decided to use MongoDB (http://www.mongodb.org/), which is a widely used, open source, NoSQL document-oriented database system developed and supported by 10gen (http://www.10gen.com/). To follow our guiding principles, we make heavy use of its replication feature, which replicates data over specified groups of servers thereby increasing data reliability, as well as efficiency. We also us the sharding feature, which distributes the data over many replica groups, thereby enabling horizontal scalability and increasing efficiency via load-balancing.

Flexibility of the system is achieved due to its NoSQL nature, since the database stores collections of generic JSON (http://www.json.org/) objects, which can represent various data-structures that might come up during the lifetime of the project. To increase efficiency we also implemented a database compression method, which compresses collections of data older than a specified date. The method automatically collects data objects coming from the same source and compresses the array of dynamic field values. To retrieve data from compressed collections of object, the method first uncompresses the data and then runs the required filter over them.

The GIRAFFPLUS Middleware Listener component. The GIRAFFPLUS Middleware Listener component is instantiated automatically as the middleware is run. It listens to all data sources connected to the middleware (sensors, other components, logs, *etc.*), properly formats the data, attaches the timestamp and stores the data in the database.

The RESTful GIRAFFPLUS Storage Web Service. The implemented RESTful (REpresentational State Transfer) web service enables secure (using certificates via SSL) and efficient access to the data stored in the database. It is implemented in Java using Jersey (https://jersey.java.net/), which is an open-source, production quality, JAX-RS Reference Implementation for building web services. To run the web service we use the Apache Tomcat (http://tomcat.apache.org/), which is an open source software implementation of the Java Servlet and JavaServer Pages technologies and enables easy scalability by using load-balancing between multiple servers.

Data between the web service and its clients is exchanged in JSON format and, if required by the client, compressed to increase the efficiency of data transfer.

The GIRAFFPLUS Middleware Storage component. We predict that at some point during the project there might be multiple Long-term Data Storage centers (LTDSC) located in various locations around Europe (and later on outside Europe) due to possible legal or performance constraints. One of the design goals for the Long-term Data Storage system was to enable partners to write data-location agnostic applications and components. Therefore the only point of contact for each application/component should be the GIRAFFPLUS Middleware and requests for data storage or retrieval should be done via the middleware. For this purpose we implemented a simple Storage component. Its configuration contains the location of the LTDSC to be used for the instantiated middleware and all requests for data storage and retrieval are done through the Storage component, which then forwards requests to the corresponding LTDSC. In this fashion we enable developers to write data-location agnostic applications and components, or, if required, to forward data requests to a specific LTDSC, which contains the requested data.

5 Data Visualization and User Interaction

The GIRAFFPLUS support infrastructure described in the previous sections guarantees two very important capabilities: (a) the long term storage of sensor data; (b) the possibility of real time connection for both information sharing and immediate alarm

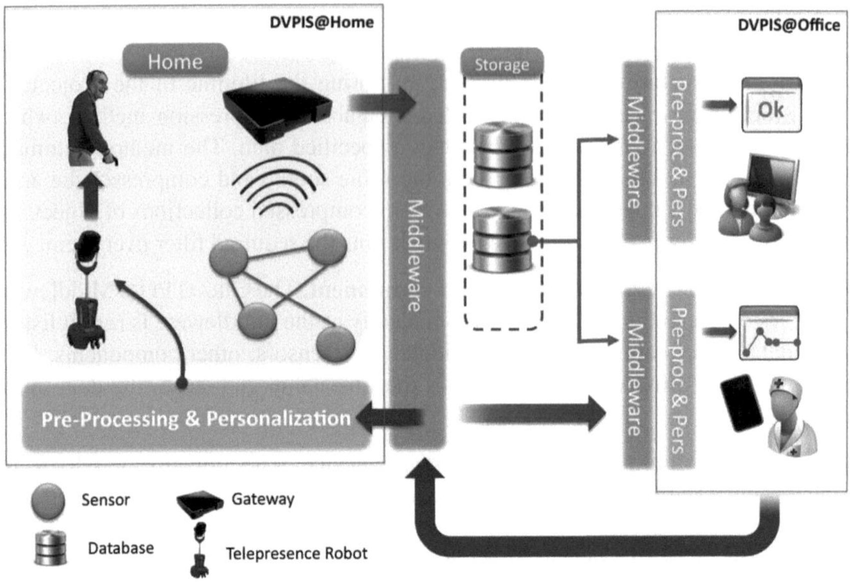

Fig. 2 Outside versus inside services through DVPIS

intervention. The work on user-oriented services has started with an analysis of the different type of people that can take advantage of the GIRAFFPLUS services. As said before, the basic subdivision between users concerns Primary (the old person in the house) versus Secondary (different people outside the house). The secondary users can be further subdivided in various groups: within GIRAFFPLUS we are particularly interested in services for (1) formal caregivers (e.g., doctors, social workers); (2 informal caregivers (e.g., relatives). The further basic question now is "how can we build useful services for such users?". Among services, one interesting distinction concerns the temporal scope we are expecting: (a) long term data analysis, for example, to observe trends for creating regular reports for different users, (b) short term reactive services, like in the case of alarms, to detect emergencies (e.g., fall detection sensors, emergency call buttons, etc.) (c) continuous asynchronous dialogues (e.g., in the case of social network channels, reminding services, etc.) (d) synchronous communication channels for conversations through the telepresence robots.

Figure 2 describes the complete scenario we have realized. Again we underscore different aspects: first of all the double directionality pursued in the complete GIRAFFPLUS service model: (a) Internal versus External Data Flow: there is a basic direction of data gathering from the house toward the external world (red arrows in the figure). Notice that we distinguish: (1) long term data storage for subsequent analysis and on-demand visualization (red arrow in upper part of the figure), and (2) direct real-time fast connections managed directly through the middleware for alarms (red arrow in lower part); (b) Bidirectional External/Internal communication: this channel allows to support the social interaction services. Initially, a basic telepresence service of the Giraff telepresence robot has been created (http://www.giraff.com/), subsequently the system has been empowered with additional services for increasing support to the primary users (blue arrow in the figure).

The Fig. 2 also identify the modules specifically called DVPIS (for Data Visualization, Personalization and Interaction Service) that have been realized to manage interaction with the different actors in an AAL scenario. In particular we have two instances of the DVPIS one for use "outside the house", called DVPIS@Office, and another dedicated to the primary user, called DVPIS@Home. Both the DVPIS modules are composed of a front-end which is a visualizer and interaction manager and a back-end that deals with pre-processing of data and personalization of services.

DVPIS@Office. The module dedicated to secondary users strongly rely on querying the data store. Additionally access to the telepresence robot via its external interface is possible within the same comprehensive layout. At present we have designed an interaction front-end that runs on a personal computer but the technological choices allow us to develop also App-style versions for other mobile platforms like tablets or smartphones. Different services are provided for formal and informal caregivers. The personalization for a doctor or a social worker takes into account the fact that this workers may connect to multiple patients at home. Hence there is an environment that manages different houses and help the user to maintain information on the different cases he/she follows. The formal caregiver may access data over time, although specialized the visualization environment allows a combination of queries that are able to generate different graphical views. Figure 3b presents a test example of the current DVPIS@Office for a doctor observing a week of blood pressure data. Figure 3a shows a user inspecting a printed report received through the GIRAFFPLUS system and at the same time has decided to query the long term data for observing temporal trends (see the PC screen). The personalization for an informal caregiver (the current target is a relative of the person at home) is far straightforward. Such a person needs a more synthetic information and just report about warning over an interval of time. At present it is possible to ask for daily reports that contain a summary of the informations of the day. We are currently working at producing a scenario similar to the one in Fig. 3a tailored for informal caregivers, with reports that contain a different perspective on the information delivered.

DVPIS@Home. The current front-end in the house for delivering information services is the telepresence robot. In fact, some user requirements gathered during the initial phase of the project suggested to avoid the introduction of further technological objects in the house. In addition, as the comprehensive design of the project includes the Giraff robot and it is endowed with an on-board computer for its basic operations, we have decided to exploit the robot also to run the local additional software services. Then, specifically for the project, the company producing the robot has synthesized a touch screen version. The DVPIS@Home is designed to take advantage of this new functionality, and somehow the telepresence robot resembles a "tablet on wheels" from the point of view of old people at home. Thus, having the robot as the front end in the house (a) allows to access the standard telepresence services of the robot (b) allows the home users to read simplified reports concerning his/her own

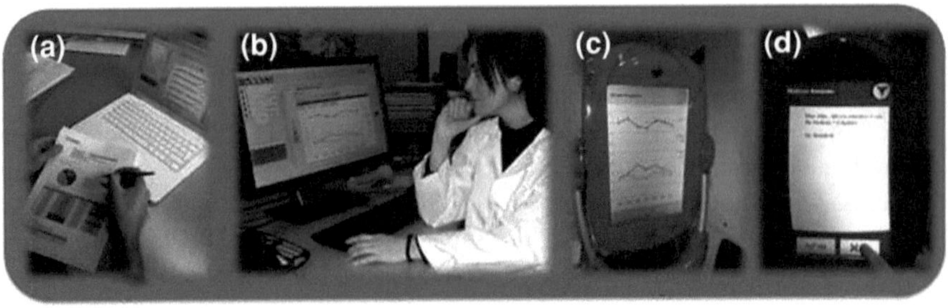

Fig. 3 The current DVPIS in operation

health status (c) enable the delivery of asynchronous messages to the user. This last functionality is currently integrated for obtaining a reminding service (a message example is shown in Fig. 3d).

The DVPIS shared dialogue space. The GiraffPlus project is in its second year. Since the first dialogue with potential home users, the issue of the dialogue "Inside/Outside" the house has been a very important one. For this reason, we are endowing the DVPIS of a capability of synchronous communication fully supported by the middleware. Using this functionality we have created a shared information space that allows a dialogue Primary/Secondary user sharing some information on the screen. The current demo of this functionality is shown in the combination of Fig. 3b and c. In fact, it is possible to see that the same information (the weekly blood pressure samples) is in front of the doctor in her office and shown to the old person at home while they can share the audio channel for discussing the assessments. In this way, we have set up an environment for flexible end-to-end information delivery on top of an AAL service.

6 Current Status

The basic path from data to user services has been first tested in our laboratories and, then, delivered as a complete system in the Rome Living Laboratory, an apartment of young colleagues of ours. Our goal is the test the technology outside the lab but without the ethical burden of testing it with fragile people. This home testing has been useful to discover weaknesses in the technology integration of the different blocks (sensors, middleware, data-store, software interaction environment). After quite an amount of work to cope with various bottlenecks and lack of robustness we obtained a stable working system. We have been able to visualize snapshots of temporal data from the apartment (see an example in Fig. 4) on a DVPIS@Office version installed in a Rome office computer, while the data store is remotely connected from Ljubljana at XLAB premises, additionally we have been able to simulate real time alarms and their respective management procedures (e.g., water leaks in the kitchen). We are

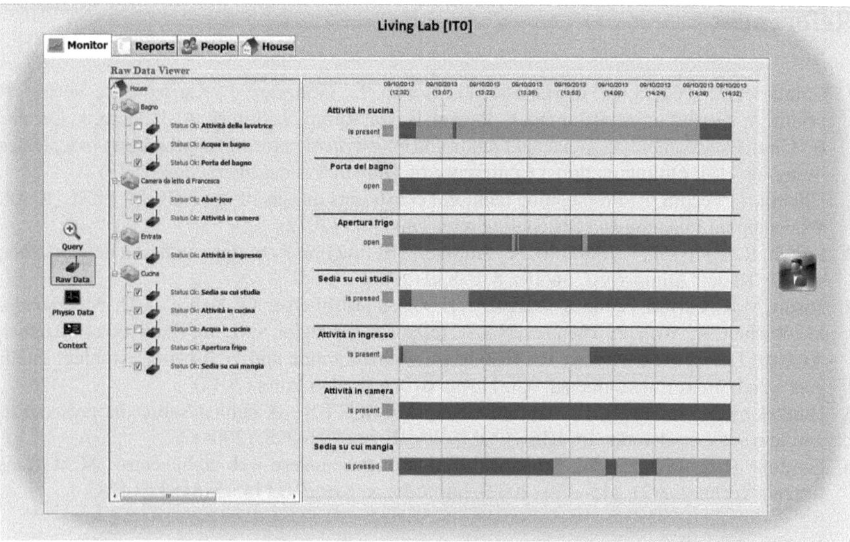

Fig. 4 A data sample from the Italian Living Lab shown on the DVPIS@Office

now monitoring the system on a longer period of time while continuously refining the functionalities of the interaction environment. Following this important testing phase we have delivered the complete GIRAFFPLUS system in six real homes (2 each in Sweden, Spain and Italy) and are currently receiving data from these test sites installations.

Going back to Fig. 2, the current live experiment concerns the path "from-Inside-to-Outside" with respect to a house. Other partners in the project have currently experimented the opposite connection ("from-Outside-to-Inside") using the Giraff robot as a media, others are testing the Context Recognition and Configuration Planning module that aims at enriching the basic gathered data with additional temporal deductions. Finally, another parallel activity is designing protocols for "safe delivering" the GIRAFFPLUS system in real test cases respecting the social and ethical expectations. In general, delivering innovative AAL technology for long term use is quite a complex and multi-faceted effort. Describing parts of this effort has been the topic of this paper.

Acknowledgments Authors are partially funded by the European Community's FP7 under the GIRAFFPLUS project (contract #288173). Francesco Furfari and Filippo Palumbo are also partially supported by FP7 under the UNIVERSAAL project (contract N.247950).

References

1. Coradeschi, S., Cesta, A., Cortellessa, G., Coraci, L., Gonzalez, J., Karlsson, L., Furfari, F., Loutfi, A., Orlandini, A., Palumbo, F., Pecora, F., von Rump, S., Stimec, A., Ullberg, J., Ostlund, B.: GiraffPlus: combining social interaction and long term monitoring for promoting independent living. In: HSI: 6th International Conference on Human System Interaction (2013)
2. Ullberg, J., Pecora, F.: Propagating temporal constraints on sets of intervals. In: PSTL: ICAPS Workshop on Planning and Scheduling with Timelines (2012)
3. Lundh, R., Karlsson, L., Saffiotti, A.: Autonomous functional configuration of a network robot system. Robot. Auton. Syst. **56**(10), 819–830 (2008)
4. Tazari, M.R., Furfari, F., Fides-Valero, Á., Hanke, S., Höftberger, O., Kehagias, D., Mosmondor, M., Wichert, R., Wolf, P.: The universAAL reference model for AAL. In: Handbook of Ambient Assisted Living—Technology for Healthcare, Rehabilitation and Well-being, Ambient Intelligence and Smart Environments, vol. 11, pp. 610–625. IOS Press (2012)
5. Hunkeler, U., Truong, H.L., Stanford-Clark, A.: MQTT-S - A publish/subscribe protocol for wireless sensor networks. In: IEEE COMSWARE, pp. 791–798. (2008)
6. Fielding, R.T., Taylor, R.N.: Principled design of the modern web architecture. ACM Trans. Internet Technol. **2**(2), 115–150 (2002). http://doi.acm.org/10.1145/514183.514185

A WSN Integrated Solution System for Technological Support to the Self-Sufficient Elderly

Paola Pierleoni, Alberto Belli, Lorenzo Palma, Michele Palmieri and Luca Pernini

Abstract The proposed system is basically an automatic dispenser of medicines which integrates functionalities of remote and real-time monitoring of the activities of the elderly in home environment. It includes the generation of alarms resulting from the omission of taking medicines and other critical events such as the permanence of the elderly in a room for a period not consistent with normal daily activities. The main elements of this system are a pair of LED smart lamps called MARCH'ingegno and Sibilla, the first one is the dispenser of medicines which embeds WSN coordinator role, located in the house of the elderly, the other is the displays of real-time alarms, located in the house of relatives or caregiver. The mobility monitoring system is composed of wireless sensor nodes distributed in the house of the elderly. MARCH'ingegno acquires information from these sensors through its RF interface, processes it and transfers the report to a remote web server together with the notes of taking medicines. The tests show that the system operates properly and the wireless sensor nodes ditributed at home provide an adequate coverage area and correct response times.

1 Introduction

This project involves the construction of an intelligent system of remote real-time assistance of elderly people living alone. With this project, Semar s.r.l. won the competition organized by the Marche region [2], "Casa intelligente per una longevitá attiva ed indipendente dell'anziano."

P. Pierleoni · A. Belli · L. Palma · L. Pernini
Information Engineering Department (DII), Universitá Politecnica delle Marche, Via Brecce Bianche 12, 60131 Ancona, Italy
e-mail: p.pierleoni@univpm.it

M. Palmieri (✉)
Semar s.r.l., Via Sardegna 5, 60022 Castelfidardo, Italy
e-mail: Michele.Palmieri@semar.biz

S. Longhi et al. (eds.), *Ambient Assisted Living*, DOI: 10.1007/978-3-319-01119-6_29,

The main elements of this system are MARCH'ingegno and Sibilla: a pair of LED smart lamps where local and global communication systems and various solutions for Ambient Assisted Living are integrated. MARCH'ingegno, is a low energy lighting and design element. It is equipped with a service of automatic dispenser for the daily dosage of medicines. In addition, it integrates a system for the mobility monitoring of the elderly in the home environment with the generation of alarms resulting from critical events such as the omission of taking medicines or the permanence in a room for a period not consistent with normal daily activities.

The system of the mobility monitoring is realized by wireless presence sensors distributed in the home environment. MARCH'ingegno acquires the information from these sensors, processes data and transfers the report to a remote web server via Internet connection. The system is capable of producing a remote alarm on the mate lamp Sibilla (placed in the house of relatives, guardian, nurse, etc.) in case of critical events. Sibilla can report critical events and provides real-time information on the mobility of the elderly through LED display. The mobility details, the taking of medicines and the related alarms are also available on a web server accessible by authorized users through a PC or Smartphone application.

The strength of the system is its scalability with a number of density of nodes and different kinds of sensing features. Therefore, we can simply extend the variety of applications for the constant care and the safety of the elderly to other applications such as vital sign monitoring, fall detection, dietary/exercise control, etc. Our system benifits from the wireless capability, the collaborative and synergic sensing task, the date processing and the efficient communication that enable reliable unattended operations.

2 Design and Implementation of the System

The goal of the present work is the development of a system of continuous aid elderly people. This section describes the basic elements of the realized system: the sensor nodes, the smart lamps MARCH'ingegno and Sibilla and the web server services.

2.1 Sensor Nodes

The mobility monitoring system consists of indoor distributed sensor nodes. Each node contains a motion sensor that detects the movements in a given coverage area and therefore the presence of the elderly in a specific room. The sensor nodes are equipped with RF, interface which transmits data to a concentrator node (located into MARCH'ingegno) that suitably elaborates the received information.

The sensor nodes are designed to be plug and play installed and non-invasive for everyday life.

2.1.1 Motion Sensor

The motion sensor includes a Passive InfraRed (PIR) sensor and a processing unit suitable for our application. In order to improve performance of PIR sensor a Fresnel's lens is added.

PIR sensor

In order to sense human motion, the PIR sensor should only be tuned to detect radiation from a human that arrives in its proximity. However, the sensing elements are responsive to radiation over a wide range. Therefore, a filter is used to limit incoming radiation in the 8–14 μm range, according to the human body radiation.

Thus, the IR rays, after passing through the filter, strike the sensing element that generates a charge. The magnitude of the charge produced is directly proportional to the amount of rays that strike on the sensing element. The charge produced by sensing element manages to generate voltages that control a Field Effect Transistor (FET) following by an impedance converter. Thus, the FET modulates the output of the PIR and transmits the output signal to a processor unit.

In this project a RE200B-P PIR sensor [3] is used. It is designed to pick up heat radiation of wavelengths in a band of about 10 μm. It contains two sensing elements connected in a voltage bucking configuration. Thanks to this configuration we can have a good compensation of environmental temperature, excellent sensitivity for small changes and reduction of signals caused by vibration and sunlight.

The output signal of the PIR generates voltage changes on the order of μV which are difficult to process, this output is amplified and suitably elaborated by a processing unit.

Processing unit

A motion sensor can be realized by just connecting the output of PIR to a processing unit. As shown in Fig. 1 the realized processing unit can be divided into three elementary blocks: amplifier, comparator and signal processing. The output of PIR sensor feeds into a two stage amplifier having signal conditioning circuits and a high gain. The amplifier is bandwidth limited to about 10 Hz to reject high frequency noise and is followed by a window comparator that responds to both the positive and negative transitions of the amplifier sensor output signal. The output of the comparator produces a 0 to Vcc transition on the signal based on the amount of IR rays that strike the surface of the PIR sensor. This signal is sent to the signal processing block which provides a digital output indicating the presence (high level) or absence of motion (low level). Thus, the movements can be detected by checking for a high signal on a single digital pin of this motion sensor.

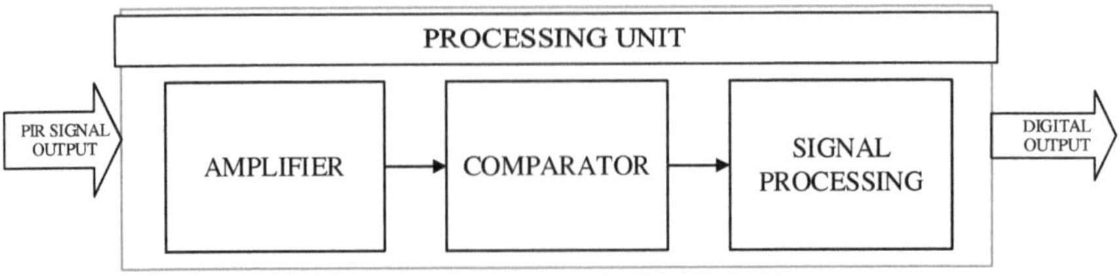

Fig. 1 Block diagram of the motion sensor processing unit

The reliability of this kind of sensor can vary with environmental conditions. The processing unit is designed to slowly adjust at the changing conditions in the environment that normally take place during the day. The motion sensor has a specific coverage area in which a moving person can be detected. The sensor coverage area varies based on of the size of the monitored environment, for this reason, the motion sensor sensitivity must be adjusted according to the distance range and visibility angle range of the sensing region.

Fresnel lens

In order to increase the coverage area of motion sensor it is necessary to focus the infrared radiation on the surface of the PIR sensor. This is accomplished using a plastic chamber called Fresnel lens. It consists of an array of curved segments that capture more infrared radiation and focus it to a relatively smaller point. In fact, the lens is designed to have its groove toward the sensing elements of PIR sensor.

The placement of the sensor beneath the Fresnel lens is crucial as it supposed to receive maximum amount of infrared radiation that comes from the lens array.

In order to focus the infrared radiation on the surface of the realized motion sensor a NL-11NH Fresnel lens [4] is added. This lens is positioned at a distance of 1.29 inches from the sensing element of PIR to direct maximum amount of infrared radiation on it.

2.1.2 RF Interface

After detecting the presence of a person in a room, each sensor node transmits data through its embedded RF interface to a concentrator located in MARCH'ingegno that processes the received information.

In this project, the XBee ZB RF Module [1] is used as RF interface. It contains a transceiver operating on frequencies of 2.4 Mhz with a power that varies depending on the models from 1 mW up to about 60 mW, an antenna or a connector for external antennas and an integrated microcontroller. Communication protocols that provide wireless connectivity to other several modules are implemented in the embedded microcontroller to create a ZigBee mesh networks. The XBee ZB RF modules are designed for high-throughput applications requiring low latency and predictable communication timing. It operates at data rate of 250 Kbps with an indoor range coverage up to 40 m. In the XBee ZB modules there is also the possibility to monitor the logic level of digital pins that are provided in their RF interfaces. The logic level of specific digital pin can be read locally and then transmitted to a remote XBee ZB RF Module.

Our system, MARCH'ingegno is equipped with XBee ZB RF Module because it has the function of network coordinator. It is the core of the Zigbee network and keeps the security keys and the network information. Thus, the RF interface of MARCH'ingegno is programmed to automatically create the mesh network and permits association of the other XBee ZB RF nodes to the WPAN.

Each other sensor node has a ZB RF Module which is configured for connection to the mesh network created by MARCH'ingegno: as soon as a sensor node authenticates itself it can transmit data within the network. Each sensor node is authenticated with the identifier of the room where it is placed. The XBee ZB RF Module is configured to immediately transmit to the coordinator a data whenever a specified digital pin changes state. The output of the motion sensor is attached to this specific digital pin.

2.1.3 Performance of Sensor Nodes

The proposed detection system was experimented in an indoor testing environment in order to evaluate its performance. The results of the tests show that each motion sensor requires a calibration time to properly function. This calibration time varies from 10 up to 60 s according to the environment in which it operates. After the calibration time, any thermal signal far below to one μW is sufficient to trigger a voltage change on the output amplifier of the motion sensor. In order to evaluate the coverage area, several tests were conducted to measure the sensor distance range and visibility angle range changing the sensitivity of the device. These tests were also carried out with the sensor without lens to evaluate the decrease in performance. The results of tests carried out with the addition of the Fresnel lens show a considerable increase of the coverage area of motion detection. Indeed, as shown in Table 1, the detection is more stable and the visibility angle range is also increased about 30° with a Fresnel lens.

As illustrated in Fig. 2, the results also show that, at the maximum sensitivity, the coverage area of the motion sensor has a range of about 6 m. So follows that the sensor node provides an adequate coverage area to monitor an entire room.

2.2 The Smart Lamps MARCH'ingegno and Sibilla

The ZigBee module placed into MARCH'ingegno receives the data sent from sensor nodes distributed in the house. In order to determine the permanence of the

Table 1 Visibility angle range of motion sensor

Sensitivity (%)	Visibility angle range with lens (Degree)	Visibility angle range without lens (Degree)
0	90	60
25	105	90
50	120	105
75	150	120
100	180	150

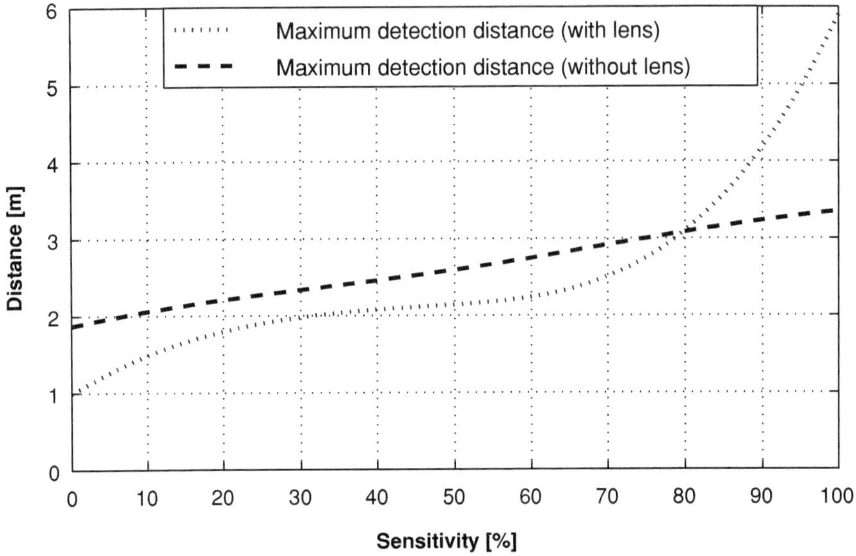

Fig. 2 Distance range of motion sensor changing the sensitivity

elderly in a specific room, the information sent from each sensor is processed by an embedded microprocessor into the smart lamp. An algorithm implemented on this microprocessor determines the residence time of the elderly in each room. Finally, these periods are compared with a specific threshold of danger to generate alarms in case of exceeding this time limit.

The danger thresholds corresponding to the specific room are obtained on the basis of a statistical survey that we have carried out. This study was conducted on a sample of 100 subjects (43 males and 57 females), aged between 70 and 87 years. The results of the study are shown in Table 2 where there is a subdivision in four typical rooms and four periods of the day.

The data resulting from this study are used as default parameters for the generation of alarms but are also remotely editable via a web server that allows to calibrate the system based on the behavior of each user.

Table 2 Percentage of time in which the subjects examined remain in every room of the house, for every time band

Time of the day	Kitchen (%)	Livingroom (%)	Bedroom (%)	Bathroom (%)
Morning (7–13)	27	39	24	10
Afternoon (13–19)	20	55	20	5
Evening (19–23)	20	50	23	7
Night (23–7)	1	4	92	3

The dispenser of medicines can be monitored through the remote web server. In fact, MARCH'ingegno is primarily an automatic daily medicine dispenser. Has several drawers that can contain a variety of different medicines. The microprocessor of MARCH'ingegno can be programmed to open the drawer at fixed daily times and contemporary to emit light signals and sound effects to remind the elderly to take the medicines. The service dispenser of medicines can be configured locally or remotely and the settings be modified whenever changing the mode of drug administration. In case of non-withdrawal of the medicines, this system is able to produce remote alarms.

The alarms and the presence data received from each sensor node are transmitted to a remote web server that allows a real time view of the elderlys mobility in the house and any potentially dangerous situation. Using the web server the presence data and alarms are immediately displayed on the Sibilla and on a Smartphone equipped with the dedicated application. In fact, the Sibilla smart lamp is programmed to continuously check the information stored in the web server and to display it. The same information could be viewed by dedicated smartphone or PC applications. The system also allows anyone who is registered and included in the list of contact numbers receive the alarm message in case of emergency.

The data exchange with the web server is allowed by the WiFi connection between the smart lamp and the home router. As shown in Fig. 3, the WiFi module integrated into MARCH'ingengo and Sibilla provides the communication between the two smart lamps and the home router. This WiFi module communicates with dedicated microprocessor via UART interface. The lamps were equipped with WiFi connectivity for not constrain their placement in proximity to a wired ADSL plug thus leaving the maximum freedom of positioning inside the dwelling.

The selected WiFi module manufactured by STMicroelectronics [5] is one of the least expensive on the market and is adequate for our needs to project a smart and inexpensive system. On this module we have implemented the WPS protocol to easily and securely connection to the domestic WiFi network . Pressing the WPS button on the domestic WiFi router it sends an 8 digits PIN code that is used for smart lamp module authentication by exchanging network name and keys set.

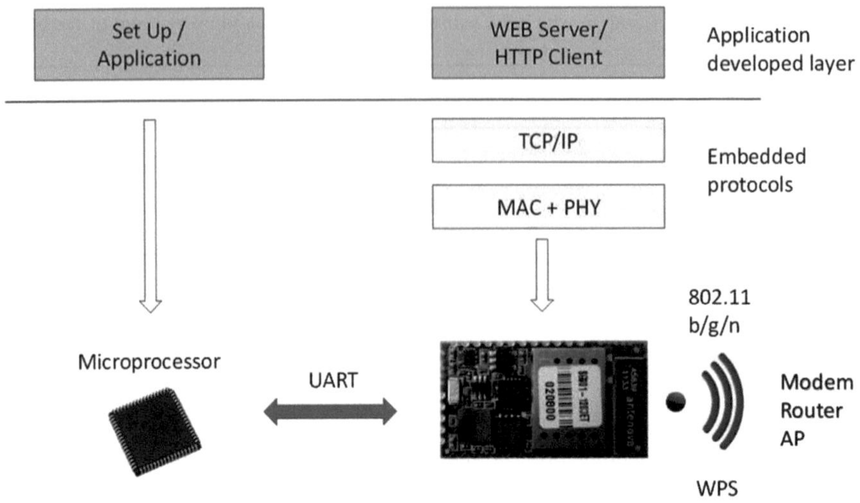

Fig. 3 Diagram of the link between the microprocessor of smart lamp and the home router via WiFi module

2.3 The Web Server

One of the most important features of the system is the capability for the authenticated remote user to manage the setting of the dispenser. It is allowed to define the conditions for opening each drawer with possible repetitions daily and duration of medicine treatment and with the inclusion of explanatory notes.

At the same time the information collected by the smart lamp MARCH'ingegno is transmitted to a remote web server via Internet to monitor the state of the elderly and his mobility inside the house. The system is developed exploiting the client-server architecture model. The client lies on MARCH'ingegno that is connected via the Internet to the mate lamp Sibilla, PC and smartphone. The diagram of the realized system is illustrated in Fig. 4.

The Microsoft Windows prototype server contains a socket based back end and a PHP based front end. The back end captures data received from the Internet, extracts the useful information and stores it into a database. The server needs to deal with many communication events because of the possible high number of clients. The server application consists of a listening socket, created at the starting of the task, and a client socket that are set up when a new client tries to connect to the server. The clients are grouped in three different types (MARCH'Ingegno, Sibilla, generic remote client and administrator) and every set can receive or transmit different types of data.

When a client calls the server, the listening socket will accept the call and a client socket is established. The front end provides services and database access, handling

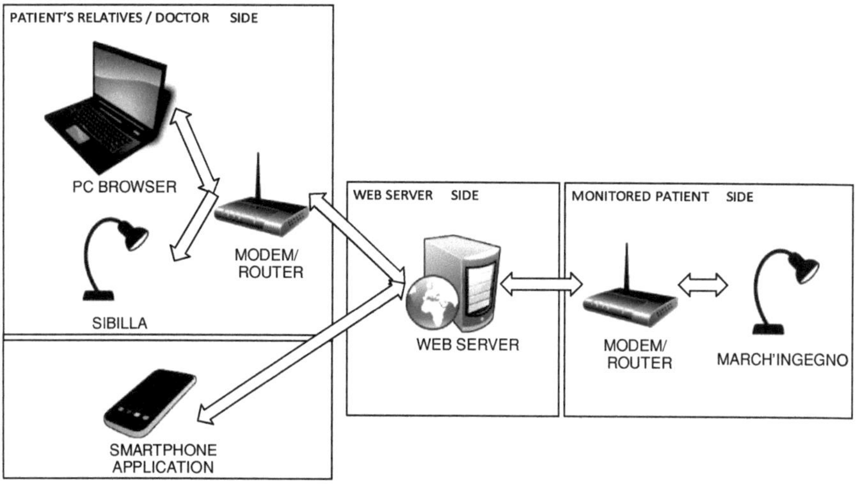

Fig. 4 Block diagram of the basic elements of the monitoring system

the data using PHP and MySql, displaying all the required information on the website, launching alarms or notifications, if necessary.

3 Conclusion

We have designed, by proposing a common object such as a LED lamp for smart lighting, an integrated multi-function system consisting of wireless sensor nodes. This WSN allows the remote setting and monitoring of a dispenser of medicines integrated into the lamp. It is also an environmental watching and alarm transmission resulting from the occurrence of abnormal events. In this way a design and everyday object becomes the main component of an AAL system integrating perfectly in a home environment without the need of invasive devices. Thanks to WiFi connectivity, RF interfaces and sensing capabilities MARCH'ingegno and Sibilla allow family members the continuous remote monitoring of the elderly, his mobility inside the house and his proper use of medicines.

References

1. Digi International: In: XBee ZB modules datasheet. http://www.digi.com/products/wireless-wired-embedded-solutions/zigbee-rf-modules/zigbee-mesh-module/xbee-zb-module
2. Marche Region (2013) "Casa intelligente per una longevitá attiva ed indipendente dell'anziano".. In: Call for the selection of project proposals aimed at developing integration platforms dedicated to active aging and to ambient assisted living (DGR 1464 del 7/11/2011)

3. Nicera Philippines Inc., In: RE200B-p datasheet. http://www.rapidonline.com/pdf/61-1466.pdf
4. Nicera Philippines Inc.: In: NL-11NH datasheet. http://www.rapidonline.com/pdf/61-1528.pdf
5. STMicroelectronics (2001) "SPWF01Sx.11". In: WiFi Module datasheet.

Part VII
Smart Housing

RITA Project: An Ambient Assisted Living Solution for Independent and Safely Living of Aging Population

Raffaele Esposito, Manuele Bonaccorsi, Dario Esposito, Massimo Filippi, Erika Rovini, Michela Aquilano, Filippo Cavallo and Paolo Dario

Abstract This paper presents the work carried out during the RITA Project, a study that focused on designing and implementing Ambient Assisted Living (AAL) services in the real context of Province of Pisa (Tuscany, Italy). The main objective of the RITA Project was to demonstrate the efficiency and the feasibility of new socio-medical services based on AAL approach. The user target of this project were elderly persons of Pisa area, living mainly alone or with a partner at their home, and their formal and informal caregivers. According to their needs new services and ICT system were developed and tested in order to improve the sense of safety of elderly people and caregivers.

1 Introduction

People older than 65 years are the fastest growing segment of the European population and they will account for a third by 2060 [1]. As aged people, most of them are affected by physical and cognitive deficits and often need family or caregivers' support for their daily activities and health monitoring and care. For these reasons the ageing of the population is raising significant issues for new sustainable economic and welfare systems in order to meet needs of elderly citizens, families and caregivers in an adequate way and to impact positively on their quality of life and quality of work/services. In order to face the cost-effectiveness of new technologies for elderly three factors have to be considered:

R. Esposito · M. Bonaccorsi · D. Esposito · M. Filippi · E. Rovini · M. Aquilano · F. Cavallo (✉) · P. Dario
The BioRobotics Institute, Scuola Superiore Sant'Anna, viale Rinaldo Piaggio 34, 56025 Pontedera, PI, Italy
e-mail: f.cavallo@sssup.it

R. Esposito
e-mail: r.esposito@sssup.it

S. Longhi et al. (eds.), *Ambient Assisted Living*, DOI: 10.1007/978-3-319-01119-6_30,

- *The increasing demand for elderly cares*
  In Europe people over 65 years will be a third of the population by 2060 and the consequence will be the increase of the demand for care due to a growing number of chronically ill people. The ageing of the society will led to an increasing demand for Nurse Practitioners (+94 % in 2025) [2] and Physician Assistants (+72 % in 2025) [3] with several implications for quality of care and for the configuration of future cost-effectiveness care delivery systems.
- *The reduction of funds for social-medical services*
  The decreasing of worker population [4] and the reduction of funds for social-medical services, due to the current economic crisis, couldn't fulfill the demand for care.
- *The growth of smart technology market*
  In the last years we are witnessing from one side to the rapid increase of "intelligence" of technologies and at the same time to the improvement of their economic accessibility among common people (i.e. computers and smartphones). According to this phenomenon it is estimated that medical electronics equipment production will increase from the $91Bn in 2011 to the $119Bn in 2017 with an average rate of 4.6 % per year [5]. Particularly the EU smart home market is estimated to grow from $1,544.3 million in 2010 to $3,267 million in 2015 [6].

On the base of these premises the AAL services and the ICT systems could be a valid solution to support the independence of elderly people and improve a sustainable healthcare system.

In the last years, several research projects have been focused on new approaches and solutions for innovating socio-medical services, based on the use of ICT devices that could improve safety and independence of elderly persons. Some examples of these projects belong to the Ambient Assisted Living Joint Programme (AAL-JP): CARE [7], DOMEO [8], HOPES [9] and CARE@HOME [10] etc. Considering this context, the Tuscany Region (Italy) funded RITA Project to develop an 'aging-in-place' model, in which elderly people live as long as possible independently and safely in their own homes and caregivers can remotely follow them. RITA aims to demonstrate the potential efficiency of socio-medical services based on an Ambient Assisted Living (AAL) approach, the interest among elderly and caregivers and the sustainability of these AAL services in Tuscany context. So this project studied and designed services and ICT technologies suited for the Tuscany context and new legal framework needed to implement these new services in Tuscany. RITA Project was not only 'elderly people centered', but it also involved all persons supporting senior citizens, as relatives, caregivers, socio-medical workers and volunteers. This document reports an overview of the technological work carried out in RITA Project for designing and testing AAL services and ICT devices.

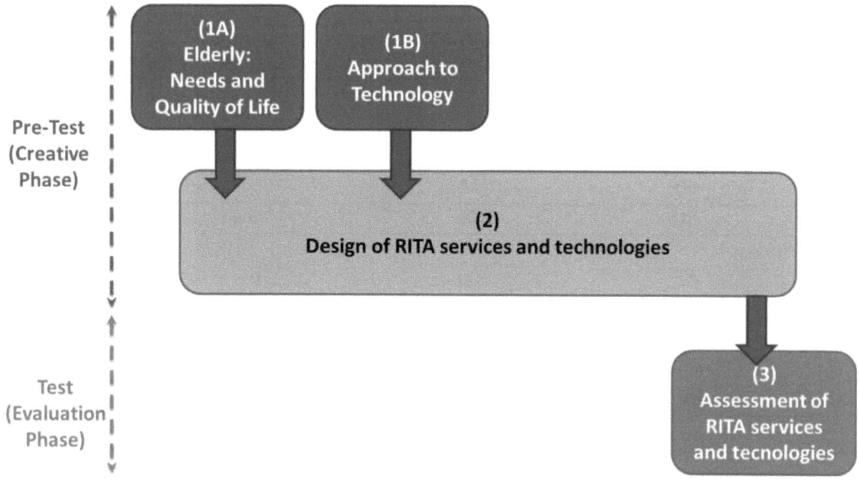

Fig. 1 RITA Project: work phases

2 Methods

RITA consisted of two parts (see Fig. 1):

Creative part included:

> *Phase 1*: Analysis of (a) elderly people's needs and quality of life, and (b) attitude towards current and future technology.
> *Phase 2*: Design of services and ICT systems.

Evaluation part included:

> *Phase 3* Assessment of developed services and ICT systems.

2.1 Phase 1 (a) Elderly Needs and Quality of life

During the Phase 1, researchers involved elderly citizens and caregivers living in the province of Pisa, Italy, through a number of focus groups and interviews in ten local end-user organizations and in some cases at elderly homes. Ad-hoc questionnaires, combining multiple choice and open questions, were used to define the profile of older persons and identify their perceived needs and quality of life. The interviewees were 211 elderly people 65+ without severe cognitive pathologies and 68 formal and informal caregivers. The details of participants are showed in Tables 1 and 2.

In general elderly users were enough satisfied about their life and their current independence; however the main aspects influencing their perception of quality of life concerned possible degeneration of their health status, safety and social life.

Table 1 Elderly people's characteristics

Gender	65–75 Years	Over 75 Years	Unspecified age	TOT
Number of elderly				
Men	24	24	4	52
Women	36	120	3	159
TOT	60	144	7	211
Gender	Min.	Max	Mean	SD
Age of elderly				
Men	65	100	76	9
Women	65	103	80	8

Table 2 Caregivers' characteristics

	Formal caregivers				Informal caregivers		
Gender	Men		Women	TOT	Men	Women	TOT
	22		33	55	5	8	16
Role	Health workers		Volunteers		Family	Private care	
	23		32		13	0	
Age	0–40	41–60	60+	ND	0–40	41–60	60+
	12	22	16	5	0	2	11

In summary from these surveys three main elderly requirements emerged:

1. increasing the attention of doctors and caregivers about their health;
2. living at their own home maintaining their autonomy and independence but in a safe and secure context;
3. carrying out social activities to reduce their sense of loneliness and general negative feelings.

Looking to caregivers point of view, two main critical points were highlighted:

1. the willing to follow more carefully the elderly without increasing their burden and having a negative impact on their quality of life;
2. the hope that technologies could help them to monitor old persons, especially during their absence, and inform them in case of possible dangerous events.

Caregivers felt that assisting elderly persons was a heavy work and its burden influenced negatively their life. At the same time, they would like to follow more efficiently the older users and for this reason they believed that current technologies could help them to assist elderly relative.

2.2 Phase 1 (b) Elderly and Technology

Fifty-three (53) older adults drew up a specific questionnaire in order to investigate their attitude about current and future technology in the context of their home, daily life, free time and healthcare. This survey showed that participants used different kind of technology; particularly they were familiar with home appliance and electronic devices for everyday use. Interestingly elderly people involved in the study declared to have a mobile phone to dial and answer call, even if only few of them were able to send SMS. Also most of them used technological devices for monitoring their health status, such as blood pressure or glycaemia, but few of them used sensor devices for monitoring the domestic environment and even less used a tele-assistance system.

About attitude towards future technology, most of elderly people reported to be interested in using technology for monitoring their health status and for controlling remotely their home when they were both inside and outside of it. Also they thought that internet and informatics were useful to keep relationship with family and friends. At last the half of the sample emphasized the benefit of a tele-gym system in keeping a good fitness.

2.3 Phase 2 Design of Services and ICT System

The new socio-medical services based on AAL approach may allow elderly people to stay in their own homes for as long as possible delaying institutional care.

Many studies and technologies [11] that can assist older adults to continue living at home with safety and independence were developed, but few of them faced the cost of the new technologies. Suryadevare and Mukhopadhyay [12] developed a wireless sensor network integrated with ZigBee modules for monitoring ADLs of elderly living alone reporting that it was a "low cost" solution. Also Lotfi et al. [13] implemented a "low cost" sensor network including movement and door entry point sensors for supporting independent living of the elderly suffering from dementia. Tomita et al. [14] conducted a study with frail elderly people, who lived alone, based on a computer with Internet access and X10-based smart home technology while van Hoof et al. [15] used a prototype of the Unattended Autonomous Surveillance system to promote the ageing in place.

In RITA Project in order to meet elderly people's need a modular and pervasive wireless sensor network was developed. Furthermore an Ambient Intelligence (AmI) Infrastructure was created for the integration of information collected from all sensors.

The modules (Fig. 2) and services implemented are described below.

General health status monitoring: the commercial wearable monitoring system BioHarness™ (Zephyr) (Fig. 2b) was used to measure physiological parameters as heart rate, respiration rate, temperature, activity and posture. The system was composed by an electronics module and an ad-hoc wearable band. Data could be transmitted directly to a PC, via Bluetooth, and viewed in real-time or logged inside the electronic module and later uploaded and transmitted to clinicians. In this way physiological data could be analyzed and monitored remotely by doctors and so elderly people could receive more attention about their health by clinicians also beyond direct medical examinations.

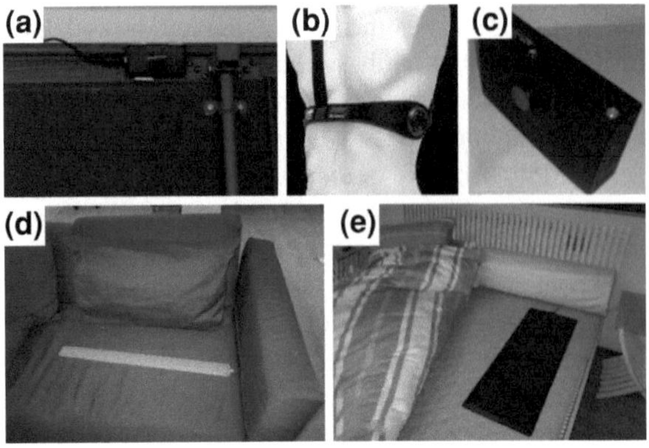

Fig. 2 Modules developed for RITA Project: **a** system for door monitoring; **b** commercial band monitoring physiological signals (BioHarness^{TM}, Zephyr); **c** localization and alarm module; **d** sofa monitoring (BOS, Recora); **e** bed monitoring (COS, Recora)

Indoor and outdoor localization: a wearable module (Fig. 2c), powered by a Lithium battery, was developed for localizing elderly and monitoring their motor activity in daily life. The system was composed by an inertial sensor (iNEMO-M1, STMicroelectronics) to monitor motor activity, a GPS receiver and GSM/GPRS module for outdoor localization (GM862-GPS, Telit), a ZigBee module (SPZB250M, STMicroelectronics) for indoor localization and the GPS and ZigBee antennas. The system was able to switch automatically from GPS (outdoor open space) to ZigBee module (indoor environment). This system could help caregivers to know always where old persons were especially during their absence. Furthermore the wearable module allowed elderly people, that have feelings of vulnerability and insecurity, to go out for their activities in safety because they knew that they could be localized with high precision in case of need.

Help request: the wearable module (Fig. 2c) was equipped with a red button that could be pushed by elderly people in case of need. Thanks to the GSM/GPRS module the caregivers received an SMS help request that informed if the elderly person was inside or outside home. In this way elderly people could maintain their autonomy and independence and the caregivers could be dismissed from a permanent assistance.

Domestic environment monitoring: a wireless indoor sensor network based on ZigBee technology was implemented and some environmental sensors were integrated into domestic spaces. In particular sensors acquired information about door opening/closing (Fig. 2a), user presence on sofa (Fig. 2d) and user presence in bed (Fig. 2e). All these modules were coordinated and managed by a PC, present inside the house and working as server. This sensor network allowed elderly people to live at home in security because unexpected environmental changes were detected and alert requests were promptly sent to the caregiver. Furthermore these sensors together with the localization module allowed to estimate elderly motor and static activities (for example, time in front TV) and recognise some activities as sleeping or napping. This information could be useful for caregivers to plan activities to maintain senior users active from the motor and social points of view.

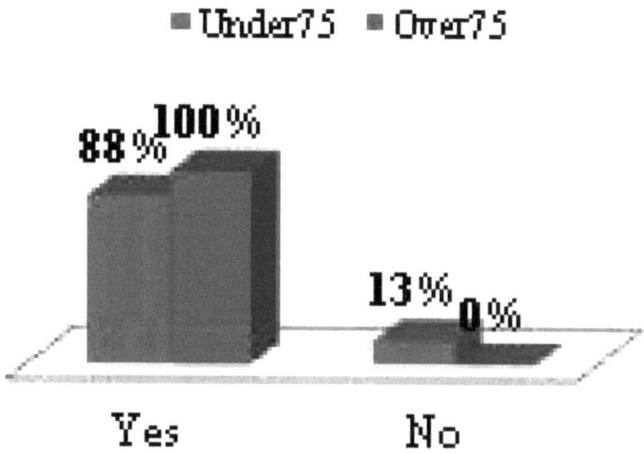

Fig. 3 Do you think that the use of RITA services and ICT system in everyday life could improve your quality of life?

Remote control interfaces: ad-hoc computer interfaces for the indoor environment monitoring and outdoor localization were developed. These components were conceived to allow caregivers to easily control elderly status. Also some information as help request, localization and door status were transmitted on caregiver mobile phone through SMS. Finally a mobile phone interface for physiological parameters was developed.

2.4 Phase 3 Experimentation of Services and ICT System

Two experimentation loops were conducted with elderly people and formal caregivers for testing RITA services and ICT system. The first loop was organized at DomoCasa Lab in Peccioli (Pisa, Italy) and involved 17 persons over 65 years old. During the test the elderly volunteers tested RITA sensors and services, described in the previous paragraphs, and then reported their feedback through a questionnaire. Data suggested that the minority of the sample thought that using the showed technology and services could have an excessively negative impact on their lifestyle, while the majority of them perceived the benefit of this technology use in everyday. At last elderly people reported that the perceived sense of safety and quality of life could improve (Fig. 3) thanks to the technology and services use.

The second loop was conducted at a residential care home for elderly with 9 formal caregivers and it unfolded with the same modalities of the first. All participants reported that the elderly people's quality of life could improve if this technology and services were implemented in everyday life. Also the caregivers thought that showed technology would increase the quality of job of socio-medical workers (Fig. 4a) and only a few of them suggested that all this would lead to a less personal relationship between caregiver and elderly person (Fig. 4b).

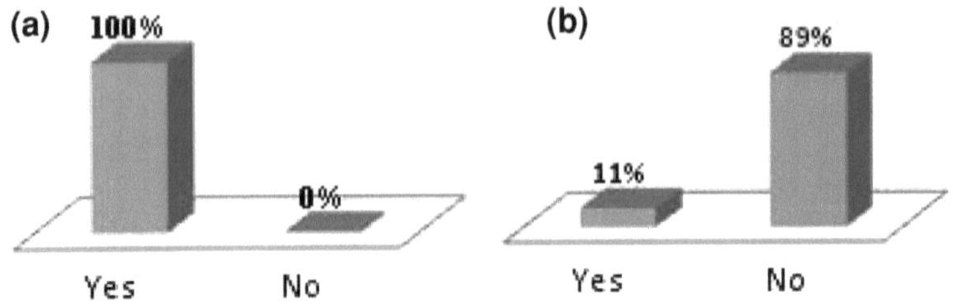

Fig. 4 **a** Do you think that the use of RITA services and ICT system in everyday life could increase the quality of job of socio-medical workers? **b** Do you think that the use of RITA services and ICT system could lead to a less personal relationship between caregiver and elderly person

3 Conclusions

As demonstrated in other studies conducted in Europe even in Tuscany, a new sustainable economic and welfare system to develop an aging in place model is indispensable.

The feasibility and utilization of new technologies for elderly is related to the perception of benefits due to their use, but the budget is also not a negligible aspect. Even if the cost-effectiveness was difficult to elicit, promoting ageing in place and facilitating older people to live independently in their homes should lead to more effective forms of care for elderly. In fact the new technologies, home and health monitoring, could reduce the burden on informal and formal caregivers. Thanks to a new socio-medical services based on AAL approach the informal caregivers could be always informed about their relative's condition reducing the necessity of their presence at home of their older relatives; this condition allows them to reduce costs due to time savings and avoidance of travelling and also to avoid to take frequently day or time off work for older persons' cares. Also the formal caregivers could be supported in their caring work offering a more focused and efficient job that could allow to caregivers to spent more time for a direct contact with elderly people.

In line with these objectives of efficient cares and with the AAL approach the RITA Project designed some possible socio-medical services, starting from the analysis of elderly persons' real needs, that are based on the use of "low-cost" ICT system. This research obtained positive and encouraging results, in fact, at the end of the project old people involved in the study said that they were interested in using the RITA technology and services in everyday life because this technology could improve the their perceived safety feeling and quality of life, not having a negative impact on their lifestyle and home environment. Moreover also the caregivers asserted that RITA services could improve the quality of job of socio-medical workers without damaging the personal relationship between caregiver and elderly assisted person.

In conclusion, RITA Project showed that a care model, in which elderly people live independently and safely in their own homes as long as possible, is feasible. Achieved results and feedbacks suggested that both elderly people and caregivers are favorably disposed to use new services and ICT systems in everyday life, if they perceive the benefits, related to the technology use, on the elderly people's quality of life and the quality of welfare services offered by social-medical workers and families.

Acknowledgments The RITA project has been supported by Regione Toscana under the POR CReO FESR 2007–2013 Programme, Asse 1 Attività 1.1 Linea di intervento D. Furthermore authors thank Società della Salute—Area pisana—which provided contacts with local associations and service providers, working with elderly people, allowing the involvement of a large number of elderly persons and caregivers in the project.

References

1. European Commission (DG ECFIN) and Economic Policy Committee (AWG): The 2012 Ageing Report: Economic and budgetary projections for the EU27 Member States (2010–2060). European Economy 2/2012 (2012)
2. Auerbach, D.I.: Will the NP workforce grow in the future?: new forecasts and implications for healthcare delivery. Med. Care **7**(7), 606–610 (2012)
3. Hooker, R.S., Cawley, J.F., Everett, C.M.: Predictive modeling the physician assistant supply: 2010–2025. Public Health Rep. **126**, 708–716 (2011)
4. Eurostat: European social statistics—edition 2013 (2013)
5. iNEMI Advancing manufacturing technology: iNEMI Technology Roadmap 2013 (2013)
6. Markets and Markets: European Smart Homes and Assisted, Living Market (2010–2015) (2011)
7. Belbachir, A.N, Litzenberger, M., Schraml, S., Hofstatter, M., Bauer, D., Schon, P., Humenberger, M., Sulzbachner, C., Lunden, T., Merne, M.: CARE: a dynamic stereo vision sensor system for fall detection. In: Proceedings of IEEE ISCAS2012 (2012). doi:10.1109/ISCAS.2012.6272141
8. Rumeau, P., Vigouroux, N., Boudet, B., Lepicard, G.: Home deployment of a doubt removal telecare service for cognitively impaired elderly people: a field deployment. In: Proceedings of IEEE CogInfoCom 2012 (2012). doi:10.1109/CogInfoCom.6422015
9. Resca, A., Spagnoletti, P.: ICT and elderly: a tool for supporting on line communities or off line communities?. In: Proceedings of Personal Health Records and patient-oriented infrastructures (2011)
10. Pommeranz, A., Fitrianie, S., Alers, H., Guldemond, N.: Care@Home: an integrated approach to care and social inclusion of elderly. In: AAL Forum 2012 (2012)
11. Morris, M.E. et al.: Smart-home technologies to assist older people to live well at home. J. Aging Sci. **1**, 1–9 (2013)
12. Suryadevara, N.K., Mukhopadhyay, S.C.: Wireless sensor network based home monitoring system for wellness determination of elderly. IEEE Sens. J. **12**, 1965–1972 (2012)
13. Lotfi, A., Langensiepen, C., Mahmoud, S.M., Akhlaghinia, M.J.: Smart homes for the elderly dementia sufferers: identification and prediction of abnormal behaviour. J. Ambient Intell. Humaniz. Comput. **3**(3), 205–218 (2012)
14. Tomita, M.R., Mann, W.C., Stanton, K., Tomita, A.D., Vidyalakshmi, S.: Use of currently available smart home technology by frail elders: process and outcomes. Top Geriatr. Rehabil. **23**, 24–34 (2007)
15. van Hoof, J., Kort, H.S., Rutten, P.G., Duijnstee, M.S.: Ageing-in-place with the use of ambient intelligence technology: perspectives of older users. Int. J. Med. Inform. **80**, 310–331 (2011)

Smart Technologies in Social Housing: Methodology and First Results of the HOST Project Experimentation Activities

Luigi Biocca, Paolo Casacci, Assunta D'Innocenzo, Davide Di Pasquale, Filippo Fabbri, Marialisa Manuzzi, Marco Padula, Nicolò Paraciani, Massimo Pistoia and Paolo L. Scala

Abstract The HOST project, funded by the Ambient Assisted Living Joint Programme, aims at the realization of ICT tools for social inclusion, that are specifically designed to be integrated in the context of social housing and to encourage direct access to a package of services by elderly and/or fragile people, residing in these structures. The project has been implementing test versions of these tools in three European countries—France, Italy and UK—with the aim of analysing the results and establishing a shared view about the features of this type of applications, as well as producing three prototypes that should be released on the market within 2 years after project's completion. Project objectives and initial results of the Italian experimentation phase are here presented.

L. Biocca · N. Paraciani (✉)
CNR-ITC Construction Technologies Institute, Area della Ricerca RM 1, Via Salaria Km 29, 300, 00015 Monterotondo Stazione, RM, Italy
e-mail: nicolo.paraciani@itc.cnr.it

L. Biocca
e-mail: luigi.biocca@itc.cnr.it

P. Casacci · M. Manuzzi
eResult s.r.l., Via Ferruzzi Serafino 167, 47521 Cesena, FC, Italy
e-mail: Paolo.Casacci@eresult.it

M. Manuzzi
e-mail: Marialisa.Manuzzi@eresult.it

A. D'Innocenzo
Associazione Abitare e Anziani, AeA, Via Nizza 154, 00198 Roma, Italy
e-mail: assunta_dinnocenzo@fastwebnet.it

D. D. Pasquale · M. Padula · P. L. Scala
CNR-ITC Construction Technologies Institute, Area della Ricerca MI 1, Via Bassini 15, 20133 Milano, Italy
e-mail: davide.dipasquale@itc.cnr.it

S. Longhi et al. (eds.), *Ambient Assisted Living*, DOI: 10.1007/978-3-319-01119-6_31,

1 Introduction

The housing environment can be viewed as an ideal framework for allowing elderly people to enhance their skills with—and approach to—Information and Communication Technologies (ICTs), by making use of dedicated digital technologies, with particular focus on their application in providing a wide set of innovative services that can be delivered at home, improving as a consequence the quality of life of users.

In this context, the European research project “HOST—Smart technologies for self-service to seniors in social housing” [1], funded by the Ambient Assisted Living (AAL) Joint Programme [2], aims at providing senior residents of social housing with a package of services delivered through the mediation of ICTs [3, 4].

The services forming this package, or framework, are specifically customized depending on the identified needs of a previously selected user sample, placing the aging person at the centre of the communication process (being at the same time provider and consumer of information) and favoring his/her participation to social life, while retaining autonomy and independence in daily life.

HOST involves three experimentation activities being carried out in three partner countries, namely France, Italy and UK. In this paper, we will focus on features and initial outcomes of the Italian case study.

2 Concept and Development of the ICT Platform

The conceptual development methodology of the platform is based on “User Centered Design” (UCD), whose purpose is to tailor the system around the user, employing a co-design procedure, which implies that designers, technicians and providers have to necessarily work in direct contact with end users in each design phase.

Selected kinds of services to be provided with the mediation of ICT devices were chosen in order to promote awareness of independent control among older consumers

M. Padula
e-mail: marco.padula@itc.cnr.it

P. L. Scala
e-mail: paolo.scala@itc.cnr.it

F. Fabbri
Université Paris-Sud, 15 Rue Georges Clemenceau, 91400 Orsay, France
e-mail: filippo.fabbri@u-psud.fr

M. Pistoia
BioResult s.r.l., Via Cerchia di San Giorgio, 145, 47521 Cesena, FC, Italy
e-mail: massimo.pistoia@bioresult.it

in selecting their own appropriate responses to personal requirements, always promoting communication with and between their service providers and the "circle of support", composed of family members and local services, both public and private [5].

A detailed list of services and features which are currently available for provision through the ICT platform developed and adopted in the Italian experimentation phase can be found in Sect. 2.1, Table 1.

The platform is constituted by two main parts: a client-side Android application that can be installed on tablets and is specifically conceived for the elderly users, and a server-side management platform, that is used both by project's technical staff and service providers and operators participating in the experimentation [6].

In Fig. 1 the system architecture is detailed.

It should be noted that the experimentation used only tablets, which were chosen by users as the preferred technology, rather than smart-TV/set-top-box and smartphones. In particular, smartphones were discarded because they were not well received by users, who found them uncomfortable due to the small touch-screen and icons.

2.1 The Android Application

The Android application is based on a simple (touch-screen) interface through which a user can browse through different categories of services and utilities provided by the system. By means of the application, users can, for example, place orders to request a home service delivery (see Table 1), indicating the time frame in which it should take place, or gain access to cultural or recreational features, as well as communicate with the people forming their own circle of support. This latter feature is, at present, the most developed and complete one and is frequently used by the elderly participants.

Contact details of people included within the circle of support are retrieved by the application through a synchronization with the central database server containing all user data, based on the user profile that the application itself is configured on. This means that each circle of support is individual to the single user and can be expanded or modified at any time, upon user's request. The two main communication features of the circle of support are: short text messages and Skype calls (voice and video calls facilitated by the application).

The User Interface (UI) is constituted by menus with large icons and symbols, as well as text descriptions, that should facilitate the identification of the desired functionality. In the previous version of the UI, menus could be browsed using left and right arrows, going back and forth through icons. The currently selected icon was placed in the middle and bordered in red. Inactive icons were smaller and bordered in gray (see Fig. 2a). In the current updated version, services and features are shown in the main menu and can be accessed by tapping on the respective icon (see Fig. 2b). In the case of services, tapping on the icon leads to a submenu which shows all service

Table 1 Services and features currently provided through the ICT system

Service/Feature	Description
Home delivery of medicines (or drugs)	Users can place an order through the app requesting the delivery of over the counter drugs, that they can select from a list (that can be customized). They can specify the date and time range in which they want the goods to be delivered
Home delivery of groceries	Users can place an order for the delivery of grocery items, that can be selected through the app from a list provided by local commercial facilities, specifying date and time range
Transportation by car/Accompanying for walks and shopping	Users can request transportation by car, which is also provided for being carried to doctors for medical examinations and appointments, or to be accompanied for walks/shopping
Company	Users can request the company of a volunteer for a certain period of time, which is specified in a way similar to the one used for deliveries
Help with managing operations/practices	Users can request help of a volunteer with managing their practices, like paying bills or rents, filling forms etc.
Booking cinema and/or theatre tickets/Entertainment	Users can book tickets for cinemas or theatres by means of links to websites that provide these facilities which are embedded in the application. They can also use applications for reading, browsing the Internet, playing games, etc., which were selected by the project team
Reminder	Gives the possibility to set a reminder, specifying a description, date and time and notification mode, that is: vocal message (that reads the description), sound notice (beeps), vibration of the device, or any combination of these
Circle of support	Users can use this feature to communicate with a customizable list of contacts, by sending text messages and performing Skype voice and video calls
Requested services	This feature enables users to see a navigable list of the services they have requested, checking their progress status
Service evaluation	Users can provide their feedback about the quality of the services being received, choosing between three scores (positive, sufficient, negative), with the possibility of writing remarks
Settings	Users can access settings in order to reset the database, in case they are experiencing synchronization problems
Home maintenance and management services	These services are planned to be implemented in the next experimentation phase. They will be provided with the cooperation of social housing operators

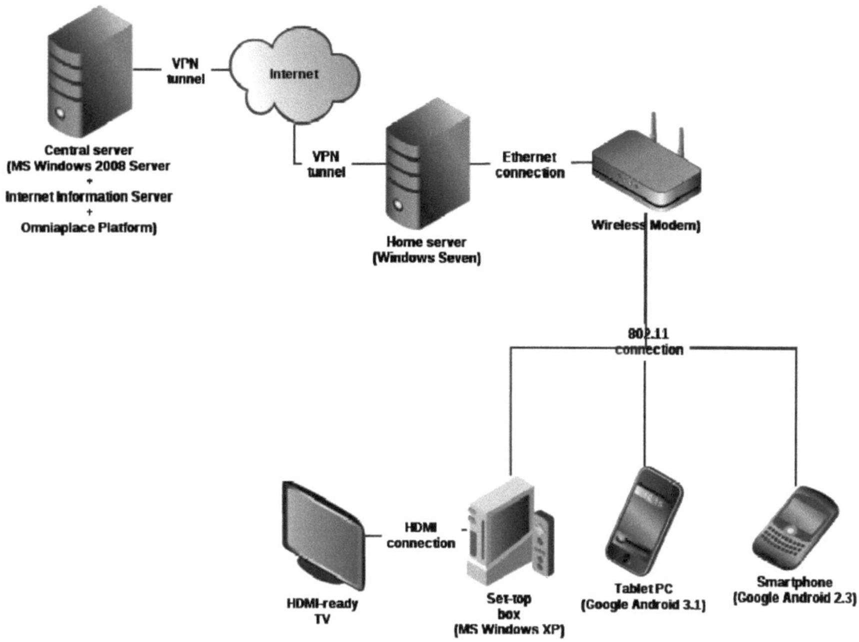

Fig. 1 System architecture

categories in one view, hence eliminating the need for arrows to browse them. These updates and changes of the UI were based on user feedback (see Sect. 4).

The interface also includes a voice synthesis engine, which is implemented to give a natural language feedback for some functionalities (in particular, for the reminder feature).

2.2 Central Server Application Interface

The server-side desktop platform gives technical staff the possibility to connect to the central server (through a web-based GUI, shown in Fig. 4), which is the core of the entire ICT system, in order to perform updates, manage user profiles and generally conduct maintenance, monitoring and support activities.

It also allows service providers to manage users' service requests, checking whether a service order has been placed, or if a delivery has been carried out properly, and responding to a request by specifying whether or not the preferences indicated by the user can be satisfied.

It is also possible for tutors to keep in contact with users by sending and receiving text messages from those belonging to the relevant circle of support. In some cases, before a service request can be processed by providers, a confirmation from tutor may be required, which they can give using this server-side platform.

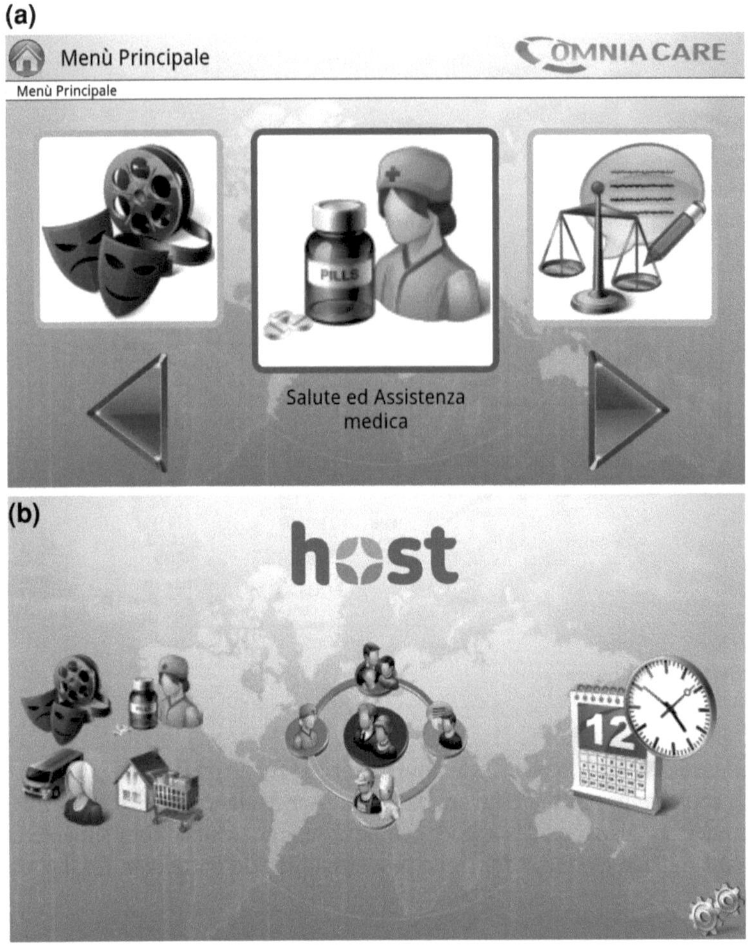

Fig. 2 **a** Main menu in the previous version (v. 1.24). **b** Main menu in the latest version (v. 1.28)

3 Experimentation Phases: Methodology

The project methodology consists in three main phases: co-design, development and implementation, experimentation.

The co-design phase was based on the organization of meetings and workshops, with the following objectives:

a. getting in touch with end users, managers responsible for housing and tutors/families;
b. identifying habits, behaviours, social relationships and final users degree of autonomy to formulate project requirements;

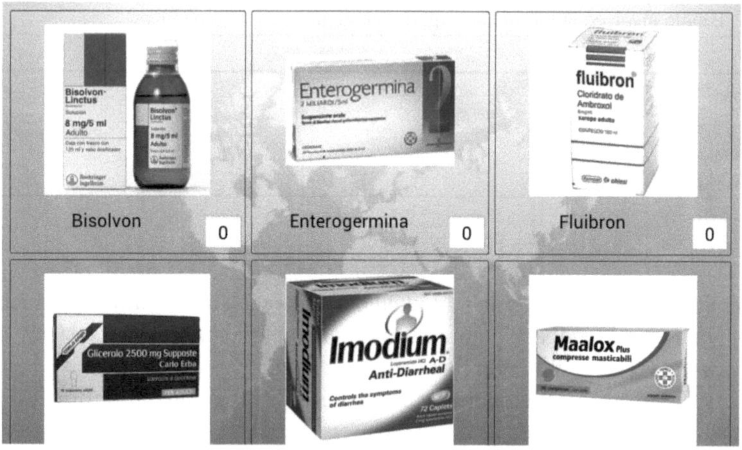

Fig. 3 An example of service request: home delivery of over the counter drugs

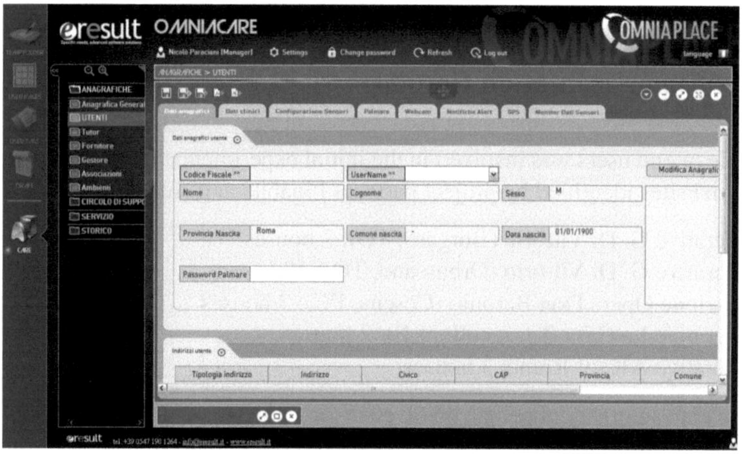

Fig. 4 Back end user interface: a view from users registry facility

c. focusing on the needs of people involved by means of individual interviews, based on a simplified questionnaire.

The results from the co-design phase were used to fine tune the system during the subsequent development and implementation phase, particularly in relation to the features of the client-side application hosted on tablet devices.

Finally, the actual experimentation phase started with a first installation step, needed to properly configure the tablet devices that were then given to users. A series of meetings with users were held in each trial site, where project staff carried out "first-level" tutoring activities in order to teach participants how to use basic functionalities of tablets and of the HOST platform.

Table 2 People involved in the Italian experimentation

Role	Number
End users	20
Tutors	6
Caregivers	6
Social workers	10

After this initial step, several individual meetings with users followed, which were aimed at collecting feedback coming from their experience with the system, as well as performing technical interventions when needed.

During the experimentation phase, which is still on-going, an informal tutoring service has been established, allowing the user to send messages requesting any kind of help directly through the communication facilities of the HOST platform (see Sect. 2).

4 First Experimentation Results

A sample of end users was involved in the initial experimentation phase, carried out in five trial sites identified by project partners Finabita and AeA [7]:

- Cooperative G. Di Vittorio (Borgaro, TO): 2 houses
- Cooperative G. Di Vittorio (Orbassano, TO): 10 houses
- Fondazione Opera Don Baronio (Cesena, FC): 3 houses
- Cooperative La Casa (Lizzanello, LE): 2 houses
- Cooperative AIC of Rome: 3 houses.

An overall summary of people involved in the Italian experimentation phase at the beginning is shown in Table 2: caregivers should be considered as both formal and informal, thus also including family members. Figure 5 shows instead the distribution of the initial user sample by age groupings.

It should be noted that during the course of the experimentation, some people left the project, mainly because of health problems or incompatibility with personal time schedules, and were replaced, where possible, by other users.

Currently, the initial results issued from this phase have shown that elderly users were rather attracted, at first approach, by new technologies, but were generally unsure about how to use them. What has been noted, in particular, is that they initially missed to see the practical usefulness of the tablet and of the HOST platform. This situation, however, improved in time through further tutoring activities. It was also found that some users had difficulties with the interface, since they were not able to easily determine which icon was currently selected, because they found it confusing, and also remarked that some font sizes were too small, especially those related to notifications shown by the application. This feedback gathered from users by project

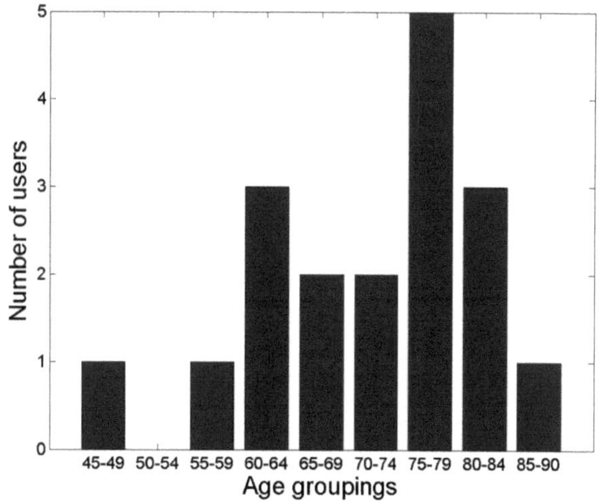

Fig. 5 Distribution of the user sample by age groupings

staff, led to further development of the application, with the release of a new version in late September 2013, whose UI reflects the modifications suggested or requested by users to make it easier (see Figs. 2b and 3), and is currently in use in a new experimental phase.

At present, the participants are mostly using the communication facilities of the circle of support and sometimes requesting home deliveries.

These activities are being monitored periodically in order to analyze the gradual evolution of the experimentation. For the next future, it is expected that more services will be available for provision and they will be therefore tested, with the involvement of both users and social volunteers. This will lead to the acquisition of a larger number of data about the actual use of the system, which will serve to better establish its effectiveness.

References

1. HOST Project Website: http://www.host-aal.eu/cms. Accessed 1 June 2013
2. Ambient Assisted Living: AAL FORUM 2011 Lecce (Active Ageing: Innovations, Market, and EU Initiatives). Italy, 26–28 Sept 2011
3. Bagüés, M.I., Bermúdez, J., Illarramendi, A., Tablado, A., Goñi, A.: Using ontologies in the development of an innovating system for elderly people tele-assistance. In: Proceedings of the 2003 International Conference on Ontologies, Databases and Applications of Semantics (ODBASE'03), 3–7 November 2013. Catania, Italy (2003)
4. Chen, H., Perich, F., Finin, T., Joshi, A.: SOUPA: standard ontology for ubiquitous and pervasive applications. In: Proceedings of the 1st Annual International Conference on Mobile and Ubiquitous Systems (MobiQuitous2004) 22–25 August 2004. Cambridge, MA, USA (2004)

5. Biocca, L., Morini, A., Padula, M., Paraciani, N.: HOST: smart technologies for personalized services to older people living in social housing, L'Edilizia, Speciale ITC-CNR 2011, extracted from structural, no 171. pp. 123–126 (2011)
6. Biocca, L., Morini, A., Padula, M., Paraciani, N., Scala, P.: HOST, smart technologies for personalized services to support independence of older tenants of social housing. In: BRAID Final Conference 28 May 2012. Prague (2012)
7. Casacci, P., D'Innocenzo, A., Fabbri, F., Pistoia, M.: HOST: smart technologies for self-service to seniors in social housing. In: Foritaal: Ambient Assisted Living, IV Forum Italiano, 18–20 June 2012. Parma (2012)

Smart Object for AAL: A Review

R. Bevilacqua, Silvia Ceccacci, M. Germani, M. Iualè, M. Mengoni and A. Papetti

Abstract In the last years, some attempts have been made to explore the use of smart objects, with the purpose of monitoring well-being and supporting independent living. However, an inventory of characteristics of smart assistive devices currently available on the market is still lacking. The aim of this study is to provide an overview of such products and in particular to analyse their functionalities in order to understand if they can be considered inclusive and if they effectively match the needs of elderly. This review can be useful for researchers and designers of Ambient Intelligent Environments, to realise where efforts should be focused in order to match the demand from the elderly people's side.

R. Bevilacqua
Direzione Scientifica, Polo Scientifico-Tecnologico, INRCA, Via Santa Margherita, 5, 60124 Ancona, Italy
e-mail: r.bevilacqua@inrca.it

S. Ceccacci (✉) · M. Germani · M. Iualè · M. Mengoni · A. Papetti
Dipartimento di Ingegneria Industriale e Scienze Matematiche, Università Politecnica delle Marche, UNIVPM, Via Brecce Bianche, 60131 Ancona, Italy
e-mail: s.ceccacci@univpm.it

M. Germani
e-mail: m.germai@univpm.it

M. Iualè
e-mail: m.iuale@univpm.it

M. Mengoni
e-mail: m.mengoni@univpm.it

A. Papetti
e-mail: a.papetti@univpm.it

S. Longhi et al. (eds.), *Ambient Assisted Living*, DOI: 10.1007/978-3-319-01119-6_32,

1 Introduction

Ambient Assisted Living is becoming one of the most important research fields on ICT-based products. Its general goal is to enable people with specific demands (e.g., disabled or elderly) to live longer in their preferred environment, by increasing their autonomy and self-confidence, by monitoring and providing care and by enhancing the security and saving resources.

In the last years, some attempts have been made to explore the use of intelligent objects, or *smart* objects, with the purpose of monitoring well-being and supporting independent living: the idea is that a collection of smart artifacts can be added to almost any physical environment in order to make it intelligent.

The notion of Smart Object (SO) describes technology-enhanced everyday objects which are equipped with sensors, memory and communication capabilities [1, 2]. Hence, they are able to capture information about their surrounding, communicate with each other and react according to previously defined rules [3], and act on behalf of user needs. Through the capability to interact with humans directly, they can help users to accomplish their tasks in new, intuitive ways [4].

Smart objects raise unique challenges and opportunities for designing interaction with intelligent systems, coping with the mostly limited interaction capabilities, exploring context information to provide a more natural interaction, helping the user to understand the behavior and capabilities of the objects.

In a broader sense, SO are able to help people in participating in their environment through adaptation, accessibility and communication. This also stresses their role in offering assistance to the elderly people's independent living. As applied to people with disabilities, a smart object can be considered as an Assistive Technology (AT), in line with the definition of AT given by Cowan and Turner-Smith [5]: "any device or system that allows an individual to perform a task that they would otherwise be unable to do, or increases the ease and safety with which the task can be performed".

Today a lot of smart artefacts are commercially available but an inventory of characteristics and functionalities of these products is lacking as well as an evaluation of their applicability to support independent living of elderly people.

Such an inventory and overview may be useful to researchers and designers of Ambient Intelligent Environments, to understand where efforts should be focused in order to match the demand from the older people side. In particular this review focuses on domestic smart products and aims:

- to provide insight into the state of art of smart objects, in order to analyse which kind of information are provided by smart products and if they provide context aware applications that are able to support elderly people's independent living.
- to understand if the available products can be considered inclusive and if they effectively match the specific needs of the older population with their functionalities.

2 The Search Strategy

In order to create an inventory of assistive smart devices commercially available, at first we have conducted a search into the network by using the keyword "smart object". This preliminary search highlighted a large improper use of the term "smart object" making it loose the initial connotation to become a marketing means (e.g., any innovative product from a technical, esthetic and communicative point of view is called "smart").

The enormous quantity of available information made the detection of proper smart objects very complicated. Consequently, as in this context the most important principle is to be specific, therefore several search strategies have been explored in order to select the most appropriate one.

The first step consisted in searching and consulting of the list of domestic objects (e.g., clothes, scales, toilets etc.) by which people come in touch every day. Also assistive products have been included: in particular we considered the assistive device typologies collected by McCreadie and Tinker [6]. As a second step, keywords as "smart", "interactive" and "inclusive" have been selected, starting from the smart object definition. Objects and keywords have been used together for the search, exploiting several combinations and search engines. Finally, both search engine tips and synonyms for the previous words have been considered.

For each found object, it has been verified if it was compliant with the requirements. The producer company site has been investigated in order to select other products related to the search topics. At the end, the search strategy resulted in 172 relevant smart devices.

3 The Classification Method

Each parameter, monitored by the collected smart objects, has been identified, analysed and grouped with the similar ones. Thus, four categories of data have been defined as follows:

- *Vital signs monitoring*: it includes parameters that allow to check the user's level of physical functioning. Vital signs (i.e., heart and breathing rate, core and skin temperature, blood pressure etc.) belongs to this category.
- *Lifestyle monitoring*: it comprises useful data to "observe" the user's daily behaviour and activity (i.e., eating habits, hygiene care, physical exercise etc.). Also data provided to the users by applications associated with smart objects, which aim to support the user's lifestyle improvement by setting goals and providing useful advises, have also been considered.
- *Mobility and falls monitoring*: it refers to parameters related to the people's indoor and outdoor mobility and safety. For instance, data collected by video cameras and falls detectors have been associated to this category.
- *Domestic environment monitoring*: this section contains data provided by sensors for the monitoring of the domestic environment, such as temperature, humidity, CO_2 level, etc.

In this way, it was possible to associate all the smart objects to the properly functional category, according to the specific data they provide.

Table 1 summarizes the results of the classification, grouping them in homogenous classes of devices. In particular objects have been classified in: Clothes (e.g., sweater, vest, etc.), *Clothing Accessories* (e.g., shoes, elastic bands, bracelets, etc.), *Dishes* (e.g., forks, glasses, etc.), *Household Devices* (e.g., video-cameras, decorative objects, etc.), *Medical Devices* (e.g., glucometer, sphygmomanometer, etc.), *Personal Care Products* (e.g., toothbrushes, toilet, etc.). Moreover, an estimation of the quantities of intelligent devices of each category has been reported.

The results show how a specific functional category is generally covered by the smart objects pertaining to a same class. Therefore, it is important to highlight the lack of transversality of the available solutions. In particular, vital signs are monitored almost exclusively by apparel and medical devices. On the other hand, Household Devices are focused in monitoring users mobility and safety and/or domestic environment (Fig. 1).

By analysing Table 1, it is possible to observe that in the context of each functional category some data (e.g., for lifestyle monitoring: water consumption, fork servings frequency, brushing duration etc.) are monitored by only one or few objects.

This can be justified considering that these data are most related to particular product functionalities which aim to support special pathologies/needs. Otherwise, most popular data (e.g., for lifestyle monitoring: weight, calories burned, body mass index, covered distance etc.) are provided by the majority of the devices.

The analysis has focused also on other aspects such as technical specifications (connectivity and compatible devices) and market information (availability and price), which are not reported in this paper. These factors are not negligible when the purpose is to define an appropriate set of smart devices to support people with specific fragilities, pathologies or needs.

4 Results

In order to evaluate the efficacy of the collected smart objects in supporting elderly people's independent living, we analysed all the functional domains of activities that can be performed in a domestic environment. It could be said that a proper inclusive environment is able to support all these activities. Firstly, the activities' domains defined in the International Classification of Functioning, Disability and Health (ICF) [7] were considered, and in particular: learning and applying knowledge (d1), general tasks and demands (d2), communication (d3), mobility (d4), self-care (d5), domestic life (d6).

The research has shown that the majority of smart objects provides support to the activities related to "Looking after one's health" (d 5700) and in particular they are focused on ensuring physical comfort, health and physical well-being by monitoring specific parameters, for the achievement of a better lifestyle.

Table 1 Correlation between categories of smart objects and functionality respectively supported

Categories functions	Data provided	Smart objects					
		Wearable		Not wearable			
		Clothes	Clothing accessories	Dishes	Household devices	Medical devices	Personal care products
Vital sings monitoring	Ankle brachial index (ABI)					•	
	Blood glucose					•••	•
	Breathing rate	•				••	
	Cardiac output (CO)					•	
	Core body temperature						
	Diastolic blood pressure					•••	•
	Electrocardiography (ECG)					•••	
	Electroencephalography (EEG)					•	
	Electromyography (EMG)					••	
	Galvanic skin response (GSR)					••	
	Heart rate	•••	••			•••	•
	Heart rate variability	••				••	
	Hematocrit					•	
	Hemoglobin					•	
	Oxygen saturation (SpO_2)		•			•••	
	Perfusion Index					•	
	Phonocardiogram					•	
	Photoplethysmography (PPG)					•	
	R-R Interval					•	
	Skin temperature		•			•••	

(Continued)

Table 1 (Continued)

Categories functions	Data provided	Smart objects					
		Wearable		Not wearable			
		Clothes	Clothing accessories	Dishes	Household devices	Medical devices	Personal care products
	Spirogram					•••	
	Systolic blood pressure					•••	•
	Urine analysis						•
Lifestyle monitoring	Activity pace						
	Activity speed					•	
	Activity time		•••				
	Activity/Movement						
	Altitude					•	
	Awoke numbering						
	Body fat						•••
	Body mass index (BMI)						•••
	Body water						••
	Body weight distribution						•
	Bone mass						••
	Brushing duration						•
	Brushing frequency						•
	Calories burned	••	•••				
	Covered distance	•	•••			•	
	Daily calorie intake (DCI)						••
	Daily medicines missing					••	
	Fork servings frequency			•			
	Fork servings interval			•			

(Continued)

Table 1 (Continued)

Categories functions	Data provided	Smart objects					
		Wearable		Not wearable			
		Clothes	Clothing accessories	Dishes	Household devices	Medical devices	Personal care products
	Last sip time			•			
	Lean mass						•
	Meal duration			•			
	Meal time			•			
	Muscle mass						••
	Posture	•	•				
	Sleep quality		•••			••	
	Sleep time		•••			•	
	Stairs climbed numbering		•				
	Steps numbering	•	•••			•	
	Visceral fat rating						•
	Water consumption			•			
	Weight						•••
Mobility and falls monitoring	Audio				•••		
	Fall detection					••	
	Motion detection				•••		
	Position (GPS)	•	•				
	Video				•••		
Domestic environment monitoring	Fertilizer				•		
	Room air quality (CO_2 level)				•		•
	Room barometric pressure				•	•	

(Continued)

Table 1 (Continued)

Categories functions	Data provided	Smart objects					
		Wearable		Not wearable			
		Clothes	Clothing acces-sories	Dishes	Household devices	Medical devices	Personal care products
	Room humidity				••	•	
	Room lighting		•		•		
	Room noise level				•		
	Room temperature				•••	•	•
	Soil moisture				•		

1 bullet means "low covered"
2 bullet means "medium covered"
3 bullet means "high covered"

Fig. 1 Smart objects categories

Self-monitoring is not a new concept in healthcare. Indeed, there are a lot of products on the market for the measurement of the vital signs, but the user's data cannot be easily exploited due to their storage just inside the device. On the other hand, smart products allow measuring, tracking and sharing data with family, caregivers and/or doctors. Furthermore, the relative apps aim at empowering users to better understand their health and manage it more effectively. It could be said that they support the health literacy of the older adults. In addition, some of them allow tracking multiple users across different devices.

As far as the several functioning categories concerned, these objects can help to maintain a balanced diet and an appropriate level of physical activity (d 5701) and by monitoring eating habits and daily routines and by measuring physical parameters such as weight, body mass index and body water. The jointed analysis of these factors allows tracking the user's behavior and suggesting him/her appropriate guidelines for maintaining a good lifestyle (i.e., right fork servings frequency, need to drink, goal to reach about the physical exercise etc.).

Maintaining one's health (d 5702) and looking after body parts are other important functions supported by intelligent devices. Indeed, they can support regular physical examinations, for example, by simply wearing smart clothes and/or accessories and/or using smart toilets. In this way, appropriate algorithms analyze the parameters and generate alerts only when necessary, in order to prevent risks and reduce the need of professional assistance.

Other products aim at ensuring the user physical comfort (d 5700), supporting the domestic environmental care and informing the users, for example, about the current temperature or lighting and suggesting the most appropriate ones.

Finally, outdoor and indoor mobility can be monitored respectively through GPS and video cameras, performing also the fall detection functionality.

As a second step, it was decided to use a different tool for the evaluation of the applicability of the selected smart objects, specifically used in the field of gerontology.

The Instrumental Activities of Daily Living [8] was selected, in order to deeply understand which SO can be used to support the most important daily activities for elderly people. In particular, the Lawton Instrumental Activities of Daily Living Scale (IADL) is an appropriate instrument to assess independent living skills. These skills are considered more complex than the basic activities of daily living as measured by the Katz Index of ADLs [9]. For this reason, IADL scale was preferred respecting to the Katz Index of ADLs, representing the most useful tool for identifying how a person is currently functioning but also for identifying his/her improvement over the time, that can be gained, for example, through the use of a technological artifact.

There are 8 domains of function measured with the Lawton IADL scale [10]. Among them, five were chosen, as they refer to the domestic environment: food preparation, housekeeping, laundering, ability to use the phone, responsibility of the own medications. Persons are scored according to their level of functioning in each category.

Just the category of responsibility of the own medication is supported by the considered SO, for what concerning for example the measurement of vital signs or reminding to take medicines. The ability to use the phone is the most controversial domain: actually there are not assistive devices to support this activity, even if smart phones and tablets are pervasive in the market of technology and entertainment. These technologies should be considered as pioneers in the inclusiveness field as they represent a potential hub of smart technologies. This is confirmed by the large presence of apps, which are provided with SO.

5 Conclusion

In an ontological point of view, much evidence show that inclusiveness is a prerequisite of any assistive technology artefacts, aimed at cooperating with people, especially if they have particular needs to face, as the elderly.

To understand if the available products can be considered inclusive and if they effectively match the needs of the elderly, we have started from the analysis of the activities, that it is expected an assistive environment should be able to support.

Using ICF, we have noticed that the available SO are not able to support all the activities, most of all the ones ascribed to the cognitive domain, such as for example, learning and applying knowledge. Nowadays, there are a lot of software programmes and apps to support these activities, already available in the market. It seems that the implementation of more complex software architectures inside the smart objects should be stressed, in order to have a new generation of products, that can be considered more intelligent and intuitive. For example, this can strongly support cognitive processes like work memory, attention and reasoning. This, can be a key issue for assuring the success of the future generation of SO.

From the analysis of the SO, using the IADL scale, it was learnt that there are not SO for truly supporting the older people in carrying on the activities reported in the IADL. The monitoring activity is the most common reported functionality of the SO, performed at light level. The reason of this difficulty in supporting the IADL activities, should be found in the interoperability of the SO, which is an issue amply treated in literature [11]. In fact, the majority of the SO includes and/or requires specific apps or software to interpret the monitored information and to provide feedbacks to the users. This requirement may represents a limit when these objects have to be interconnected each other, to create an integrated assistive environment targeted on the specific user's needs. Indeed, it is necessary to develop an "intelligent management tool", essential to collect, elaborate and exploit the data generated by several objects.

Enhancing the SO interoperability can lead to the support of more complex activities, such as IADL. In line with this, also the interoperability can be considered a prerequisite of the system inclusiveness, allowing also the system flexibility and customization.

Currently, it could be said that the SO are not inclusive, because (a) they are not able to answer to a specific problem, but just to collect information on some parameters; (b) they are not planned to be integrated in more complex systems, that can describe the extended concept of "environment".

Nowadays, the SO can be divided in system-oriented, importunate smartness and people-oriented, empowering smartness [12]. In the first case, smart objects can take certain self-directed actions based on previously collected information, so that the space would be active, in many cases even proactive. In the other case, smart objects empower users to make decisions and take mature and responsible actions.

Through the implementation of the software architectures of the SO, also inside more complex platform, this subdivision can be overcome, creating a new generation of products, that can be context-aware oriented, and finally becoming inclusive.

References

1. Ferguson, G.T.: Have your objects call my objects. Harvard Bus. Rev. **80**(6), 138–143 (2003)
2. Gellersen, H.W., Schmidt, A., Beigl, M.: Adding some smartness to devices and everyday things. In: Proceedings of WMCSA 2000–3rd IEEE Workshop on Mobile Computing Systems and Applications, December 2000. IEEE Computer Society, Monterrey, USA (2000)
3. Ziefle, M., Rocker, C.: Acceptance of pervasive healthcare systems: a comparison of different implementation concepts, pervasive computing technologies for healthcare (PervasiveHealth). In: 4th International Conference on-NO PERMISSIONS, 22–25 March 2010, pp. 1–6 (2010)
4. Bohn, J., Coroama, V., Langheinrich, M., Mattern, F., Rohs, M.: Living in a world of smart everyday objects-social, economic, and ethical implications, human and ecological risk assessment, vol. 10, pp. 763–785. Taylor & Francis, New York (2004)
5. Cowan, D., Turner-Smith, A.R.: The role of assistive technology in alternative models of care for older people. In: Tinker, A., et al. (eds.) Alternative Models of Care for Older People research, vol. 2, pp. 325–346. The Royal Commission on Long Term Care, The Stationery Office, London (1999)
6. McCreadie, C., Tinker, A.: The acceptability of assistive technology to older people. Ageing Soc. **25**, 91–110 (2005)
7. World Health Organization: International classification of functioning, disability and health. World Health Organization, Geneva (2001)
8. Lawton, M.P., Brody, E.M.: Assessment of older people: self-maintaining and instrumental activities of daily living. Gerontologist **9**(3), 179–186 (1969)
9. Katz, S., Down, T.D., Cash, H.R., et al.: Progress in the development of the index of ADL. Gerontologist **10**, 20–30 (1970)
10. Lawton, M.P., Moss, M., Fulcomer, M., Kleban, M.H.: A research and service oriented multi-level assessment instrument. J. Gerontol. **37**, 91–99 (1982)
11. Perumal, T., Ramli, A.R., Leong, C.Y., Mansor, S., Samsudin, K.: Interoperability for smart home environment using web services. Int. J. Smart Home **2**(4), 1–14 (2008)
12. Streitz, N., et al.: Designing smart artifacts for smart environments. Computer **38**(3), 41–49 (2005)

Living in the Living Lab! Adapting Two Model Domotic Apartments for Experimentation in Autonomous Living in a Context of Residential Use

Massimiliano Malavasi, Roberta Agusto, F. Marcella Ioele, Stefano Martinuzzi, Maria Rosaria Motolese and Matteo Rimondini

Abstract The development of Ambient Assisted Living (AAL) is clearly heavily dependent on a capacity to respond to the needs of potential users. Research and development based on User Centered Design methodologies must be accompanied by testing with users in contexts which are as close as possible to those encountered in everyday life. The whole process should conclude with an adequate means of measuring the outcomes of experimental solutions. The characteristics of the place of experimentation and of the methodologies employed thus appear to be particularly critical. This article describes a project path involving the realization and adaptation of two experimental domotic apartments (and related methodologies of use) for a Living Lab dedicated to experimentation in the use of assistive technologies and AAL. The two apartments concerned are to be found in the Corte Roncati complex in Bologna.

M. Malavasi (✉) · R. Agusto · F. M. Ioele · M. Rimondini
Centro Regionale Ausili, Bologna, Italy
e-mail: mmalavasi@ausilioteca.org

R. Agusto
e-mail: ragusto@ausilioteca.org

F. M. Ioele
e-mail: fioele@ausilioteca.org

M. Rimondini
e-mail: mrimondini@ausilioteca.org

S. Martinuzzi · M. R. Motolese
CAAD Bologna, Bologna, Italy
e-mail: smartinuzzi@ausilioteca.org

M. R. Motolese
e-mail: mmotolese@ausilioteca.org

S. Longhi et al. (eds.), *Ambient Assisted Living*, DOI: 10.1007/978-3-319-01119-6_33,

1 Experimenting with AAL and Quality of Life Solutions in Home Environments

Research and development based on User Centred Design, experimentation in contexts that are as close as possible to real life and real users, and the development of adequate methodologies for the measurement of outcomes could all prove to be crucial factors in the development of the AAL (Ambient Assisted Living) sector, a sector still relatively immature in some respects.

These same aspects are cited as extremely critical at the international level too [1–3] and it is thus imperative that resources and activities be dedicated to their development. Experimenting with innovative technologies in everyday life environments and measuring the outcomes of interventions is, however, an extremely complex business and one that requires preliminary and ongoing evaluation of many contexts. The environments used for this kind of experimentation should therefore be designed with extreme care, in accordance with adequate methodologies.

2 Corte Roncati's Experimental Domotic Apartments

The Corte Roncati Regional Technological Center in Bologna, which has been operative since 2007, was set up in order to bring together under one roof a number of highly specialized centers concerned with neuro-motor and cognitive disabilities and with finding ways of promoting the autonomy and security of the disabled.

These centres are supported by a number of operative structures, among them two experimental domotic apartments (hereafter, ADS 1 and 2). They were designed by an interdisciplinary team drawing on the participation of all the centres. The primary object in creating the ADSs was to provide a means of responding to the various needs of people with disabilities or difficulties associated with frailty.

ADS 1 is equipped with solutions aimed above all at people with severe motor disabilities, and is designed to promote the greatest possible degree of autonomy. ADS 2 is rather more specialized. It is designed for the needs of people with reduced sensorimotor/cognitive efficiency or with mild motor disabilities, including situations of difficulty associated with frailty and old age.

Both were equipped with a technological infrastructure that can with justification be called state of the art even when compared to the latest solutions at an international level. In order to ensure users complete control and conditions of guaranteed safety, even in cases of extremely severe disabilities, the designers opted for an open expandable domotic architecture which allows for the incorporation of a range of different solutions, especially with regard to interfaces between the user and the environment. The user can in fact use various methods to access the building's functions: buttons with ergonomic features, traditional or high accessibility remote controls, voice control, remote control via a specific module integrated in the wheelchair, or portable devices such as multifunction tablets and smartphones. Among the built in

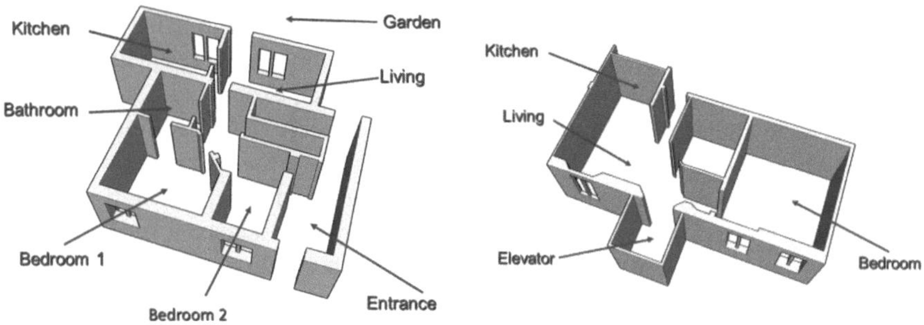

Fig. 1 3D Plans of the two ADS of Corte Roncati

functions there are various types of specialized motors designed to increase accessibility, systems for the automatic control of lighting and air conditioning, access rfid units which allow for hands-free entrance, remote controlled sound systems and video surveillance, solutions for the safety of users and systems that facilitate communication both inside and outside the home. The system of real-time supervision can handle complex scenarios and can automatically make decisions linked to comfort and safety, as well as create functional profiles based on specific needs.

In order to ensure the full enjoyment of the apartments on the part of the user and care giver in conditions of maximum autonomy and safety, a complete set of furnishings, solutions for accessibility and aids are incorporated with the technological infrastructure (Fig. 1).

3 A Tool to Support the Nascent "Ausilioteca Living Lab" and the Pursuit of Autonomy Backed by Local Services

The apartments were initially expected to serve as a permanent exhibition of AT solutions, as an instrument to support educational and training activities, as a place where people with disabilities and their families/caregivers could try out solutions and finally as a test bed for new assistive technologies and for AAL.

Despite being equipped with all the features necessary for everyday life, it was not expected that the ADSs would be occupied on a residential basis for 24 h and testing activities were limited to daylight hours. It was only in 2012, after 5 years of use on a daily basis, that the possibility of extending its operational role with new types of activities was taken up.

This was triggered by a desire to develop User Centred Design activities on an ongoing basis and to create a permanent community of users and professionals who could work side by side utilizing the new but well validated Living Lab approach.

A further factor was a realization that emerged in the course of trying to better integrate Corte Roncati's specialised centres with those working to provide support for people with disabilities in their quest for greater autonomy at regional level: namely, that the achievement of greater autonomy could be furthered by residential experience. Obviously, in order to maximize the utility of such experiences, it is essential to trial as wide a selection of aids and solutions as possible. It hence seemed evident that the two ADSs would provide the best possible context for such experimentation. In keeping with the principles that had guided their initial design, a multidisciplinary participatory process was activated in order to improve and adapt the facilities of the apartments and to define the procedures and protocols for their use.

4 The Process of Adaptation

4.1 Methodology

Among other things, the process of adaptation was guided by the International Classification of Functioning, Disability and Health (ICF) [4] and by Design For All [5]. According to the bio-psycho-social approach proposed in particular by the ICF, disability can be defined as a multidimensional phenomenon resulting from the interaction between the person and the physical and social environment [6]. In the light of these considerations and in order to empower the person with disabilities in his/her interactions with the physical environment, we looked for products, even in the mainstream non-specialist market, that matched our criteria of accessible, ergonomic design and which were at one and the same time aesthetically pleasing, useful and affordable (many products are not eligible for National Health System subsidies).

4.2 An Accessible Domestic Environment

The ADS apartments in Corte Roncati are the product of the restructing of part of what used to be the Roncati psychiatric hospital in the centre of Bologna.

In designing the apartments, windows and the distribution of the inner space were adapted to the constraints of the pre-existing architecture. Inner spaces respect the position of load-bearing walls which thus defined the dimensions of the new environments. In the case of ADS 1, the front door opens on to a living area which in turn provides access to a kitchen area; a central space provides access to the bathroom and bedroom zones, which are directly connected by a door that allows for the passage a mobile overhead hoist for movement from the bed to the sanitary facilities (toilet and shower). There is also accommodation for a carer. In the case of ADS 2, there is an open space living area that includes a kitchen; a small hallway provides access to the bathroom and to an adjacent bedroom.

Spaces in the ADS apartments were designed to provide enough room to ensure access to all with convenient circulation routes and areas of maneuver for wheelchair users. Interior fixtures were chosen for maximum ease of use with swing, sliding or motorized doors preferred. The same went for the exterior: motorized or easily opened doors were chosen which could be controlled by the home automation system. Front doors lock electronically and can be opened using a rfid badge. In the case of ADS 1 the front door was motorized.

The same prioritizing of accessibility, ease of use and versatility guided the choice of sanitary fittings in the bathrooms (wall-mounted toilets, washbasins accessible from a seated position, toilets with a built-in bidet function, adjustable showers for personal hygiene, floor-level showers, etc.), as well as of furniture and accessories, encouraging the choice of products and components with immediately recognizable functions:

- in the bathroom (mirrors flush with the level of the wash basin, showers with double curtains for use both sitting and standing, easy to move cabinets on wheels, etc.);
- in the bedroom (electronically controlled bed articulated in three places and adjustable in height, mobile hoist with IR remote control which reaches the bathroom area, walk-in wardrobe with variable heights for hanging clothes);
- in the kitchen (motorized cupboards, height-adjustable work surfaces; mobile wall-mounted ovens with foldaway doors, fridges and freezers with drawers, wall-hung storage units, accessible sinks including a "portable" one, accessible hotplates and faucets, pull-out pantry containers, trolleys, accessible table/work surface, hinged to allow for repositioning).

The furnishing accessories for the ADS apartments were chosen from those available in the mainstream market or from among available aids and innovative products. They fell into the following broad categories: products which allowed improved handling and grip, object stabilizers, facilitators for the activities of drinking and eating and "ergonomic" objects for leisure and domestic activities (Fig. 2).

4.3 An "Adaptable" Technology

The technologies used in the ADS apartments, based on open standards, have facilitated the addition of new functions and the expansion of existing ones (Fig. 3).

ADS 1, originally designed to accommodate the activities of a single user with severe disabilities, was redesigned so that it could be used simultaneously by two users accompanied by an assistant. The domotic functions were reprogrammed so as to make the principal environments multifunctional (for example the living area can now become a bedroom). For the same reasons, automatic functions (such as continuously regulated lighting activated by presence sensors and environmental brightness sensors) were modified to make it possible for users to deactivate them easily. New sub-scenarios were created which permitted independent control of ADS functions since users may be engaged in activities in different rooms.

Fewer changes were necessary in the case of ADS 2 as it had already been designed to host two people (e.g. an elderly couple). Here too it was to necessary to reprogram

Fig. 2 Interior views of the two ADS

the domotic functions of the day area so that it could become a sleeping area for a carer during the night.

In both units, the duration of automatically activated lighting in the bathrooms was extended in order to ensure that the lighting was sufficient for hygiene procedures; the positioning and performance of the sensors were found to be fundamental factors with regard to detecting movement. The two ADS apartments were provided with the means to communicate with each other. The phones provided are of various types, including a cordless with an accessible keyboard, a single radio button activated phone (with scan mode) and a simplified phone with a memory to which images can be attached to facilitate use. Moreover local and remote control was expanded through the use of tablets and smartphones. In both apartments it was made possible to deactivate the video surveillance system based on IP type cameras.

5 The Experimental Trials

5.1 Methodology

During the phase in which the testing protocols were being defined, it seemed important to be able to measure changes in both a "direct" way, with the tests that could be given to individual users at the beginning, middle and end of their experience of

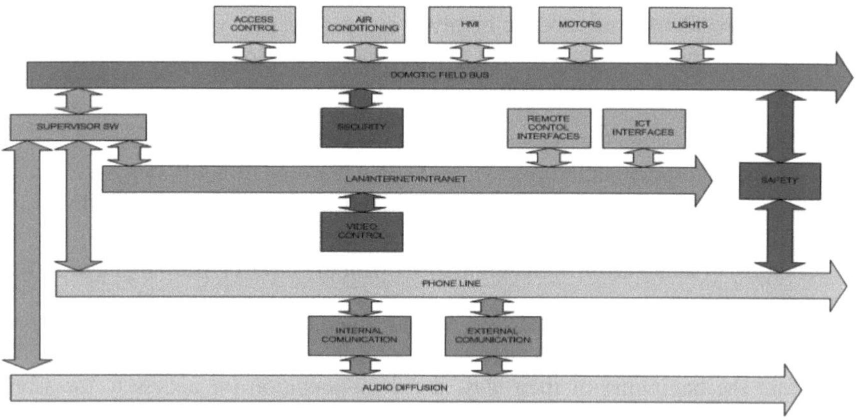

Fig. 3 The technological architecture of the two ADS

the apartments, and in an "indirect" manner through their operators of reference, as well as through a "logbook" where in a free, narrative and iconographic way users could make judgments and express their needs or other requirements with regard to the solutions offered by the apartments.

When it came to selecting the instruments to be used for assessment purposes, the choices made by the team supervising the ADSs were determined by a method of work that focused not so much on the clinical characteristics of the disability (important as they are from the point of view of customizing technologies in situations of progressive debilitation) as on the functional characteristics of the user (what the person is able to do and what he/she would like to do with the aid of simple or complex technological solutions).

5.2 The First Residential Experiences

The first experimental autonomy trials in the Corte Roncati ADS apartments planned a total of nine weekends beginning in February 2013, and concluding in December, its stated objectives are:

- "Activities inside the apartment for the development of skills and competencies aimed at achieving as much autonomy as possible and a life lived independently of the context of the family of origin, with the adequate and competent assistance of a facilitator";
- "activity in the external environment for the development of autonomy in an urban context."

The first residential experience involved a group of four young adults from the province of Bologna, selected for this experience by their local Health Services. The group consisted of two young women and two young men:

- A. 21 year old (sequelae of cerebral palsy, uses a manual wheelchair);
- B. 19 year old (personality disorder and slight learning difficulties);
- C. 19 year old (spastic quadriplegia and poor vision, uses a walker aid indoors and a manual wheelchair outside);
- D. 31 years (sequelae of right hemiparesis with underdeveloped language skills).

The girls stayed in ADS 1 and were accompanied by an educator who was known to them; the boys were accompanied by their teacher and stayed in ADS 2.

Before the beginning of their stay, the documentation for access to the Corte Roncati aids assessment process was completed and a meeting was organized with the four young people and the educators involved. During the preliminary meeting they were shown the apartments in which they would be staying and they met the Regional Centre for AT team that would be supervising them during the experimental trial.

It was clear from the first meeting that the four young people had different characteristics and objectives; the group appeared to get along well together but the autonomy that would be possible for each of them would be different. They were well motivated and enthusiastic about participating in the project for which they had been recommended by their operators. They did not appeared particularly concerned by level of technology they would encounter in the ADS apartments, even though it was not usual for them.

Expectations were high and the fear of failure seemed, for some, outweighed by the fear of success and of having to deal with a level of autonomy they had never experienced before.

Given the nature of the experience of this first group it was realized that finding suitable instruments to measure changes attributable to residence in the ADSs would be no simple matter, especially in the absence of similar experiments for comparison.

Two validated tests were identified which address specific targets and are capable of in some way of measuring a change:

1. DASH, Disability of the Arm, Shoulder and Hand (Italian Version), which measures a person's ability to use the upper limb in different fields [7];
2. The IPPA, Individual Prioritized Problem Assessment, a tool for measuring the effectiveness of assistive technology [8].

While these tests were not specifically designed for this kind of experimentation, they seemed appropriate to the collection of the information we needed. Both were administered, and deliberately so, without a rigorous assessment of the individual's motor skills or functional skills as users, as would normally be required by a specific program for the improvement of work skills.

A rereading of this first experience from a qualitative point of view is awaited as well as the possibility of obtaining data of a quantitative nature thanks to these tools that were originally intended to measure subjective and objective aspects of changes linked to assistive solutions and improvements following their adoption.

The team has also developed a semi-structured interview with open-ended questions that are designed to detect both satisfaction with the tools and dissatisfaction with respect to tools and wherever possible, what changes there are.

6 Conclusions

The Corte Roncati ADSs provide a context in which people can experiment and develop new more autonomous patterns of behavior which can then be introduced and tested in everyday personal life.

During the periods of stay planned by the Living Lab, choices concerning technological systems, assistive technology and designs, fittings and furnishing accessories will be subject to testing and evaluation. Thanks to the experience of the user-residents, there will be changes in the collocation of these elements, new products will be introduce, others will be integrated, and those already present will be fine-tuned.

By analyzing the data gathered from the experimental trials in the ADSs, it should be possible to obtain valuable information that will make solutions for autonomy more effective and responsive to the subjective needs and that will lead to improvements in the ADSs themselves and how they are used.

References

1. Brandt, A., Salminen, A.L.: Systematic review: activity outcomes of environmental control systems and smart home technology. In: AAATE Conference Proceedings (2009)
2. Brandt, Å., Samuelsson, K., Töytäri, O., Salminen, A.-L.: Activity and participation, quality of life and user satisfaction outcomes of environmental control systems and smart home technology: a systematic review. Disabil. Rehabil. Assistive Technol. **6**(3), 189–206 (2011). doi:10.3109/17483107.2010.532286
3. Martin, S., Kelly, G., Kernohan, W.G., McCreight, B., Nugent, C.: Smart home technologies for health and social care support. Cochrane Database Syst. Rev. (4) (2008). doi:10.1002/14651858.cd006412.pub2
4. http://www.who.int/classifications/icf
5. http://www.designforall.it
6. OMS: ICF Classificazione Internazionale del Funzionamento, della Disabilità e della salute. Eriksson, p. 191 (2004)
7. Institute for Work & Health (IWH): GLOBE, Gruppo di Lavoro per l'Ortopedia Basata sulle prove di Efficacia (2003)
8. Wessels, R., Persson, J., Lorentsen, Ø., Andrich, R., Ferrario, M., Oortwijn, W., VanBeekum, T., Brodin, H., de Witte, L.: IPPA: individually prioritised problem. Assessment Technol. Disabil. **14**(3), pp. 141–145 (2002). ISSN: 1868-643x
9. D.M. 27/8/1999: Regolamento recante norme per le prestazioni di assistenza protesica erogabili nell'ambito del S.S.N: modalità di erogazione e tariffe, G.U. 27/9/1999

Preliminary Findings of the AALIANCE2 Ambient Assisted Living Roadmap

Alessandra Moschetti, Laura Fiorini, Michela Aquilano, Filippo Cavallo and Paolo Dario

Abstract The AALIANCE2 Project, funded by the European Commission's ICT Programme within the 7th Framework Programme, aims at identifying the research priorities in the Ambient Assisted Living (AAL) field in Europe and worldwide for the next decades. One of the main objectives of this Project is the development of an AAL Roadmap and Strategic Research Agenda (SRA) that, starting from the needs of the elderly and caregivers, describes the possible next generation of AAL service scenarios, the necessary key enabling technologies (KETs) and the technological, legal and economic requirements necessary for the implementation of the proposed AAL systems. Some of these new AAL scenarios show how technologies, such as robotics and ICT solutions, could be used in senior citizens' daily life activities to maintain their independence and to stay healthy and active in society. At the same time, other scenarios propose new approaches and solutions for caregivers to efficiently support old persons and optimize their work. The Roadmap and the Strategic Research Agenda finally present the future technological challenges to developing the proposed service solutions. This paper provides a short overview of the preliminary version of the AALIANCE2 Roadmap.

1 Introduction

The population in Europe is rapidly ageing due to increased life expectancy and low birth rates [1]. The growth in the number of elderly people also means that the number of "retirees" will get higher, reaching the point in 2060 where there will be

A. Moschetti · L. Fiorini · M. Aquilano · F. Cavallo (✉) · P. Dario
The BioRobotics Institute, Scuola Superiore Sant'Anna, Viale Rinaldo Piaggio 34, 56025 Pontedera, PI, Italy
e-mail: f.cavallo@sssup.it

A. Moschetti
e-mail: a.moschetti@sssup.it

S. Longhi et al. (eds.), *Ambient Assisted Living*, DOI: 10.1007/978-3-319-01119-6_34,

almost one person over 65 for every two persons of working age in Europe [1]. This increase is causing a wider need for services and more responsibility for society both from an economic and a social point of view.

The current concern of most worldwide governments is that the growth in retirees could affect public finances, increasing costs for the society. Moreover, during the past years the European Union, as well as other countries worldwide, has been facing an economic crisis that is involving all parts of the society. So it is necessary to help people stay as healthy and active as possible and give them the opportunity to remain involved in society and in work for longer.

In order to face the current socioeconomic context, a new approach is necessary, and in this context the Ambient Assisted Living (AAL) model is gaining a foothold. The main characteristic of the AAL approach is the use of ICT technological solutions to improve the quality of life of older persons and to optimise socio-medical services. In the past years several AAL services have been developed following this approach. According to this vision, AAL could support elderly people in staying independent and active and help them enjoy more years of healthy life. Technologies could indeed monitor their health, assist them in executing daily life activities and support them in maintaining social contacts and being involved in community life. The current studies carried out in this field show the great potentialities and the possible benefits of AAL in the society; however, there are still technological, social and political barriers delaying the exploitation of AAL in communities. For this reason the European Community is financing the AALIANCE2 Project, a Coordination Action funded within the ICT Programme of the 7th European Framework Programme, aiming at identifying challenges and research priorities in the AAL field over the next decades that could allow and facilitate AAL exploitation and deployment of the actual services. In particular, one of the main objectives of AALIANCE2 is development of the AAL Roadmap and Strategic Research Agenda. This paper shows the preliminary results of the AAL Roadmap developed up to now in the context of this project.

2 AALIANCE2 and Methods

The aim of the AALIANCE2 Project is to study and investigate the technological innovations needed in the next years for the real deployment and exploitation of AAL services in the society.

The AAL Roadmap and the Strategic Research Agenda written during the Project will propose new potential and necessary AAL scenarios and the technologies that should be developed to provide these new services to elderly persons and to formal and informal caregivers. This work took as its starting point the AALIANCE AAL Roadmap, a document developed in 2010 during the AALIANCE Project [2]. The workflow followed during the elaboration of the roadmap is shown in Fig. 1.

The first phase was devoted to the identification of actual necessities of senior persons and caregivers. Starting from this information new scenarios of AAL services,

Fig. 1 Stream used in the development of the AALIANCE2 AAL Roadmap and the SRA

effective for satisfying real user needs, were conceived. Finally, the innovations and challenges for the five enabling technologies (acting, reasoning, sensing, communicating and interacting) necessary for developing the new AAL service scenarios were studied and presented. In the following paragraphs the main results of these three work stages reported in the first preliminary version of the AAL Roadmap are briefly presented.

2.1 Stakeholders' Needs

The AAL Stakeholders were identified and divided into four groups, which are:

- Primary Stakeholders: old people and informal caregivers;
- Secondary Stakeholders: organisations offering services;
- Tertiary Stakeholders: organisations supplying goods and services;
- Quaternary Stakeholders: organisations related to the economic and legal context of AAL.

The identification of stakeholder necessities was the first step of the work and was necessary in order to conceive the key AAL service scenarios that were really implementable and effective in the society. These needs were identified starting from the experience of AALIANCE2 partners, representatives of the four kinds of stakeholders and an analysis of the state of the art. Moreover, some questionnaires were also developed and filled in by external experts in order to collect their opinions about the needs of AAL users and to find suitable solutions for all stakeholders. The choice of AAL solutions also took into consideration the AAL market, because the AALIANCE2 Project aims at suggesting services and systems that can be commercialized.

Analysing the elderly persons' point of view, it emerged that they want to maintain their independence, autonomy and dignity without, however, being excluded and uncared for by society. They want to remain in their own home, but at the same time they want to feel safe there; in other words, if something happens to them, they want to be able to easily get in contact with someone who can help them, or receive help directly from the "environment". Moreover, old persons are at high risk of being isolated, due to their weakening health and to territory. Indeed, Europe presents a variegated territory made of metropolitan district and wide rural areas. These geographical differences influence access to services and could then increase the isolation of older persons due to their increased difficulty in moving.

Furthermore, the economic crisis strongly influences the lives of most of the elderly population increasing their risk of poverty and social exclusion—as shown in the Eurostat studies [3]—and consequently worsening their health and quality of life [4].

The AAL approach should help elderly people stay healthy as long as possible, allowing them to keep on living their own life and helping them to be involved in society and to continue working if they wish to. Senior subjects feel they are citizens who may still be useful in society, so they would like to be more included and involved in community life. This is important not only for them but also for society itself, which can still benefit from their experiences and capabilities.

Moreover, AAL should help informal caregivers in managing elderly people, reducing the stress and also the amount of work related to that. AAL devices would allow them to monitor older persons more easily, as well as to maintain contact with experts who could support them in making decisions and in evaluating the best behaviour to adopt. This kind of help could also be useful for formal caregivers. Often these professionals spend short periods of time with each of their senior users because they have to follow several people. If AAL devices could optimise their work, make it more efficient and also carry out part of it autonomously, they could spend more time with the elderly people and focus their efforts on more important aspects.

Furthermore, in analysing the needs of all stakeholders, the importance emerged of considering the economic and legal contexts, factors strongly influencing AAL exploitation and deployment in society. In particular, the economic crisis is having a negative influence on various countries, which are making cuts both to health care and to socio-medical services, while many older persons do not have the economic wherewithal to buy these services privately. For these reasons it is important to identify both new AAL solutions that could reduce the costs of health and socio-medical carers, e.g. to reduce hospitalization, and also alternative forms of finance for these services.

2.2 Service Scenario

The scenarios described in the current version of the AALIANCE2 Roadmap are identified on the basis of three main service areas:

- *Prevention*, that is, action to reduce or eliminate the onset, causes, complications or recurrence of disease;
- *Compensation and Support*, which concerns elderly people with physical or cognitive impairments who need help in their daily activities;
- *Independent and Active Ageing*, which aims to extend healthy life expectancy and Quality of Life for all people as they age, including those who are frail, disabled and in need of care.

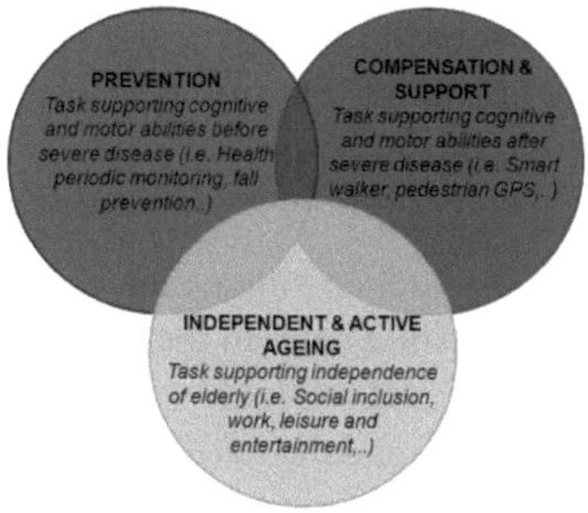

Fig. 2 The AALIANCE2 macro-areas: prevention, compensation, support, independent and active ageing

Fig. 3 Possible effects on quality of life

These areas are at the same time complementary and overlapping, as shown in Fig. 2.

These aspects are strongly related to the Quality of Life (QoL) of senior citizens. Some research [5, 6] has asserted that the perception of QoL is conditioned mainly by five factors: (physical) health, (psychosocial) welfare, social contacts, activities and living environment. The way in which each of them influences QoL depends on personal experiences and life [5, 6]. The three AALIANCE2 service areas are strongly correlated to these aspects; therefore, innovative AAL services related to these areas could positively influence the QoL of elderly people. Figure 3 shows how the QoL of elderly people could be influenced by Prevention, Compensation and Support and Independent and Active Ageing.

In particular, the light blue line of Fig. 3 represents the QoL of a person who is not provided with any AAL technologies and services. It would decrease after a certain age due to disabilities and morbidity related to age. This line would decrease later (dark blue curve) if people perform a series of preventative actions that help delay or better managing diseases, in order to reduce or eliminate morbidity. When, coupled with prevention, compensation and support actions are carried out, the line could decrease even more slowly (red curve). As a matter of fact, ICT solutions could help patients do desired activities, making them perceive a higher QoL. Eventually, maintaining independence and remaining active, continuing to live their own life, increases the perceived Quality of Life (yellow curve).

Thanks to the workshops carried out during the AALIANCE2 Project, several AAL service scenarios were identified, describing some possible care services implementing ICT and robotic solutions to help elderly people in their daily life.

Some examples of scenarios are briefly listed, as follows:

- *Healthy living*: an intelligent kitchen suggesting recipes for healthy meals, also considering what is going to expire soon in the fridge;
- *Falls*: advanced environmental and wearable sensors linked to intelligent systems could prevent possible falls during daily activities;
- *Prevention for health*: motor and cognitive games carried out with smart tools to maintain elderly people's active and healthy lives, to support rehabilitation and prevent cognitive diseases;
- *Senior citizens at work*: considering the actual socioeconomic context, it is important to help people to keep working longer, and this can be done by supporting them in learning new skills and also in doing their daily work thanks to smart tools;
- *Support in driving*: smart cars proving more information about the traffic and street status and helping the elderly to drive safely;
- *Sensing emotion*: by understanding emotions, formal and informal caregivers could better understand how the elderly people are feeling and how best to help them;
- *Assistance at home*: a robotic assistant that helps older persons to carry out the most complex and dangerous activities of daily life in caring for him/herself and the house;
- *Keeping social contacts*: age-friendly environment that enables easier access to services, so to make people more involved in the society. Moreover radio and television can be used to help people interact with others and create topic discussion channels.

2.3 Enabling Technologies

Starting from these scenarios, some technological components and innovations necessary for the development of these services were identified. AAL systems are typically composed of sensors, components that process the sensor data and derive conclusions, human–machine interface components and actors that execute all the actions. Finally all these components have to communicate with each other. These enabling technologies were then grouped into:

- Sensing
- Reasoning
- Acting
- Interacting
- Communicating.

Sensors are necessary in AAL systems in order to perceive what is happening. New sensors need to be developed in order to design new technologies that could improve elderly people and caregivers' life. New environmental sensors should be developed in order to reduce their impact in people's life. Besides vision sensors, a new generation of acoustic and smell sensors should be designed. These kind of sensors could in fact help informal and formal caregivers to take care of elderly people, by adding to the vision system other devices that could help them to understand what is going on. Moreover, it is important to improve the personal sensors, increasing acceptability and reducing invasiveness. That means that new wearable and implantable sensors should be developed and all these new sensors should be designed taking into account the challenge of power management and energy harvesting.

In order to deal with the great amount of data coming from different sensors, it is necessary to have good communication networks (both wired and wireless) and communication protocols. It is thus necessary to have communication standards and interoperability between systems and components. The main developments have to be made in the fields of body area or personal area network, local area or home network and wide area network, in order to allow communication between more systems and components. Furthermore, cloud computing is also considered a promising technology that could improve AAL services. In particular, it allows for managing data and supporting all the devices collaborating together. In this network of devices that can be useful for elderly people, robots must also be considered. Wearable robots and service robots could support elderly people in their lives, helping them feel more independent and safe. Moreover, cloud robotics is an enabling technology that integrates different agents together to improve collaboration and provide useful and high quality services to citizens.

If all these AAL systems are to be part of everyday life, it is important that people can interact easily with them. Efforts have to be made to make this interaction "friendly" and usable to encourage users in using AAL systems. New interaction systems should be developed, based for example on spatial interfaces, such as 3D movement-tracker or gesture-based interaction, or on sensorial interfaces.

Furthermore people should feel safe with the new AAL systems, considering them as an advantage rather than an additional problem or cause of stress. For this reason systems should be dependable and should also be able to automatically manage any possible system failure, so to prevent damage or help the user to manage it. New challenges in process systems include the capability to recognize emergency situations as well as user habits, so as to identify uncommon situations. Moreover, the systems could be able to learn from the users' habits, so as to change with them and remain up to date with users' needs.

It is important to make people feel comfortable with the technologies. Acceptability is a fundamental aspect in AAL because it conditions the real use of these new devices by elderly persons and makes them feel the system is part of their home or themselves. Only by satisfying this precondition AAL technologies can find a place in the real market and be commercialized.

In the end it is important to underline that technology should be an "enabler"—something that helps elderly people feel better and helps formal and informal caregivers in their work; but technology per se is useless without the human components provided by the experience of the caregiver and from his/her relationship with older person. AAL systems should thus help to do time-consuming chores, allowing family and caregivers to spend more time with the elderly persons.

3 Conclusion

The AAL Roadmap developed in the first period of the AALIANCE2 Project aims at finding suitable new service and technological solutions in the AAL field, starting from stakeholders' needs, that will allow old persons to remain healthy and active for as long as possible. In this way senior citizens can remain involved in society, continuing to be part of it and working for as long as they want. Moreover, one of the objectives of this project is the identification of challenges both for the AAL research and for the industry in the coming years.

This paper reports a brief summary of the first version of the AAL Roadmap. In the next months the AALIANCE2 AAL Roadmap and Strategic Research Agenda will be updated in order to refine the contents, considering new technologies and also feedback from the different experts and stakeholders involved during the project.

Acknowledgments The research leading to these results has received funding from the European Community's Seventh Framework Programme (FP7/2007-2013) under grant agreement num. 288705 (AALIANCE2 Project).

References

1. Special Eurobarometer 378: Active ageing. http://www.ec.europa.eu/public_opinion/archives/ebs/ebs_378_en.pdf (2012)
2. van der Broek, G., Cavallo, F., Wehrmann, C.: AALIANCE Ambient Assisted Living Roadmap, vol. 6. IOS Press, Amsterdamk (2010)
3. Eurostat: People at risk of poverty or social exclusion. http://www.epp.eurostat.ec.europa.eu/statistics_explained/index.php/People_at_risk_of_poverty_or_social_exclusion (2013)
4. AGE Platform Europe: Older people also suffer because of the crisis (2012)
5. Bowling, A., Gabriel, Z., Dykes, J., Dowding, L.M., Evans O., Fleissig A., Sutton S.: Let's ask them: a national survey of definitions of quality of life and its enhancement among people aged 65 and over. J. Aging Hum. Dev. **56**(4), 269–306 (2003)
6. Puts, M.T.E., Shekary, N., Widdershoven, G., Heldens, J., Lips, P., Deeg, D.J.H.: What does quality of life mean to older frail and non frail community-dwelling adults in the Netherlands? J. Qual. Life Res. **16**, 263–277 (2007). doi:10.1007/s11136-006-9121-0

Part VIII
Social Implications and Age-Friendly Smart Cities

The Third Age in the National Health System: A Proposal for Increasing the Spending Effectiveness in Health Care for the Elderly

Maurizio Ciaschini, Monica De Angelis, Rosita Pretaroli, Francesca Severini and Claudio Socci

Abstract The allocation among Institutional Sectors of Health Care Expenditure for the Elderly can be designed in a way that new processes can be activated that stimulate total output and overall employment. The impacts of these processes as they emerge from several policy scenarios in all phases of income circular flow within the economic system, can be determined with reference to the model of computable general equilibrium (CGE) calibrated on the Social Accounting Matrix (SAM) data set. The HCEE is placed within the economic flows of a bi-regional SAM for Italian economy which provides the data set for a CGE model where the hypothesis of involuntary unemployment is also considered. Two policy scenarios are then simulated to evaluate the impact of the policy reform in HCEE on total output and unemployment. Under the first scenario the Regional Government reduces the expenditure in integrated home care assistance (ADI) in order to directly increase the demand of private residential health care (ARA) services; under the second scenario the Regional Government transfers the resources saved by reducing ADI to the Households in order to stimulate their demand of ARA.

Paragraphs 1 and 2 have been authored by M. De Angelis

M. Ciaschini · R. Pretaroli · F. Severini · C. Socci
Department of Economics and Law, University of Macerata,
P.za Strambi 1, 62100 Macerata, Italy
e mail: ciasco@unimc.it

R. Pretaroli
e-mail: pretaroli@unimc.it

F. Severini
e-mail: francesca.severini@unimc.it

C. Socci
e-mail: socci_claudio@unimc.it

M. De Angelis (✉)
Department of Management, University Polytechnic of Marche,
P.le Martelli 1, 60121 Ancona, Italy
e-mail: m.deangelis@univpm.it

S. Longhi et al. (eds.), *Ambient Assisted Living*, DOI: 10.1007/978-3-319-01119-6_35,

1 Introduction: Crisis and Health Systems

Several European countries (and others) are experiencing a deep economic and financial crisis which has a direct impact on health care. In particular, the current situation is characterized by three simultaneous dynamics: the recession or, in some cases, the slowdown in the economy, which usually is accompanied by a reduction in tax revenue because of some taxes automatic shock absorber role; growth of diseases related to unemployment and risk of increased social inequalities and, consequently, of the "social gradient" in health; policies of reduction of public health spending—in terms of both current and investment expenditure—in order to reduce the public finance deficits. The economic and financial crisis is still ongoing and no one can predict with certainty its duration, its magnitude, its effects and its final outcomes. However, there is no doubt that the adverse economic situation and the consequent public finances contraints are putting the health system of many countries under severe strain; in addition, these countries have been facing—amongst other scenarios—a strong aging of the population. Despite experience of previous crises and studies on the effects caused by them, the widespread feeling among policy-makers and experts is that different countries are taking short-term approaches to face the challenges posed by the current situation, without adequate criteria for building an effective warning system to indicate trends and risks in a timely manner, e.g. with regard to state of health of the population and the issue inequalities; as well for suggesting the steps to be taken in order to fight or slow down the negative effects of the crisis on health systems, at least by considering the close relationship existing between the health of population and socio-economic welfare. This relationship reflects Virchow's paradigm according to which any cause amending socioeconomic status affects the health status of the population (or classes of this) [9, 22].

However it is worth remembering that the word crisis also means opportunity[1] and if the issue of an aging population is crucial for modern health systems, there is no doubt that any analysis having the purpose of identifying new developments for modern health systems should strongly consider this component. Moreover, the active aging (AHA) can promote long-term economic sustainability of health systems, whereas the non-self-sufficient ageing could help generating income and consequently employment. In recent years the focus has been placed on the elderly, and in particular on AHA showed that for elderly population we have to imagine new supply chains of products, services, support, tools, changing markets and rules in order to create a welfare system suitable for the conditions of aged future society. Therefore, the elders can be regarded as the reason for rethinking the market, the products and the services and as the driving force for doing so and for building an

[1] According to a John Fitzgerald Kennedy's aphorism 'the word crisis', if written in Chinese, is composed of two characters. One represents danger and the other represents opportunity. In fact, in Chinese, the word opportunity is translated 'ji hui' whereas the word crisis is translated 'wei ji'. We can note that the same symbol 'ji' is being used: the word 'crisis' is composed of two characters which mean respectively danger and opportunity. The underlying rationale is that the crisis is to take a very deep meaning, because sometimes a situation can be an opportunity (so something positive), or, in the opposite case, a risk, a danger (meaning something negative).

environment as close as possible to their needs; they can also be considered as wealth -creating people both for themselves and for the wider community, thereby allowing a greater sustainability of the welfare system.

2 An Institutional Framework: The Direction Taken by Italian NHS

At the outset, and in order to frame any type of instrument of economic policy we should ask which direction is following the health care system of reference: in this case, the Italian NHS. The events which have affected the Italian NHS since its institution in 1978 with the Law n. 833/1978, lead us to imagine an expansion of its duties, for example in terms of protection of rights to healthcare, of care arrangements and of levels of care; as well as a trend in restraint on the side of service on offer, for example with regard to spending levels or the capillarity of the hospitals. These are phenomena that accompany cyclothymic evolutions (involution) of the legal framework that leave the previous regulations without a proper implementation. It is not easy to identify the direction taken by the Italian NHS in recent years. So far, more or less essential reforms concerning, for example, organizational profiles or guarantee of the right to health, have been following each other [10] thereby creating difficulties both in the process of building new rules and in their implementation phase due to the lacking—in either case—of the proper conception and "maturation" period.

As in all industrialized countries, in more recent times there has been a huge growth in demand for health services and, consequently, for the above-mentioned situation of crisis, there has been the development of policies aimed at rationalizing services offer rationalization has taken various forms and has involved multiple dimensions: from personnel management and equipment purchases to the location of hospitals). The combination of social and institutional needs has become more and more articulated: e.g., budget constraints and the reduction of public spending collide with the intrinsic costs of technological progress of medicine which increases the expectations (although sometimes not reasonably reliable) of the patients and the progressive aging of the population. These are issues that need to be added to the inherent complexity of health systems but which will increasingly affect the future healthcare policies.

In the Italian case, this complexity certainly emerges when you try to figure out which direction the health care system will take with the implementation of Article 119 of the Constitution, as amended by the constitutional reform of 2001, which introduces the so-called fiscal federalism: the introduction first of the delegated legislation, and subsequently of implementing decrees relating to fiscal federalism, raised since the beginning the important question of how and how much the health sector will be affected in the future. With regard to the most recent past, and in particular on the basis of the steps taken in the last two decades, we can distinguish a deep regionalization of Italian NHS, and it is likely that the need for a "participated" health service, that is shared with all stakeholders, will become increasingly important. Those who study health care systems, in fact, cannot fail to consider the utmost importance that demands of citizens and change in their needs have in a modern society: an aging population will express very different needs if compared to a society in which the number of workers and young people is high. Strengthening certain services rather than others may mean improving the NHS thanks to feedbacks, by calibrating and connoting it in the sense of greater appropriateness.

Despite the complexity of the issues entailed, in an effort to understand the path taken by the Italian NHS there is no doubt that the satisfaction of the user has fulfilled the role of directional beacon of different health care policies in recent years. Although the Constitution reserves to the national level the definition of essential service levels, it simultaneously makes Regions in charge of promoting effective actions aimed at improving the appropriateness of cares, the efficiency of their supply, and the uniform access for citizens, by finalizing the use of resources on enhancing the quality and effectiveness of services and on ensuring that the health services be evenly used all over national territory. In the backdrop of a financial and economic framework as difficult as the current one, the emphasis on the quality of services may take a very important role. To pursue these objectives, then, it would seem obvious to think that the scarcity of resources push towards greater concentration on care services and support existing services, by trying to improve their performance. Each operation, however, cannot absolutely ignore that key factor concerning aging of the population. It is also impossible to ignore the general feature that indeed marks most of the public administration sectors and therefore healthcare administration too: the relationship between "public" and "private". A relationship which has different facets and which is mentioned here in the sense of "use of private law" and "correlation between the public sphere and the private sphere".

With regard to the first variant, it is worth recalling that the mentioned law n. 833 of 1978—that is a law which sets the rules and principles of a public service and designes a system based on typical administrative and traditional patterns—welcomes this trend. It should also be stressed that the use of private law in systems falling within the scope of administrative law is not just about the rules relating to contracts or the management of cares, but concernes something else, e.g. the relationships with all users. With respect to the second variant, however, it can be noted that the typical models of the private sphere included in NHS are not, today, easily assessable also because the gradual implementation of Law n. 833/1978 has seen the emergence of conflicting signals: at first, the path showed a strong acceleration towards a competitive semi-liberalized system; subsequently, there were sudden revirement, with detriment of consistency and robustness of the health care system. The instability is also derived from composite regulations, both national and regional, from the judgments of the Constitutional Court, from the contributions of the Italian Competition Authority, which in several occasions has called for structural interventions aimed at promoting competition. Many reforms have occurred, triggered also by the enormous impact that political, institutional and economic frame work has been generating on the healthcare system: hence the great attention paid to limiting costs, to efficiency and to accountability; the introduction and the application of the subsidiarity principle, the adoption of specific controls aimed at assessing the result, the emphasis on competitiveness, etc. Therefore there was no lack of measures, but, maybe, an effective control of the extent to which the introduced reforms (outcomes) had been leading to the desired achievements (real effects) was insufficient. For this check to be effectively carried out, it would be important to adopt legislative drafting techniques, i.e. a tool capable of connecting law and economics.[2] There is the tendency to assimilate increasingly the organization of healthcare to a business organization, by giving autonomy and responsibility to managers, by accepting organizational flexibility, by adopting the analytical accounting, etc.: however, this has not resulted in an undisputed improvement of the relationship between users' expectations and health services offered. In other words, what has been done so far does not seem to give satisfactory answers to the issue of a change of services offered in NHS in light of aging population.

The correlation between the public sphere and the private sphere seems to have accelerated in the Nineties, with the emerging of an articulated system for regional services, differentiated on the basis of policies concerning, for example, the services provided by public or private bodies or the possibility for the users to choose the provider for the health care service. In fact, with the legislative decrees n. 502/1992, so -called "Health care reform bis" and n. 299/1999, so-called "Health care reform ter", there is a contraction of the previous system based on partnership agreements which was unfit for a modern heath care administration, a greater openness to business schemes in order to improve organizational and managerial arrangements. The reform of Title V of the Constitution, with the introduction of the principle of horizontal subsidiarity, goes on to strengthen the process of transformation of the existing institutional models: the national level is assigned a "lighter" role (duties of supervision and control; definition of essential service levels); whereas public health organizations and private bodies are entrusted to perform even more consistent management tasks on various types of care and assistance. Therefore, in this new framework, the facet of the public-private relationship in health care, which is being referred to, can find unusual and innovative opportunities for its consolidation: for instance, in theory, the period in which the position of private actors is bound to be complementary to the role of the public sector could come to an end, especially if private bodies show they can better meet the needs expressed by users. In other words, the regional competence to regulate the organization and arrangements of healthcare services—whilst respecting the constitutional rules providing for an appropriate and universal national health service, in which the right to health is a fundamental right-may theoretically allow each Region to build its own unique healthcare services offer, always within the framework of certain nationally pre-defined standards of care, treatment and quality of outcomes: Regions showing a very significant rate of aging population may propose an organizational model and an overall set of services specifically addressed to the elderly population.

[2] On the subject, please consult the national and international bibliography on the website http://www.osservatorioair.it/air/letteratura-air/valutazione-economica-ed-air/.

3 Elderly's Health Care Expenditure: SAM and Model

Production of health services for elderly which has shown a continuous increase of its weight within health expenditure, has a role within a national economy mainly with regard to its potential to enhance income national [5]. In this paper an effort is made to simulate scenarios of reform of the composition of such expenditure for the Italian economy where health expenses for elderly are distinguished into: integrated home care (ADI), private residential health care (ARA) and public residential health care (ARP) services [4, 6–8]. Significant impacts on income of institutional sectors, which are the owners of the different types of incomes emerging in the whole economy, are to be expected when the destination of expenditure on such services is modified. This happens because of the existence of interdependencies both between each of tree typologies of services (ADI, ARA and ARP) and the whole economy [20]. In particular we reproduce a reduction in regional expenses for ADI that is compensated by an increase in the regional demand for ARA.

For this aim we develop a CGE model that assesses the impact of a regional reform of health care expenditure for the elderly on the main macroeconomic variables. This analysis can be carried out by using a database that integrates the health flows for elderly within the phases of generation and distribution of income in a multisectoral and bi-regional framework [3]. For this purpose, the CGE is based on the bi-regional (North-Centre and South-Isles) Social Accounting Matrix (SAM) for Italy [15] which describes the income circular flow for the Italian economy (year 2003), in terms of intra-regional and inter-regional flows [18]. The CGE model is considered as a suitable tool to examine the economic implications of an exogenous shock when prices change since it considers all the relationships among the agents in the economic system [2, 11]. It gives an integrate representation of the income circular flow [3] and allows to evaluate the direct and indirect effects of a policy on macroeconomic variables (GDP change, unemployment rate) and on income distribution between Institutional Sectors [16]. Even though the model does not describe the reality objectively [12], it provides an interpretation of the economy given certain basic assumptions [1, 13, 17] that influence the results. The model is calibrated on the bi-regional Social Accounting Matrix (SAM) for Italy (year 2003), which represents the benchmark data set and the income circular flow for the whole economy, in terms of intra-regional and inter-regional flows [14, 19].

The SAM considers an open economy with two regions (North-Centre and South-Isles), 16 commodities,[3] 2 components of value added (labour and capital), 2 private Institutional Sectors (Households and Firms) and 7 public Institutional Sectors (Regional Government, District Government and five Municipal Government[4]) in

---

[3] 1. Products of agriculture, 2. Energy products, 3. Metal and non metal ore, 4. Non metallic mineral products, 5. Chemical products, 6. Mechanics, 7. Transport equipment, 8. Food products and beverages, 9. Textile, 10. Other manufacturing products, 11. Construction work, 12. Trade, 13. Transport, 14. Financial services and Insurance, 15. Private services, 16. Government services.

[4] The municipal Government is disaggregate in 5 classes according to the official statistics on population. In particular we consider: (i) Municipality 1, less than 5,000 inhabitants; (ii) Municipality 2,

each region [16]. The health commodity can be included in this scheme through the integration of the production and the income national accounts [17] with the economic flows related to ADI, ARA and ARP. A whole detailed description of the model and the SAM features is presented in [8, 14, 15].

3.1 The Policy Reform in Health Care Expenditure for Elderly: Implications and Results

In this paper we simulate two different scenarios. In the first one (SR), the Regional Government in each macro-region (North-Centre and South-Isles) reduces the expenditure for ADI in order to increase the demand for ARA. The amount of resources moved is around 7.43 millions of euro in the South-Isles (corresponding to 2 % of health care expenditure in ARA) and 15.7 millions of euro in the North-Centre (corresponding to 1.60 % of health care expenditure in ARA). In the second scenario (SF), the regional government in each macro-region reduces the expenditure in ADI and transfers the freed resources to Households who are bound to use the funds to demand ARA. The amount of this manoeuvre is approximately 8 million of euro in the South-Isles and 21 millions of euro in the North-Centre (corresponding to the 2 % of Households expenditure in ARA in each region).

In both scenarios, the manoeuvre does not imply new resources and is performed in order to assess the effects in terms of total output, prices and unemployment of a change in the typology of elderly health care services (first scenario—*SR*) or a reallocation in Households demand for elderly health care services financed by the regional government (second scenario—*SF*).

The simulations aim to quantify the direct and indirect impacts of a different allocation of health care expenditure for the elderly on the phases of the income circular flow. The results are of direct and indirect type and are expressed as percentage changes from the *benchmark* represented by the SAM. In the first scenario *SR* the increase in health care expenditure for elderly in ARA directly affects the total output of ARA services that increases in North-Centre and South-Isles (Fig. 1). The same policy generates an indirect effect on the other commodities total output since their productions are strictly connected. The goods that are affected the most by the manoeuvre are: "Energy products", "Non metallic mineral products", "Food and beverages" and "Construction". The increase in ARA and other commodities total output is balanced by a decrease in ADI total output as a direct consequence of the regional government cut in ADI demand. Moreover the effect on ADI mitigate the effects of the policy on the other commodities total output.

The second scenario *SF* assumes that the Households increase the demand for ARA using the transfers from the regional government (see Fig. 2. The impact of the

---

inhabitants included between 5,001 and 15,000; (iii) Municipality 3, inhabitants between 15,001 and 30,000; (iv) Municipality 4, inhabitants between 30,001 and 60,000; (v) Municipality 5, more than 60,000 inhabitants.

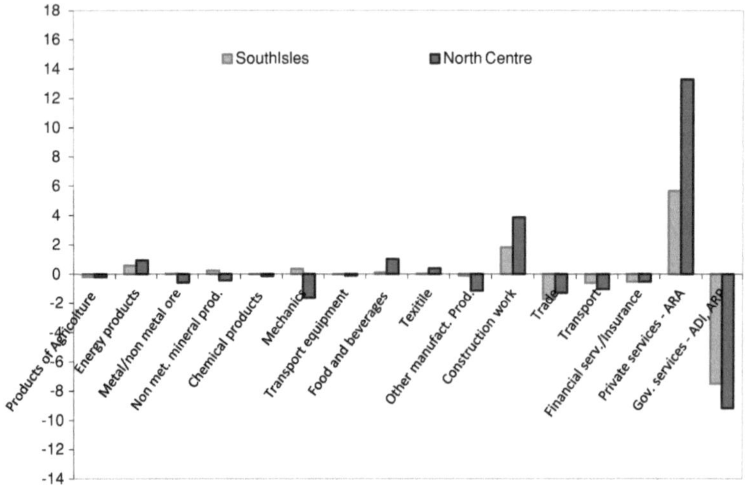

Fig. 1 Total output change (millions of euro)—SR scenario

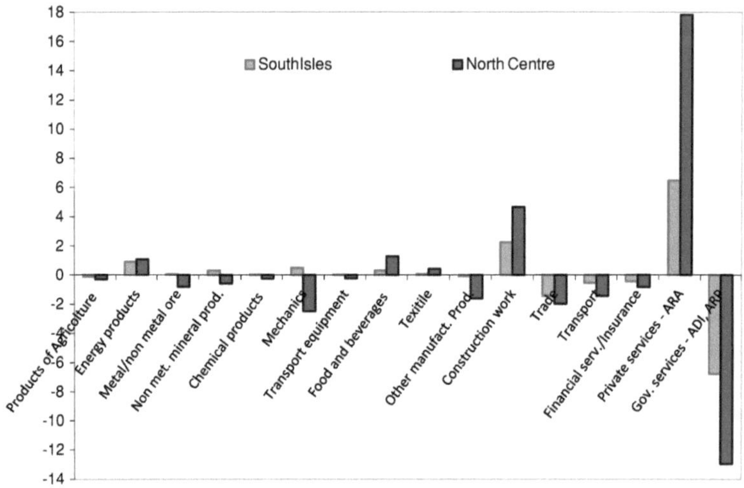

Fig. 2 Total output change (millions of euro)—SF scenario

policy on total output is similar to the previous scenario although the output changes are more accentuated in both regions. Indeed, the resources used in this scenario are higher than the previous.

Moving from total output to final prices results, the increase in the demand of ARA produces a generalized reduction in the price of commodities in both regions, as showed by the Figs. 3 and 4. According to the theoretical assumption of the CGE model on the perfect competitiveness of commodity markets, when a commodity supply increases, given the demand, the prices reduces. Therefore the results show that

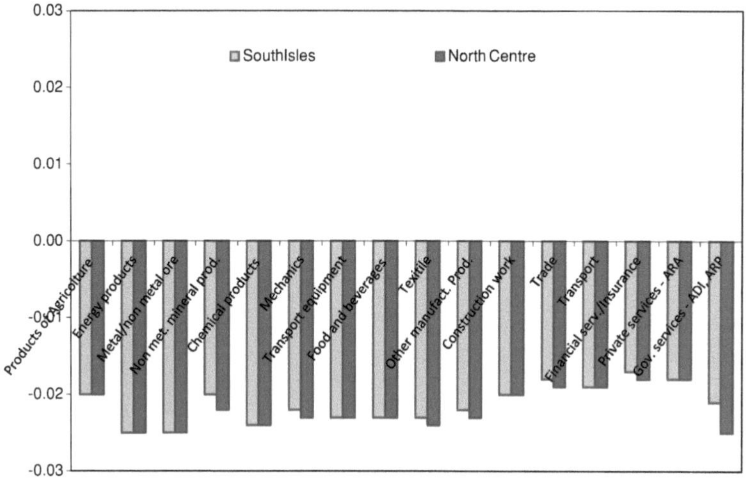

Fig. 3 Percentage change in commodity price—SR scenario

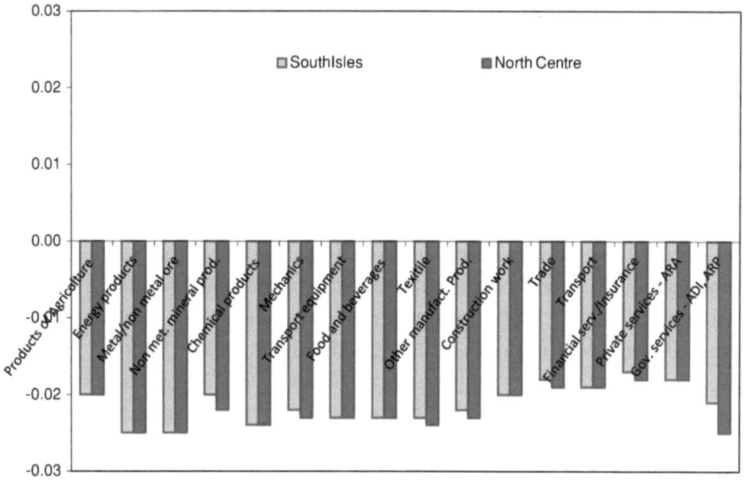

Fig. 4 Percentage change in commodity price—SF scenario

the more the total output increases, the more the price of that commodity decreases. This effect is observable especially in North-Centre region for "Energy products", "Non metallic mineral products", "Food and beverages" and "Construction".

The reduction in prices is observable even in all the other commodities in both regions since in the general equilibrium model, the reduction in the price of certain goods affects indirectly the demand for intermediate goods and then the production and formation of final prices of other goods. To be more specific in the second scenario *SF* the percentage change of prices are higher than in scenario *SR* because the amount

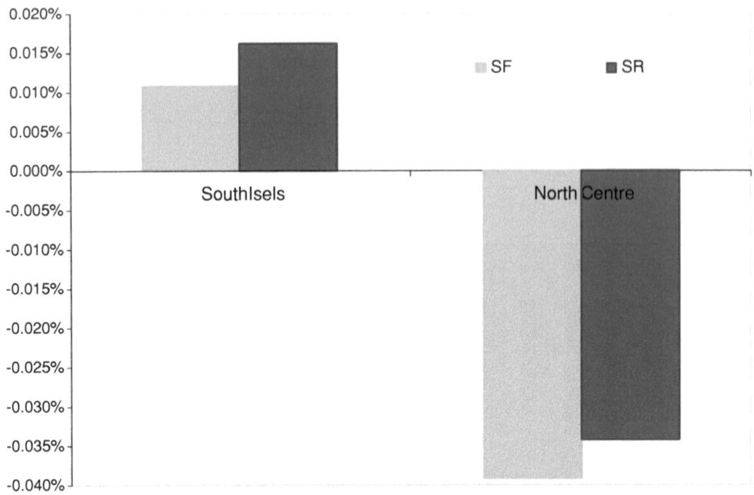

Fig. 5 Percentage change in unemployment rate

of the manoeuvre is higher. A prominent advantage of the model is related to the assumption of market imperfection for the labor market that consents to evaluate the impact of the health care expenditure policy reform on unemployment in each region. Considering a wage function within the system of behavioral equations, allows to model a non-competitive labor market that admits involuntary unemployment.

As showed by the Fig. 5, the impact of the policy on unemployment is extremely different within the regions in both scenarios. In particular, in the North-Centre region the unemployment rate reduces in both scenarios. The higher demand for ARA (the policy objective in both scenarios) generates an increase in total output and then stimulate the demand for intermediate goods and primary factors. In the North-Centre region the unemployment rate decrease by 0.034 % in *SR* and 0.039 % in *SF*. Otherwise in the South-Isles region the increase in the demand of ARA, generate an increase in total output but does not affect the demand of labour. Therefore the unemployment rate increases by 0.016 % in *SR* and by 0.015 % in *SF*. This result, which is found in both scenarios, highlights the differences between the two regions as regard to the technology used in the production processes and the primary factors demand.

3.2 Sensitivity Analysis of the Results

The results showed in the previous section are conditioned to the assumption made in the model as regard to some parameters and the elasticity of substitution in the production function. Therefore, in order to strengthen the validity of the simulations' outcomes previously described, it is worth to integrate the study with the sensitivity analysis for the elasticity of substitution between primary factors in the value added aggregate. To be more specific, in the nested production function, when combining labour and capital in order to generate the value added aggregate, we assumed a CES technology with elasticity of substitution equal to 0.5218 [21]. The sensitivity analysis is carried out assuming two alternative parameters for the elasticity: $sigma_l = 0.6262$ (the original parameter increased by 20 %) and $sigma_l = 0.4174$ (the original parameter decreased by 20 %). Then the simulations are repeated with these new parameters.

The results of this analysis are compared with the results of the simulations *SR* and *SF* as regard to the change of total output (Table 1), the change in the price of commodities (Table 2) and change in unemployment rate (Table 3).

The sensitivity analysis shows that the results of the simulations are robust since all variables' changes are similar in signs and absolute value even when assuming a different elasticity of substitution. As for the unemployment rate in particular (Table 3) the analysis confirms the opposite result in the Nord-Centre and the South-Isles. Given the production function, the elasticity of substitution between labour and capital could have been decisive in terms of unemployment change: the results of the sensitivity analysis do not confirm this statement.

4 Conclusion

The regional competence to regulate the organization and arrangements of healthcare services in principle allow each Region to build its own unique healthcare offer, within the framework of nationally predefined standards of care, treatment and quality of outcomes. Regions showing a significant rate of aging population need to rethink the organizational model and propose an overall set of services specifically addressed to the elderly population. However restructuring health care expenditure for the elderly can also provide an opportunity to the policy maker of promoting the private residential health care assistance while stimulating production and reducing unemployment rate.

The type of analysis that has been performed in this paper, in terms of general economic equilibrium, on the data set provided by Social Accounting Matrix (SAM) quantifies the direct and indirect impacts of the health care policy reform in all the phases of the income circular flow. The disaggregation in the SAM of the economic flows related to the integrated home care (ADI), the private residential health care (ARA) and the public residential health care (ARP) services, allowed to evaluate the impact in the economic system of a health care policy reform oriented to stimulate the private expenditure in health care.

For Italy the effects of an increase in the final demand of ARA services rather than ADI services were studied determining the disaggregate numerical impacts in the North-Centre and South-Isles regions. This was done through the design of two policy scenarios: in the first the regional government reduces the demand for ADI

Table 1 Change in total output (millions of Euro)

	Scenario SF						Scenario SR					
	Sigma = 0.5218		Sigma = 0.6262		Sigma = 0.4174		Sigma = 0.5218		Sigma = 0.6262		Sigma = 0.4174	
	South Isles	North Centre	South Isles	North Centre	South Isles	North Centre	South Isles	North Centre	South Isles	North Centre	South Isles	North Centre
ADI	−0.02	−0.05	−0.02	−0.06	−0.02	−0.05	−0.02	−0.04	−0.02	−0.04	−0.02	−0.03
ARA	0.05	0.14	0.05	0.15	0.05	0.14	0.05	0.11	0.05	0.11	0.05	0.10
ARP	−0.03	−0.20	−0.03	−0.22	−0.03	−0.19	−0.03	−0.14	−0.04	−0.15	−0.03	−0.13
Products of agriculture	−0.13	−0.30	−0.15	−0.26	−0.10	−0.34	−0.19	−0.21	−0.22	−0.19	−0.17	−0.24
Energy products	0.92	1.09	0.71	0.99	1.20	1.23	0.57	0.93	0.41	0.85	0.80	1.03
Metal/non metal ore	0.05	−0.83	0.05	−0.67	0.06	−1.05	0.03	−0.60	0.03	−0.47	0.04	−0.77
Non met. mineral prod.	0.30	−0.62	0.27	−0.50	0.34	−0.78	0.24	−0.44	0.21	−0.35	0.27	−0.57
Chemical products	0.02	−0.26	−0.00	−0.19	0.06	−0.37	−0.03	−0.15	−0.05	−0.09	0.00	−0.23
Mechanics	0.48	−2.51	0.43	−1.86	0.57	−3.37	0.35	−1.66	0.30	−1.15	0.41	−2.33
Transport equipment	0.01	−0.24	0.00	−0.16	0.03	−0.35	−0.03	−0.12	−0.04	−0.05	−0.01	−0.20
Food and beverages	0.30	1.27	0.19	1.07	0.45	1.54	0.11	1.02	0.02	0.87	0.23	1.24
Texitile	0.07	0.42	0.03	0.37	0.11	0.48	0.01	0.37	−0.02	0.34	0.05	0.42
Other manufact. prod.	−0.07	−1.63	−0.08	−1.28	−0.06	−2.13	−0.11	−1.14	−0.12	−0.86	−0.10	−1.54
Construction work	2.23	4.65	2.02	4.24	2.55	5.24	1.82	3.86	1.65	3.55	2.07	4.33
Trade	−1.45	−1.99	−1.46	−1.67	−1.50	−2.40	−1.71	−1.29	−1.73	−1.04	−1.74	−1.62
Transport	−0.54	−1.46	−0.53	−1.21	−0.57	−1.79	−0.61	−1.03	−0.60	−0.84	−0.63	−1.29
Financial serv./Insurance	−0.45	−0.82	−0.47	−0.73	−0.43	−0.96	−0.52	−0.52	−0.54	−0.45	−0.50	−0.63
Private services—ARA	6.44	17.81	6.39	18.20	6.46	17.29	5.66	13.28	5.61	13.58	5.69	12.87
Gov. services—ADI, ARP	−6.79	−12.96	−7.36	−13.76	−6.02	−11.98	−7.51	−9.16	−7.95	−9.77	−6.90	−8.43

Table 2 Percentage change in commodity prices

	SF						SR					
	Sigma = 0.5218		Sigma = 0.6262		Sigma = 0.4174		Sigma = 0.5218		Sigma = 0.6262		Sigma = 0.4174	
	South Isles	North Centre	South Isles	North Centre	South Isles	North Centre	South Isles	North Centre	South Isles	North Centre	South Isles	North Centre
Products of Agriculture	−0.020	−0.020	−0.020	−0.020	−0.021	−0.021	−0.019	−0.019	−0.019	−0.019	−0.020	−0.020
Energy products	−0.025	−0.025	−0.024	−0.024	−0.026	−0.026	−0.023	−0.023	−0.022	−0.022	−0.024	−0.024
Metal/non metal ore	−0.025	−0.025	−0.024	−0.025	−0.027	−0.027	−0.023	−0.023	−0.022	−0.023	−0.024	−0.025
Non met. mineral prod.	−0.020	−0.022	−0.020	−0.021	−0.021	−0.023	−0.019	−0.021	−0.019	−0.020	−0.020	−0.022
Chemical products	−0.024	−0.024	−0.023	−0.024	−0.025	−0.026	−0.022	−0.023	−0.021	−0.022	−0.023	−0.024
Mechanics	−0.022	−0.023	−0.022	−0.022	−0.023	−0.024	−0.020	−0.021	−0.020	−0.021	−0.021	−0.023
Transport equipment	−0.023	−0.023	−0.022	−0.023	−0.024	−0.025	−0.021	−0.022	−0.021	−0.021	−0.022	−0.023
Food and beverages	−0.023	−0.023	−0.022	−0.023	−0.024	−0.025	−0.021	−0.022	−0.021	−0.021	−0.022	−0.023
Texitile	−0.023	−0.024	−0.023	−0.023	−0.025	−0.025	−0.022	−0.022	−0.021	−0.022	−0.023	−0.024
Other manufact. prod.	−0.022	−0.023	−0.022	−0.023	−0.024	−0.024	−0.021	−0.022	−0.021	−0.021	−0.022	−0.023
Construction work	−0.020	−0.020	−0.020	−0.020	−0.020	−0.021	−0.018	−0.020	−0.019	−0.020	−0.019	−0.020
Trade	−0.018	−0.019	−0.019	−0.019	−0.019	−0.020	−0.018	−0.019	−0.018	−0.019	−0.018	−0.019
Transport	−0.019	−0.019	−0.019	−0.019	−0.019	−0.020	−0.018	−0.019	−0.018	−0.019	−0.018	−0.019
Financial serv./Insurance	−0.017	−0.018	−0.018	−0.018	−0.017	−0.019	−0.017	−0.018	−0.017	−0.018	−0.017	−0.018
Private services—ARA	−0.018	−0.018	−0.019	−0.018	−0.019	−0.018	−0.018	−0.018	−0.018	−0.018	−0.018	−0.018
Gov. services—ADI, ARP	−0.021	−0.025	−0.021	−0.025	−0.023	−0.026	−0.019	−0.023	−0.019	−0.023	−0.021	−0.024

Table 3 Regional unemployment rate

	SF						SR					
	Sigma = 0.5218		Sigma = 0.6262		Sigma = 0.4174		Sigma = 0.5218		Sigma = 0.6262		Sigma = 0.4174	
	South Isles	North Centre	South Isles	North Centre	South Isles	North Centre	South Isles	North Centre	South Isles	North Centre	South Isles	North Centre
Unemployment	17.702	4.558	17.702	4.558	17.700	4.560	17.703	4.558	17.703	4.558	17.702	4.558

and increases the demand of ARA (the manoeuvre is budget balanced); in the second scenario the regional government reduces the demand for ADI and transfers the saved resources to the Households to increase the demand of ARA. Both scenarios are designed to stimulate the production of private residential health care services but with a different amount of resources since the Households' and regional government's spending for ARA are markedly different.

In the whole the simulation results are rather similar under both scenarios, however they differ significantly when comparing them at the regional levels. The policies implemented stimulate the total output of ARA services and total output of all the other commodities linked to this production in both scenarios and regions. Although the impact on total output is quite different among commodities, it is possible to observe a general reduction in all goods' price as a consequence of the increase in health care services supply.

An interesting aspect of the analysis is given by the effects of the health care policy for the elderly on unemployment. Given a non competitive labour market hypothesis and a wage setting function resulting from the bargaining between Unions and Firms, the model allows to observe the different impact of the policy on regional unemployment rate. In both simulations the unemployment rate decreases in North-Centre region and increases in South-Isles. This confirms the difference in technology between the regions especially as regard to the health care commodity. The validity of these results has been tested with a sensitivity analysis on the elasticity of substitution between labour and capital in the production function. The analysis confirms the robustness of the simulations' results. It is possible to assert that for the North-Centre region, the regional government resources are able to stimulate employment when they are used to promote private residential health care services either directly or indirectly by financing Households expenditure in health care services.

References

1. Bacharach, M.: The role of verstehen in economic theory. Ricerche Economiche **43**(1–2), 129–50 (1989)
2. Bohringer, C., Bhoeters, S., Feil, M.: Taxation and unemployment: an applied general equilibrium approach. Econ. Model. **22**, 81–108 (2005)
3. Ciaschini, M., Socci, C.: Bi-regional SAM linkages: a modified backward and forward dispersion approach. RURDS 19(3), 234–255 (2007) (Wilye-Blackwell). ISSN: 0917–0553, doi:10.1111/j.1467-940X.2007.00138.x.
4. Ciaschini, M., Pretaroli, R., Socci, C.: Contenimento della spesa corrente e spesa in conto capitale nel sistema sanitario pubblico, Tendenze Nuove, n.2/09, Il Mulino (2009). ISSN: 1010-1010
5. Ciaschini, M., Pretaroli, R., Severini, F., Socci, C.: Health services as a key sector for income and employment change, Bulletin of the Transylvania University of Brasov, 3(52). Social Sciences - Law, Series VII (2010)
6. Ciaschini, M., Pretaroli, R., Severini, F., Socci, C.: La spesa per anziani in una prospettiva di crescita economica. Tendenze Nuove **5**(2011), 435–462 (2011)
7. Ciaschini, M., Pretaroli, R., Severini, F., Socci, C.: La produzione di servizi per anziani non autosufficienti nell'economia italiana: un'analisi di dispersione. Tendenze Nuove **5**(2011), 463–476 (2011)

8. Ciaschini, M., Pretaroli, R., Severini, F., Socci, C.: L'impatto della spesa per gli anziani sull'occupazione attraverso un modello computazionale. Tendenze Nuove **5**(2011), 477–89 (2011)
9. DeWatt, D.A., Pincus, T.: The legacies of Rudolf Virchow: cellular medicine in the 20th century and social medicine in the 21st century. In: IMAJ, vol. 5, in particular pp. 395–397 (2003)
10. De Angelis M.: Introduzione, in La complessitá dei sistemi sanitari. Problematiche giuridiche del caso italiano, Bologna (2012)
11. Fossati, A.: Equilibrio Economico Generale e Simulazioni. FrancoAngeli, Milano (1991)
12. Grassini, M.: Rowing along the computable general equilibrium mainstream. In: Input-Output and General Equilibrium: Data, Modeling and Policy Analysis. ECOMOD 2–4 Settembre 2004. Free University of Brussels (2004)
13. Hicks, J.: Rational Behavior: Observation or Assumption?. New York University Press, New York (1986)
14. Pretaroli, R., Severini, F.: Assessing the effect of a value added tax policy on wine sectors. Enometrica Int. Rev. EAWE **2**(2), 41–1 (2009)
15. Pretaroli, R., Severini, F.: Dalla sam al modello computazionale cge. Ministero dell'Interno. Modelli Computazionali per la valutazione dell'impatto del Federalismo Fiscale sulla finanza locale, Roma, pp. 77–100 (2008)
16. Pretaroli, R., Socci, C.: Produzione e distribuzione del reddito in una sam biregionale. Ministero dell'Interno. Modelli Computazionali per la valutazione dell'impatto del Federalismo Fiscale sulla finanza locale, Roma, pp. 29–76 (2008)
17. Rubinstein, A.: Dilemmas of an economic theorist. Presidential Address. Econometric Society, Madrid (2004)
18. Socci, C.: Distribuzione del reddito e analisi delle politiche economiche per la regione Marche. Collana della facolt di Giurisprudenza, Universit di Macerata. Dott. A. Giuffr Editore 118, 6–160. ISBN: 88-14-11259-2
19. Socci, C.: La costruzione di SAM regionali: aspetti teorici e applicativi una SAM biregionale per le Marche. In: A. FOSSATI E R. TARGETTI LENTI, Matrici regionali di contabilit sociale e analisi di politiche economiche: il caso della Liguria, Toscana e Marche, 365.311:123–156, FrancoAngeli, Milano (2004) ISBN/ISSN: 88-464-5816-8
20. Trabucchi, M.: Ospedale e territorio: una problematica sempre nuova. Tendenze nuove **1**, 5–7 (2010)
21. Van der Werf, E.: Production Function for Climate Policy Modeling: An Empirical Analysis. FEEM Working Papers n.50 (2007)
22. Virchow, R.C.: Report on the Typhus Epidemic in Upper Silesia. 1848. Am. J. Public Health **96**, (12) (2006)

Active Ageing and Public Space: The Creative City 3.0

F. Bronzini, N. Barbini, G. Marinelli, M. Pantaloni and E. Giordano

Abstract The article presents a study for a "creativity incubator" in the Torrette district, Ancona, Italy, next to the Regional Hospital Centre. The urban-territorial project involves the creation of multifunctional public spaces to increase the spaces and operational relations between the social-healthcare structures and the public space of the district. The primary element of innovation introduced by the project is an integrated vision between different disciplinary fields, in order to define an original urban planning approach to the theme of active-creative ageing. Is it feasible to reduce the impact of the public spending in the development process by increasing the quality of life? The resilient city, which adapts to change (of citizens), ensures better conditions of psycho-physical wellbeing (and thus for the health of individuals) and may help the autonomy of the elderly population, propelling the induced processes of active ageing even through the careful planning of public urban spaces.

1 New Relations Between Urban Regeneration and Active Ageing

According to recent demographic statistics, Italy is one of the countries with the highest ageing indexes in the world. In this special index, our country ranks second in Europe after Germany and at a world level only Japan has a higher life expectancy than ours.

In the national ageing index, Marche Region ranks second after Liguria. In recent years, the elderly population in Marche Region has in fact increased progressively showing an increase of 1.8 in the period 2002–2012; despite the fact that this index

F. Bronzini · G. Marinelli (✉) · M. Pantaloni · E. Giordano
SIMAU Department, Polytechnic University of Marche Region (UNIVPM), Ancona, Italy
e-mail: g.marinelli@univpm.it

N. Barbini
Italian National Institute of Health and Science on Aging (INRCA), Ancona, Italy

S. Longhi et al. (eds.), *Ambient Assisted Living*, DOI: 10.1007/978-3-319-01119-6_36,

in the same period decreased for the city of Ancona, the percentage is in any case higher than the national average of 147.2 % and the regional average of 170.2 % [14].

As amply demonstrated by various of studies, the progressive ageing of the population has an impact on public spending, debt and *deficit*, consumptions, investments, productivity, the labour market, and more generally on the development process [20].

Despite the growing rate of elderly citizens in our country, the focus to date has been above all on reducing public spending (above all pensions), while qualitative aspects such as the way in which an elderly person lives urban spaces and the city and how relationships between the city (and more specifically outdoor spaces) and individuals affect psycho-physical and social health, have been neglected [22].

The radical reconsideration of the relationship between development and ageing would lead to improved labour integration and social-cultural inclusion of elderly people, encouraging a different attitude towards ageing [16].

The World Health Organization has developed the concept of ≪active ageing≫, intended as a process, applicable both at an individual and community level (≪mass active ageing≫), whose aim is to attain the maximum physical, mental, social and economic potential of the elderly. The term ≪active≫ refers to the tangible social, economic, cultural and spiritual participation of elderly people in the life of the city [23].

Active ageing is substantially the exploitation of diffuse relationality that ≪fills in≫ the time freed by the activities of the adult age and is closely linked to the new activities and commitments that the person takes on. While in past decades policies for the elderly focused above on all their needs, with the passage to the concept of "active ageing" there has been a change of paradigm: the focus is now also on the rights of the elderly and the possibility of their tangible participation in community life.

In this sense, an elderly person is recognised as a potential resource, in that the same has "experience", physical and intellectual skills that must be protected and valued, rather than as simply a person to whom a specific healthcare intervention is addressed.

These skills depend on many different factors, but, above all, from an appropriate quality of life [6].

2 Active Ageing 3.0: Creativity in the Ageing Process

Starting from the presupposition that active ageing is something which is developed in our daily lives and that active ageing means keeping an active role in society at a social and cultural level, the next step to take would be that of encouraging a ≪creative ageing≫ process in which ageing assumes decidedly innovative characteristics both at an individual and "community" level [13].

Creativity intended as a dynamic process which, together with senility, may reach new goals, is often tied to experiences encountered previously and is the result of achieving a certain interior equilibrium, security and confidence in oneself. It is also a symptom of freedom and independence from social pressures [9].

Creative ageing is a multi-dimensioned process which demands its own rules and orientations. It is necessary for example to acquire a certain mental flexibility, free emotionality, supportive and sensitive open mindedness to others and improving one's ability to have very close relations with others. In addition to interior prerequisites, *creative ageing* also requires ***≪creative places≫***, namely places where elderly people may develop their creativity autonomously, and which are able to stimulate them to develop relationships, finding new solutions to their requirements. All this is apparently determining to keep the brain fit, which in turn is fundamental for preserving cognitive functions and physical autonomy [10].

In order to develop creativity—through which favourable conditions are created to express oneself in an original manner and to express thoughts and skills—it is necessary to think of urban spaces which act as "incubators of creativity".

If the need for ≪creative places≫ has now become a need openly expressed by individuals of the modern society, it is even more important for elderly people who must preserve their health, intended as psycho-physical and social wellbeing for as long as possible in the period of the so-called "4th age".

In this sense the project for the use of public spaces of the city develops the idea of non-sectorial social and economic sustainability, but with an impact on the community as a whole.

3 Creative Ageing and Public Space as an Incubator of Social Quality

An urban-territorial project, currently being developed by the Urban Planning Division of the SIMAU Department, Marche Polytechnic University, with the support of the Epidemiological Centre of the Ancona INRCA (Italian Institute of Rest and Care for the Elderly), involves the study of a "creativity incubator" in the Torrette district, Ancona Municipality, next to the Regional Hospital Centre.

The Urban Planning Division of the Department of Science and Engineering of Materials, Environment and Urban Planning (SIMAU) has for years developed research work on the themes of accessibility, the slow city and urban resilience, drawing attention to weak users in a sole holistic vision of the city: "the friendly city".

The project, which is strongly integrated to the urban fabric of the district, involves the creation of multifunctional public spaces.

The Torrette area, which is identified in the Programmatic Document for the New Ancona Urban Plan, is a district that has substantial criticalities under many aspects—excessive mono-functionality, problems of accessibility, marginality with respect to the large infrastructures—and has been included in priority urban requalification projects [2].

The Urban Plan includes interventions for the regeneration of weak fabrics through the integration of various functions, and the development of a system of public spaces to strengthen the transverse connections to the historical seafront. The Project of the "waterfront of excellences" specifically includes the organisation of a new mix in the hospital area and the creation of an appropriate functional mix in the Torrette area, in order to integrate the existing building fabric with an offer of services, that currently do not exist, to support the Hospital, on the external area of its perimeter.

In this sense, the localisation project of the ***incubator of creativity*** for the implementation of comprehensive processes introduces urban values and contents that may become a booster of the regeneration processes in Torrette district as a whole, which might in the future be characterised as the ***district of creative ageing***, in line with the guidelines identified by the new Structural Plan of the city.

The primary element of innovation introduced by the project is an integrated vision between different disciplinary fields (urban planning and urban regeneration—addressed to the individual within a community—and the health sector, where the individual is the central point of interest) in order to define an original urban planning approach to the theme of active-creative ageing.

The programme of "incubator of creativity" interventions includes the development of social and cultural activities, which are already present in the hospital structure, but that may be extended to a district dimension, thus becoming an opportunity of sociality and integration, leisure, recreation and healthcare services for the city.

It may include integration with lodgings that provide free or subsidized accommodation for the families of inpatients, where elderly people with invaluable professional skills acquired within hospitals, as well as members of voluntary associations for families with family members who have particular ageing pathologies, will have an environment that helps to attenuate their social hardship [18].

The inclusion of specific innovative activities and therapies will imply the organisation of indoor and outdoor spaces, premises for creative laboratories and spaces for specific occupational therapies and functions (such as the AAT-Animal Assisted Therapy or the Ausilioteca—Aid Lab) integrated to businesses that serve the hospital structures and district. The new spaces make it possible to increase the mix of functions and opportunities for integration between different social clusters, elderly people, students, service staff, children, and persons with handicaps, encouraging the implementation of participatory processes, responsibilization of the community and phenomena of major social inclusion in the district [15]. In this sense, the "organisational management aspect" of the hospital, intended as a structure of different functions, services and spaces, over and above healthcare, emergency or welfare functions, is projected outwards and contributes actively to the regeneration of the urban space.

Going beyond the "specialised mono-functional boundary" of the hospital centre allows the structure to develop the public functions as whole, which are present in the structure, and that are fundamental to attaining the new standards of service quality offered to users who, temporarily and for various reasons, are forced to "live" in or close to the hospital structure.

The new spaces strongly linked to the urban fabric of the district represent a public space which is easily accessible to the population of "active elderly" residents, who, will perceive major security in the environment thanks to the presence of medical staff and the surveillance system of the hospital structure, thus helping them to use services autonomously, multiplying the possibilities of social integration between the elderly and stimulating inter-generational exchanges with users and personnel of the structure.

The project, which is structured in a sequence of open and built-up spaces, steers the regeneration processes of the city and offers solutions to the growing demand of an increasing urban population that is gradually ageing.

The positive impact of the strategic-organisations actions of the project help active ageing, increasing the quality of life standards of the elderly, by ensuring major access to places for all users, helping slow mobility and motor exercise, physical and social wellbeing and appropriate conditions of indoor–outdoor comfort for laboratory and therapeutic activities that encourage use of these new public spaces, that may be compared to inclusive urban microcosms.

Another positive impact is that of encouraging integration between generations [24].

Steering the project choices of urban design planning according to a holistic vision represents a possible scenario and new opportunities to rationalize public resources and reduce health costs.

The **resilient city**, which adapts to change (of citizens), ensures better conditions of psycho-physical wellbeing (and thus for the health of individuals) and may help the autonomy of the elderly population, propelling the induced processes of active ageing even through the careful planning of public urban spaces [7].

4 Torrette "Incubator of Creativity": Project Aspects and Possible Implementation Scenarios

The project is structured in five strongly integrated initiatives with progressive scenarios to increase the spaces and operational relations between the social-healthcare structures and the public space of the district.

It also provides major opportunity to obtain a feedback on the efficiency of the urban regeneration initiatives, waste control, flexibility in the management of the implementation programme through actions of sequential growth (Fig. 1).

The initiatives, described in the brief summary sheets indicated below, that give a description of the implementation actions and experts involved, are as follows:

1. The multi-functional hospital Garden (Table 1);
2. *"The Red Carpet"*: urban transport providing new access to the district (Table 2);
3. Regeneration of the district and integration with the hospital functions (Table 3);
4. The new Walk from the Hospital Centre to the Waterfront (Table 4);
5. Requalification of the Torrette waterfront: the sea made to measure for the elderly (Table 5).

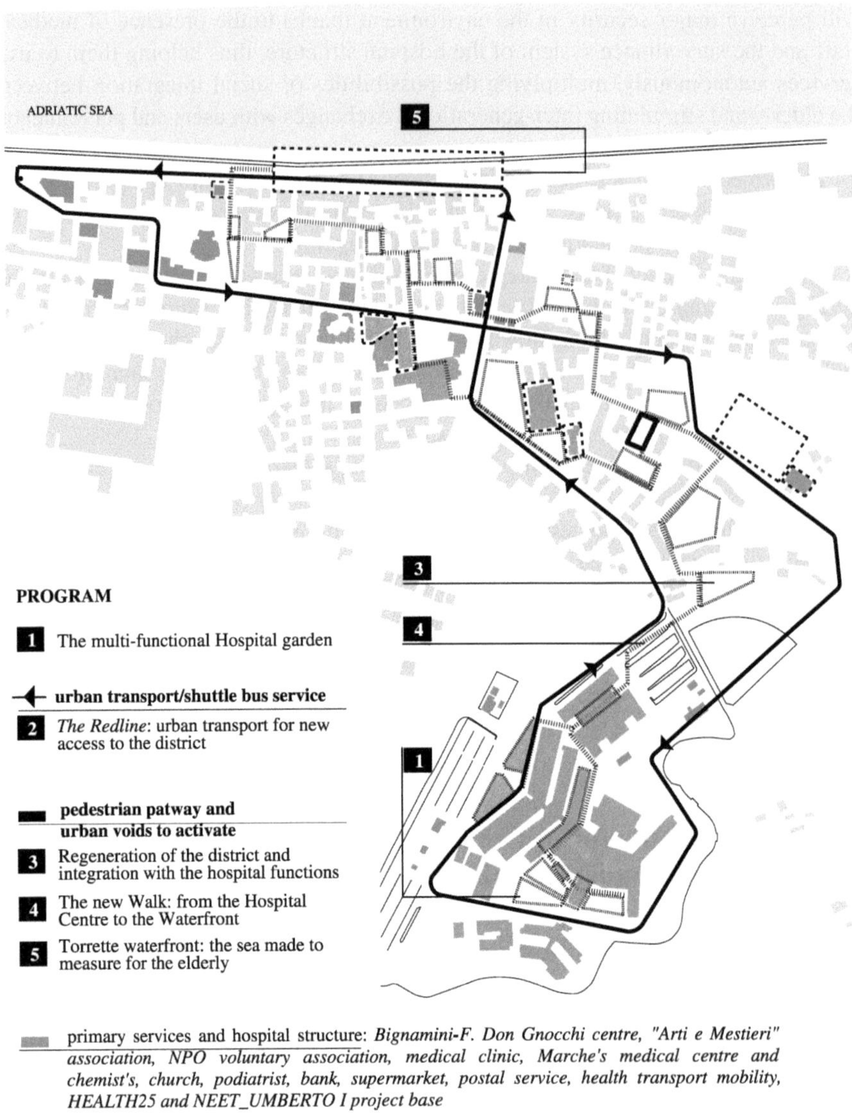

Fig. 1 Urban scale diagram with public service system, activation of urban voids and refunctionalisation of disused containers

Table 1 The multi-functional hospital garden

Actions

Construct a **system of public spaces outside** the hospital structure, implementing a mix of functions tied to leisure, psycho-physical wellbeing, sociality and inclusivity, such as

- Areas with botanical species that stimulate various sensorial capabilities [1]
- Life path consisting of floor exercises and the use of gym equipment [19]
- Garden Therapy [21]
- Accessible playgrounds to stimulate interaction between the elderly, adults and children;
- Places for relaxation, refreshments and toilets
- Spaces for information and management safety
- Teaching activities such as guided visits to urban nurseries and small botanical gardens associated to environmental education activities

Possible implementation scenarios

- Rationalisation of spaces in the hospital structure: construction of courtyards, parking areas, green areas, volumes to recover, disused areas to refunctionalise.

Experts and operators involved

- Civil service volunteers for the information service, management of toilets and catering spaces (possible implementation of the structure already operative in the hospital but which is not sufficient for the demand)
- Security guards responsible for safety, nurses and physiotherapists of the hospital
- University students of the Faculty of Agriculture, through use of the 150 paid hours system
- Personnel experienced in caring for plant to help gardening activities, 1 self-sufficient elderly person a day with experience in agriculture

Table 2 "The red carpet": urban transport for new access to the district

Actions

Definition of a route in predefined stages, served by a **specific shuttle bus** service for elderly people which guarantees facilitated access [4] to

- Services of primary necessity (chemist, supermarket)
- Spaces for socialisation and free time already active in the district
- New creative urban spaces activated in the existing urban fabric (new)
- Torrette hospital-university centre: the new "incubator of creativity"
- Torrette waterfront: the sea made to measure for the elderly

Capacity: 8 seats with platform for persons with mobility handicap

Frequency of the service: 35 min, route length: 4 km

Possible implementation scenarios

- Expansion of the protocol of agreement signed by Ancona Municipality, Umberto I, Lancisi, ASL 7 and Conerobus for active inclusion of the dedicated transport service in the urban/non-urban transport to the Hospital
- Collaboration agreement with the Marche Region Civil Service for personnel to drive/manage the dedicated vehicle and transport service for passengers with mobility handicap

Experts and operators involved

- 2 part-time participating students (2 persons for 2 daily shifts)

2 civil service volunteers for 2 daily shifts (8–13, 15–19 on weekdays)

- Elderly persons with experience in this field (retired bus drivers)
- Conerobus personnel who work in the Ancona C.R.A.S structure, with specific agreement

Table 3 Regeneration of the district and integration with the hospital functions

Actions
Activation of urban spaces dedicated to recreational-game activities which encourage social integration and stimulate interaction and creativity
– Containers for new therapeutic functions: Animal Assisted Therapy [11], Music Therapy, Art Therapy, Horticultural Therapy as a cure through interaction with nature [3]
– Auxiliary hospital structure of **Ausilioteca** as centre for information, assistance and support to users with mobility, sensory handicap. The structure will also have spaces for physiotherapy and rehabilitation [5]
Possible implementation scenarios
– Temporary use and management of disused containers and areas to refunctionalise, new public spaces to restore to the city (subsidised lease or lease agreed by Ancona Municipality)
– Direct and indirect interventions by means of urban planning tools such as ecological-environmental **compensation** and **equalisation**
– Involvement of the recreational artistic-cultural association Auser "Yesterday's young people" (that operates in Torrette district) for the integration of auxiliary hospital activities in the annual cultural programme
– Extension of the student's collaboration agreement to activities related to *UNIVPM* services—*part-time* collaboration (150 h) for the management of the local resources required to carry out the new activities
Experts and operators involved
– Volunteers registered with "Yesterday's young people", self-sufficient elderly people who provide their professional know-how
– Qualified personnel for specialised therapies (veterinary, physiotherapist, technical personnel for Ausiloteca)
– 2 civil service volunteers to ensure that activities are carried out safely
– 2 *UNIVPM* participating students-150 h for the management of premises used for programme activities

Each initiative consists of specific development actions, whose sustainability and implementation methods are verified through:

- a control of the level of involvement and social inclusion actions (by monitoring the levels of responsibilization of local associations, young students partnered with university and hospital institutions, the levels of involvement of elderly people who may provide their *know-how*),
- a control of the economic-financial feasibility by:
  - a comparative evaluation of the rate of use of public resources in the medium-term with respect to a reduction of the public social security expenditure, health-care and social-welfare costs to support non-active elderly users;
  - an evaluation of the cost of municipal policies for the re-functionalisation of disused public spaces, even through tax handles: incentives and agreed leases for inoperative commercial activities and premises;
  - evaluation of costs/benefits on municipal maintenance costs for public urban spaces;
  - evaluation of costs/benefits on the cost for public and private urban mobility for the social-welfare structures involved in the project.

Table 4 The new walk from the hospital centre to the waterfront

Actions
Development of the **pedestrian pathway** (slope no more than 5 %) connecting the hospital centre and Torrette railway station with access to the beach [12]
Intervention method
– Securing road kerbs so that they are of suitable dimensions for pedestrians with mobility handicap
– Ensuring the recognition of routes through *street design* and graphic interventions and use of shockproof paving
– Introduction of urban furniture that may be used by all users, (seats and areas for short stops, continuous line of handrails, flooring for the blind indicating changes of direction, services) [8]
Possible implementation scenarios
– Direct or indirect interventions through urban planning tools such as ecological-environmental **equalisation** and **compensation** [2]
Experts and operators involved
– Ancona Municipality, that includes in strategies for open space network and public or private in-between green areas, areas with expired restrictions for services and facilities, fringe and urban completion areas to consolidate the structure of the public city
– Private parties who collaborate with the municipal administration and actively participate in implementing the design of the new, sustainable and inclusive public city

Table 5 Requalification of the Torrette waterfront: the sea made to measure for the elderly

Actions
Requalification of the Torrette coast by constructing new spaces that may be used by otherwise—abled users [17]. The new urban functions may trigger new opportunities of longitudinal connection to the historical waterfront of Ancona
Intervention method
– Cleaning and securing the section of beach in front of Torrette station
– Construction of two salt **water pools** that may be used by persons with handicaps or mobility handicap via small floating wooden jetties. Access will be guaranteed thanks to the installation of hydraulically connected lifting platforms which function thanks to the pressure of water without electricity, allowing users to enter the water in complete safety to do exercises and rehabilitation activities
Possible implementation scenarios
– Project co-financed by Ancona Municipality—INRCA, Italian Institute of Rest and Care for the elderly
– Tourist operators
– Social-welfare structures (Bignamini-F. Don Gnocchi centre, Avulss and NPO voluntary Associations)
Experts and operators involved
– 2 lifeguards on duty together
– 1 Physiotherapists for rehabilitation exercises (one for each activity)
– 1 volunteer to clean the beach
– 1 lifeguard assistant to ensure safety on the beach, or as support to the security guards of the hospital structure

References

1. Algesi, A.: Gardening therapy: a hobby that cures. In: Diagnosis & Therapy. Retorbido (PV). Available via DIALOG: http://www.det.it/aprile08-4.html (2008)
2. Ancona Municipality: Priority projects for urban requalification. The new urban plan, vol. 4, pp. 40–51. Ancona (2008)
3. AUSL: A garden for garden therapy, cure yourself by caring for plants. Rome. Available via DIALOG: http://archive.forumpa.it/forumpa2004/sanita/cdrom/home/progetto/208.html (2003)
4. Baster, N.: It's my choice safer mobility for an ageing population. Parliamentary advisory council for transport safety. Decade of action for road safety 2011–2020. Cluta House, London (2012)
5. Bitelli, C., Desideri, L., Motolese, M.R.: Aids and social costs. It is worthwhile to propose assisted therapies "well". Bologna (2012)
6. Bresolin, F., Minello, A., Gallucci, M.: Development of the quality of life and creative longevity. In: Social Updates, vol. 02, pp. 119–128. Milan (2007)
7. Bronzini, F., Bedini, M.A., Sampaolesi, S.: The friendly city of Robero Busi. University Press, Ancona (2011)
8. Busi, R., Ventura, V.: Living and walking in cities—safety of vulnerable road users. In: Acts of the Fifth Conference of Brescia. University of Brescia, Brescia (1998)
9. Cesabianchi, M., Cristini C., Cesabianchi, G.: Ageing: corporeality and creativity. Enna (2002)
10. Cesabianchi, M.: Creative ageing. Consorzio-Nettuno, Rome (2002)
11. Cirulli, F., Alleva, E.: Zooanthropology of the third millennium: rebuilding the ancient animal–man and man–animal relationship for psychological rehabilitation: pet-therapy in Italy. In: Battaglia, L. (ed.) Topics in Bioethics: Journal of the Italian Institute of Bioethics. Dignity. The New Frontier of the Animalism, pp. 167–174. Department of Behavioural Neurosciences, Higher Health Institute of Bioethics, Name Edizioni, Genova (2007)
12. Dell'Amico, M.: The city of the future? Made to measure for the "smart" elderly. Available via DIALOG: http://life.wired.it/news/2012/09/27/aged-friendly-city-anziani-ryuta-kawashina-u-universita-milano-bicocca.html (2012)
13. Florida, R.: The rise of the new creative class. Mondadori, Milan (2003)
14. Istat: We in Italy, 100 statistics to understand the Country we live in 2013. National statistics system, Rome (2013)
15. Marino, P.: Auxiliary non-conventional therapies for the elderly. Available via DIALOG: http://www.pascalmarino.eu/Risorse%20it.htm (2007)
16. McMichael, A.J.: The urban environment and health in a world of increasing globalization: issues for developing countries. Bull. World Health Organ. **78**(9), 2000 (2000)
17. Morini, A.: Usable cities, possible solutions on a study of 33 metropoles. Assistance to the elderly, pp. 15–17. National Research Council (2008)
18. Nannini, V.: The habitat of the elderly person: the European framework and the Dutch experience. In: Act of the Conference: A Documentation Center for the Excellence of Services for the Elderly, 30 November 2006. ICIE Building Cooperative Institute, Bologna. Available via DIALOG: http://www.centrorubbi.it/scheda.asp?num=364&searchfor=L%92habitat%20dell%92anziano:%20il%20quadro%20europeo%20e%20l%92esperienza%20olandese&searchtipo=1&cat=10 (2006)
19. Pizzoni, M.: Making a point on physical exercise and longevity. Journal Club. 10 August 2008 (2012)
20. Salinas, U., Mastrorocco, N.: On ageing of the population: notes and considerations. Bari. Avaliable via DIALOG: http://www.dip-statistica.uniba.it/html/docenti/salinas/dispensainvecchiamento.pdf (2004)
21. Segesta Group: Segesta informs, smiles and emotions with garden therapy. Segesta Group Assistance and Health Care, Milan, March 2012 (2010)
22. Villani, D.: Urban planning and ageing: searching for the invisible man. Places of Care **2**(2004), 9–11 (2004)

23. World Health Organization: Active ageing: a political frame work. Geneva, Ginevra. Available via DIALOG: http://www.euro.who.int/document/hea/eactagepolframe.pdf (2002)
24. World Health Organization: Global age-friendly cities: a guide. Ginevra (2007)

“Smart” Social Housing Design: Methods, Tools and Innovation in the Assisted Living Architectural Project

Fausto Pugnaloni and Davide Di Fabio

Abstract The Architectural plan used for social housing is usually targeted to ensure accessibility to basic service of living throughout the definition of interior design and building solutions dedicated to disabled people. Nowadays, thanks to a new concept of disability and major number of interested people, the performance of this type of building have to grow up: buildings have to be accessible as well as “smart”, be able not only to satisfy help requests but to anticipate them during the management of the house. From this point of view, the architectural planner should rule the classic tools of the plan as well as manage and organize the resources of building automation, working in a “meta-architectural” dimension. Even describing some study cases, this paper want to examine the change of methods and tools of architectural planning used for the assisted living, re-defining the concept of social housing, from helpful to smart architecture.

1 Introduction

The longer life expectancy in the western society has changed the needs of the population with respect to many of the social services they are provided with. Many of us will be getting through different age groups and experiencing continual changes of their physical and health status. As a result the population’s relation to one of the most important essential goods, the house, has changed as well.

Although it is related to the “immovable good” concept, the house system should be subject to continuous mutations: firstly in order not to loose its economical value

F. Pugnaloni (✉) · D. Di Fabio
Civil and Building Engineering and Architecture department DICEA, Faculty of Engineering, Università Politecnica delle Marche, via di Brecce Bianche, Ancona, Italy
e-mail: f.pugnaloni@univpm.it

D. Di Fabio
e-mail: d.difabio@univpm.it

S. Longhi et al. (eds.), *Ambient Assisted Living*, DOI: 10.1007/978-3-319-01119-6_37,

and secondly so as to respond to the need of the people who live in it. Therefore the house is an "immovable" good that needs to be "flexible" at the same time. As far as the social housing is particularly concerned, the flexibility concept has been differently interpreted causing contrasting results. Flexibility is often defined as the capacity of the system to respond to the needs of different users that alternatively share the same living area and it is not the answer to the change of the needs of a person in their life cycle. In this respect it is necessary to develop an "house system" that is able to change accordingly to those needs. In this effort are surely included as necessary but not sufficient tools technological equipment such as home robotics [1] and furniture and accessibility solutions for disability [2].

Specifically, the building system will adapt to the moving needs picture of the individual by changing the available space and replacing some building elements if necessary.

2 Notes on History of Architecture

In the History of Building this way of understanding the Architecture was fundamental to some experimentations occurred in the second postwar period. *The Japanese Metabolist, for instance, reacted to the pressure of the overpopulation at the end of the 1950s by putting forward steady growing mega structures suitable for including joint elements: their living cells, as in the work of Noriaki Kurokawa, would become little prefabricated buildings connected to a huge helical skyscraper or, as in the projects of Kiyonari Kikutake, would be fixed as sea shells to the internal and external surfaces of enormous cylinders floating in and above the sea* [3].

An architecture able to grow and adapt to the environment and the human evolution as the cellular system in the human body. As a matter of fact, this human experience is limited by a rhetoric avant-garde with a few confirmations: except from the Sky House planned by Kitutake in 1958 and the Nagakin Tower by Kurokawa formed by capsules and build in the Tokio Ginza, a very few ideas of the Metabolists were realized (see Fig. 1). From a tightly technological point of view it is important to mention Jean Prouve, a Engineer-Artisan of genius, who understands his buildings and furniture as a system of parts that can be assembled or unassembled in some sort of a real scale Meccano [4]. In Prouve's work the dry and modular building solution is the basis of his planning and building activity: it gives birth to expandable, prefabricated and easy to construct as well as easy to maintain buildings (see Fig. 2).

As far as the recent building solution is concerned, the projects of Leo Kaufmann, the inventor of "system 3", are worth mentioning [5]. Based on prefabricated service modules and the next addition of living volumes, system 3 is a building system that puts forward vertically and horizontally expandable living units which are also easy to be carried to the building site through a container. Supported by the Xlam technology this system is able to combine the standardization and customization of the living unit.

Fig. 1 Tokyo: picture of the Nagakin Capsule Tower. *Source* Ref. [3]

3 Definition of New Living Models

The DARDUS-DICEA work group has long been working on the project of new living and social architectural models able to satisfy the contemporary living needs. Over the last years clean signs of the need to return to study, research on and pay attention to the basics component parts of the living caused by the needs of the users have appeared.

Consequently there is the opportunity to abandon the understanding of the house as a pure object (as a planning, production, building etc. object) for a vision of the house as a need therefore starting from the condition of the living needs and evaluating their developments in order to subsequently find the most suitable solutions to satisfy them. Secondly there is the need to not consider the typological residential models as the final result, that cannot be improved, of a "meta-planning" activity. It is verified, for instance, that exists a new drive to a more intense and extended use of the house with additional activities other than the traditional ones derived from the work activity system, production and assisted living.

The clear need for personalization also leads to the qualitative and quantitative clearing of the limits typical of an old understanding of standard (more or less valid for a generic as well as nonexistent average user). Based on the relative analysis, a general picture of the new users that are to be associated with a new living model, has taken shape. As for the basic shared models (see Sect. 4 for a close examination), each model adds a peculiar element which satisfies the specific needs.

4 Basic Modules: Planning, Building and Transportation

The planning philosophy in use is to plan a unique architectural-building system able to interpret all the model which came out. From this point of view the planning reference is surely the "system 3" mentioned above, a prefabricated system

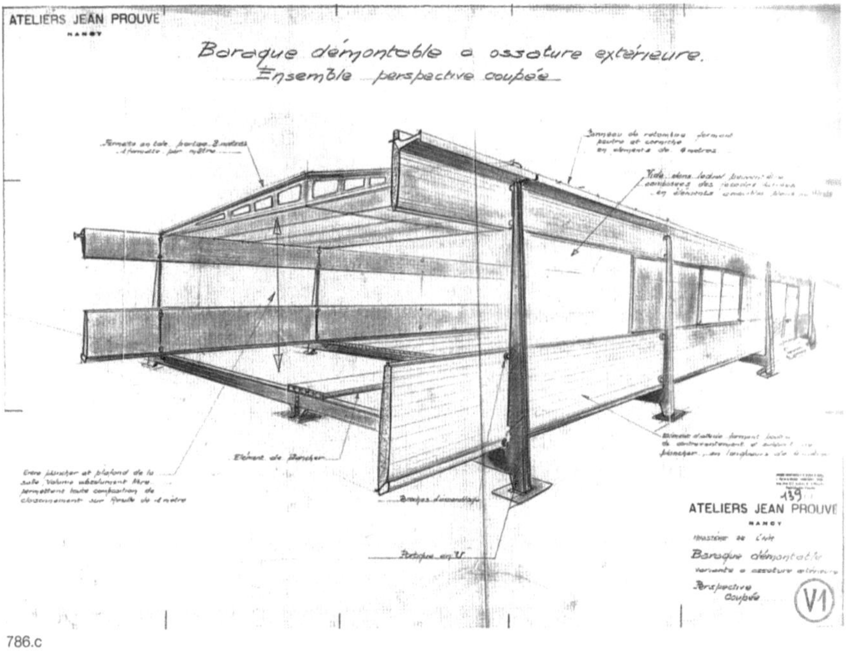

Fig. 2 Design for domountable barrack unit, variant with external structure frame. *Source* Ref. [4]

able at the same time to face the different needs of the contemporary living. This reference was however revised and implemented by the work group in the research project "Kithaus plus" [6]. Subsequently a new model was planned which includes the common equipment for residential buildings.

As far as flexibility and customization are concerned, the user can additionally configure this module with specific sub modules (see Fig. 3). Another shared element is the insertion of the so called "living" modules that contain spaces whose functionality is not previously defined but rather during their use. A therefore flexible room use typical of the present living way in continuous evolution. They are therefore empty spaces with as less as possible technical equipment. From a social housing use perspective this system gives the possibility to prefabricate its components so as to reduce the realization costs and control in any case the overall quality of the manufactures.

For this reason the "service" module is pre-configured while the "living" modules are realized on site. In order to make this process easy, the living modules use a boom that makes it possible to package them in advance and recover volume afterwards (see Fig. 4). With this device and thank to a thorough dimensional study, the system "neutral" module + "living" modules can be entirely transported in one or more containers depending on the complexity of the final living module (see. Fig. 5).

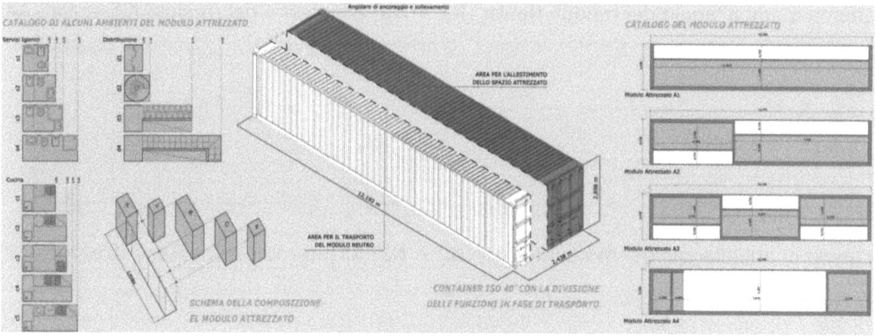

Fig. 3 Service module and sub modules for its implementation. *Source* Ref. [7]

Fig. 4 Functioning principle of the "living" module. *Source* Ref. [7]

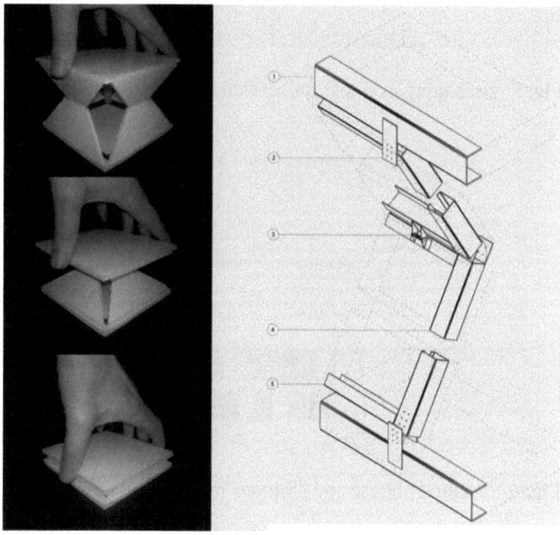

5 "Assisted House" Hypothesis for the Assisted Living

Having defined the basic elements of the architectural-building system, it is time to delve into its applications in the various living models [7]. The possible configurations are primarily obtained from the aggregation of the "living" modules around the "service" module.

Figure 6 depicts the layouts of some models developed on the basis of the analysis made. As far as the "assisted house" is specifically concerned, the system is developed starting from one "service" module to which various "living" modules can be added, at the beginning or with the passing of time, and configured as rooms for disabled people. As the users grows in number, the assisted house can be provided with another

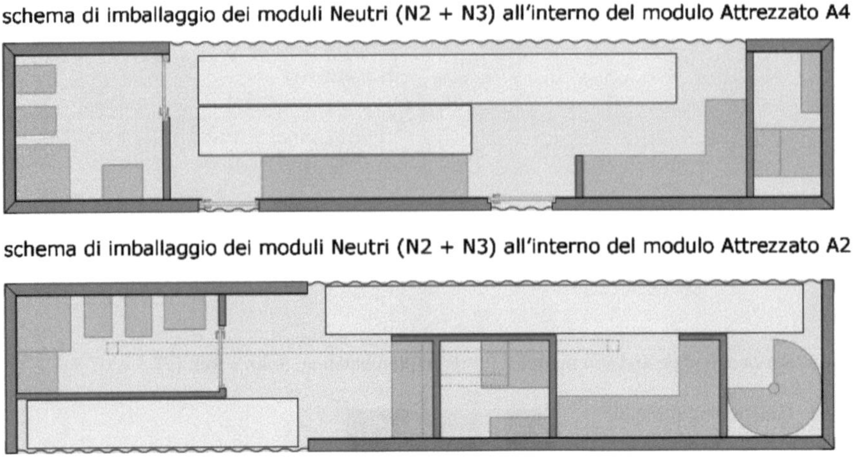

Fig. 5 Packaging example for service and living module in the container module. *Source* Ref. [7]

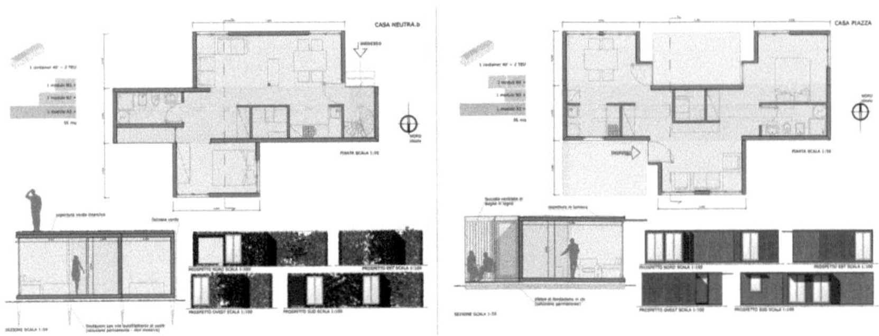

Fig. 6 "Neutral" house and "Square" house. *Source* Ref. [7]

service module as well as another "living" module which is to be fitted out as a little medical consulting room for the first aid or the normal health care for disabled people (see Fig. 7).

The building integration of the extensions is achievable thanks to the dry steel building system with prefabricated plugging panels. These can be removed in the interconnection points to new volumes. As a matter of fact, the modularity of the system is reflected in the urban planning: by aggregating different types of house in the plan, it is possible to get little one- or multipurpose communities (see Fig. 8). In this case, by aggregating different "assisted houses" it is possible to obtain shared areas such as green protected patios which represent an additional functionality for the users.

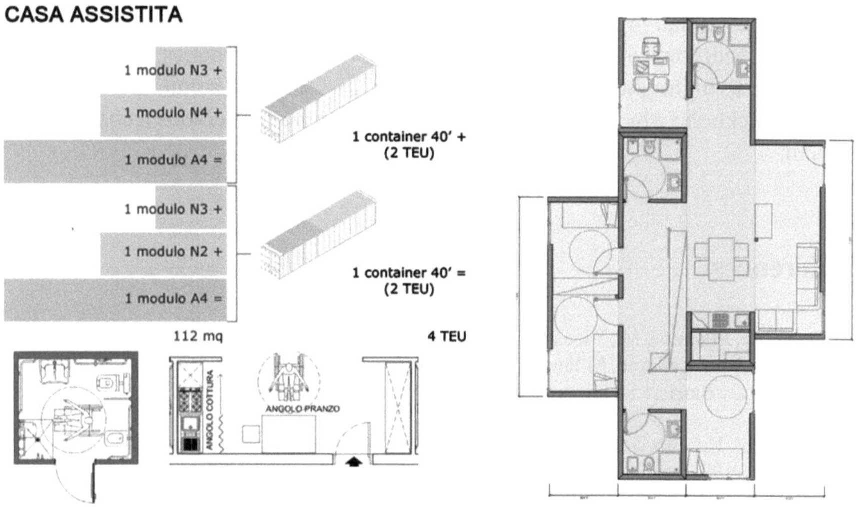

Fig. 7 Layout "Assested" house. *Source* Ref. [7]

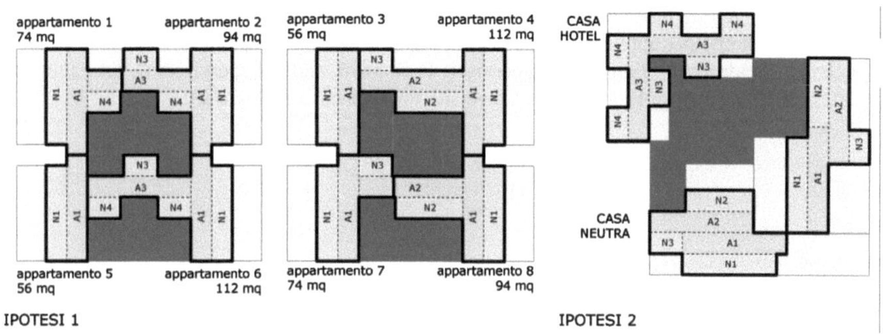

Fig. 8 Hypothesis for assisted community. *Source* Ref. [7]

6 Conclusions

It is evident that this experimentation is not only concerned with the assisted living, rather in general with the new ways of living. However, this model can be an interesting close examination which can be applied in our countries and, for example, in the developing countries as well. As a matter of fact, thanks to the prefabricated components, it is thinkable that they are produced in our territory and then transported to the other critical areas.

Although the research has primarily technological implications rather than architectural or functional ones, it will evolve to the implementation of multi story buildings which make possible, in according to the logic that has been explained so far, to achieve the needed flexibility of the house and at the same time a greater living density. It is important to point out that the invented system has to integrated with the most traditional internal design solutions for the assisted living (for example by fitting out bedrooms and bathrooms) and innovative home robotics equipment. The purpose is to highlight the architectural project as the living solution maker: a sort of solid backbone which is to be connected to the necessary nervous receptors of the system.

References

1. Zanna, G., Malavasi, M., Vaccari, G.: Manuale illustrato per la domotica ad uso sociale, maggio, Tecniche Nuove editore, Milano (2009)
2. Alberto, C.: Barriere architettoniche e tutela dei portatori di handicap, Il Sole 24 Ore, Milano (2008)
3. Frampton, K.: Storia dell'architettura moderna, Giugno Zanichelli editore, pp. 333–334. Bologna, (1993)
4. Sulzer, P.: Jean Prouvé Highlights 1917–1944, p. 146. Birkhauser- Publishers for Architecture, Germany (2002)
5. http://www.moma.org/explore/multimedia/videos/13/214
6. Pugnaloni, F., Di Fabio, D., Principi, P., Fioretti, R.: Proceedings of 1ST International Conference on Susteinable Intelligent Manufacturing, pp.601–608. IST Press, Lisboa (2011)
7. "moduli abitativi per nuovi modi di abitare", graduation thesis in Building-Architecture Engineering, Graduate: G. Franchini, Supervisor: P. Bonvini, added Supervisor: D. Di Fabio, AY 2012–2013, Faculty of Engineering, Università Politecnica delle Marche).

On the Design of Intelligent Buildings for Ambient Assisted Living

Berardo Naticchia and Alberto Giretti

Abstract This contribution concerns the integration of Ambient Assisted Living and, more in general, of automation technologies in the standard building performance frame. To date there is a mismatch between what users expect from an intelligent built environment and what the suppliers are able to deliver. This is mainly because the potential advances introduced by intelligent building organisms do not encompass the whole building life cycle. In order to achieve real effectiveness, the automation technologies must be harmonized with the building performance framework that drives the traditional building design science.

1 Introduction

To date there is a mismatch between what users expect from an intelligent environment (IE) and what the suppliers are able to deliver. One of the main reasons for this mismatch is that the intelligent environment has generally been defined in terms of its technologies, rather than in terms of the end users' goals. If the users are subservient to the technologies, this usually leads to situations where the technology is inappropriate for their needs. Hence, trying to define the intelligence of an IE, we can argue that it should be more than the technologies it uses and that it is probably better described by listing performance criteria than by a list of high-tech installations. AAL represents one of the most advanced application of the general concept of IE. An IE is basically a space made of physical objects with embedded information and communication technologies, creating a genuine user-friendly and highly

B. Naticchia · A. Giretti (✉)
Dipartimento di Ingegneria Civile Edile e Architettura (DICEA), Facoltà di Ingegneria, Università Politecnica delle Marche, via Brecce Bianche, 60131 Ancona, Italy
e-mail: a.giretti@univpm.it

B. Naticchia
e-mail: b.naticchia@univpm.it

S. Longhi et al. (eds.), *Ambient Assisted Living*, DOI: 10.1007/978-3-319-01119-6_38,

interactive ambient, where the embedded automation and the advanced computation is essentially invisible to the user. ICTs disappear as they become embedded into physical objects, infrastructures, and the surroundings in which people live, travel, and work. Despite AAL and IE are traditionally grown within the automation and distributed computing disciplines, their conceptualization and technologies are widely shared with those matured within the building technology disciplines, concerning the development of Intelligent Buildings (IB). In 1995, a CIB Working Group W098 report, Intelligent and Responsive Buildings, stated: "*An intelligent building is a dynamic and responsive architecture that provides every occupant with productive, cost effective and environmentally approved conditions through a continuous interaction among its four basic elements: places (fabric; structure; facilities); processes (automation; control; systems;) people (services; users) and management (maintenance; performance) and the interrelation between them.*" This definition generalizes and, to some extents, operationalizes in terms of building technology the IE vision. In both perspectives, the most fundamental part is human. The user decides upon the mind force of the building against which machines have to act. Nevertheless, the IB perspectives differs from the IE vision because its scope is much broader, involving all the basic performances that building must provide to their inhabitants. This has the main consequence that the IB vision embodies a systemic thinking. Systemic thinking is not just thinking in terms of systems, but in terms of system goals toward which all individuals, sub-systems, structures, methods, procurement, standards and measures, etc. are organized. If the system-wide goal is performance, then all system parts must be oriented around that goal. This involves AAL as well. In a systemic approach, whatever technology is added to the 'building organism', it must be harmonized with all overall building performance framework. This represents indeed a challenge for both the AAL and the Building Science disciplines. In fact, on one hand the scope of AAL is focused on the development of specific high-level ICT technologies that are rarely conceived, designed and deployed within the hosting building performance framework. On the other hand, the implications of systemic thinking are huge for the construction industry, which has still a commodity industry approach, characterized by fragmentation and "bottom-up" logic sufficient for producing a given commodity at the lowest cost.

This contribution focuses on the integration of AAL and, more in general, of automation technologies within the building performance framework and, on the other side, about the challenges that this new perspective opens in the systemic approach perspective of building science.

2 The Building Performance Paradigm

A building is a system that consists of materials, components (assemblies, equipment), sub-systems that interact with physical phenomena in the process of providing an intended level of performance to its immediate occupants and societal stakeholders. Standard building performance classes involve, among others, structural stability, fire resistance, efficient energy use, durability, inhabitants health and safety, indoor air quality, thermal comfort, etc. A building performance is the quality of a building's operation when measured against a reference standard concerning one or more requirements. Achieving the desired grade of building performance through design and construction is massively complex. The main reason of building complexity raises form the fact that buildings usually involve a large number of specific and competing classes of performance, each regulated by many legislative norms, economic and financial constraints, operational and construction issues. Therefore, contemporary building science is a broad body of knowledge that is concerned with the full life cycle of buildings, including policy (codes and standards), planning, design, construction, commissioning, facilities management, forensics and rehabilitation, restoration and retrofit, preservation and conservation, demolition (deconstruction) and recycling. Contemporary building science attempts to work with models of the building as a system, and to apply computational techniques to the effective solution of performance optimization each and every time a building is being designed and built. Optimizing performance goes beyond compatibility between the structure, enclosure, interior, and services. It involves the assessment of economic, social, and environmental parameters so that performance targets are attained affordably within the skill capacity of the industry. The building as a system approach requires designers to explicitly and consciously consider the interactions between the primary elements comprising the system: building enclosure (building envelope system), inhabitants (humans, animals, and/or plants, etc.), building services (electrical/mechanical systems), site, with its landscape and services infrastructure and external environment (weather and micro-climate). The harmonization of these elements is the key to well-performing buildings. An attempt to construct a general, extensible schema of building behavior applicable to the whole system, as well as its constituent elements, is presented in Fig. 1. It may be noted that by linking a number of simple schemata, complex behavior may be described, if not quantitatively modeled. In the example phenomena depicted in Fig. 1, the direct response of the building enclosure to the temperature, air pressure, and humidity difference between the indoor and outdoor environments results in heat loss and air leakage. The indirect response influences thermal comfort and indoor air quality if the enclosure provides low effective thermal resistance (excessive thermal bridging and/or insufficient insulation) and the condensation of moisture promotes mold growth. Unfortunately, computational methods suiting the performance assessment requirement of contemporary building science have not emerged so far. The integrative process of design will remain an exclusively human task.

At present, the most significant building science move toward the performance paradigm revolves around the fundamental transition towards Performance Based Contracting (PBC). PBC is about buying performance, not transactional goods and services. PBCs delineate outcome performance goals, ensure that responsibilities are assigned and facilitate the overall life-cycle management of building reliability. To achieve the PBC target, two technological steps appear necessary.

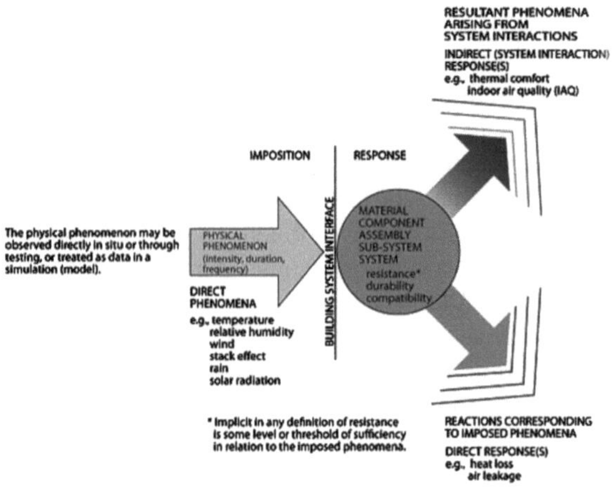

Fig. 1 Schema describing physical building behavior (courtesy of National Institute of Building Science http://www.wbdg.org/

- *Performance Oriented Design*—the first one concerns design, which must move towards a totally integrated performance oriented process.
- *Validated Building Operation*—the second one concerns the building construction and operation, that must be able to implement the quality of the processes, so that the users' needs can be met throughout the building life-cycle.

Whatever the application field will be, either Smart Energy or Ambient Assisted Living, the application of the automation technology to the building industry should be integrated into the overall production cycle, providing support to every phase of the production and operation.

3 Performance Oriented Design

Performance Oriented Design (POD) emphasizes performance as a driving concept for design that helps reconsolidate form and function into a synergetic relation with the dynamics of natural, cultural and social environments. In so doing, POD locates performative capacity in the spatial and material organization of the architecture, in the applied technologies, in the human subject and in the environment through the dynamic interaction between these domains. POD exploits the inherently performance oriented nature of design by emphasizing the effort employed in the arrangement of the processes and of the information that are necessary to suit complex performance settings. *Universal Design* or *Design for All* [1] are examples of complex performance settings for the building industry, since they involve the building from its early conception to the latest applied automation technology. The management of such complexity requires an in depth rethinking of the design process and of its information processing.

Computational thinking offers a paradigm that have been proved extremely useful to cope with the complexity of large performance settings in the building design process. The object-oriented design (OOD) is one of the computational thinking concept that is mostly affecting the design and construction industry from both a methodological and a procedural viewpoint. On one hand, the object-oriented approach naturally induces 'systemic thinking' through a system-component approach to building design. On the other hand, it is at the basis of the Building Information Modelling (BIM) [2] one of the new leading information technologies of the building sector. On the same line, Component-based software engineering (CBSE), a branch of software engineering, emphasizes the separation of concerns in respect of the wide-ranging functionality available throughout a given software system. It is a reuse-based approach to defining, implementing and composing loosely coupled independent components into systems. An individual software component is a software package, a web service, a web resource, or a module that encapsulates a set of related functions (or data). All system processes are placed into separate components so that all of the data and functions inside each component are semantically related (just as with the contents of classes). Because of this principle, it is often said that components are modular and cohesive. With regard to system-wide coordination, components communicate with each other via interfaces. When a component offers services to the rest of the system, it adopts a provided interface that specifies the services that other components can utilize, and how they can do so. This interface can be seen as a signature of the component—the client does not need to know about the inner workings of the component (implementation) in order to make use of it. This principle results in components referred to as encapsulated. The application of the system-component paradigm produces high benefits in the development of a new building conception, for two main reasons:

- the system-component paradigm maps naturally well into the physical building objects that are inherently made of hierarchies of components;
- the interface and the encapsulation concepts, which regulate the access to the component behavior and consequently the interaction between components, provide an high degree of robustness to the resulting system in any performance domains and in any phases of the construction process.

Figure 2 shows a simple example that encompasses requirements derived from the AAL domain and from the building physics (i.e. energy efficiency) domains. It concerns a new window technology, which integrates a switchable liquid shielding system in order to make windows and glazed facades dynamically adaptable, in terms of visual and thermal properties, to external conditions.

In both configurations, the window maintains its transparency but can act as a shield to sunlight, when put either in the low solar transmittance state, or as a clear transparent envelope, when operated in its high solar transmittance configuration. Visually impaired people can take advantage by sunlit environments provided by these windows without compromising the energy performance of the building. The component interface schema is shown on the right side of Fig. 2. Making explicit its functions in a structured way for each performance domain, the window component

Fig. 2 Functional schema of the reversible liquid shading window (*left*), the two test cells during the outdoor experimental campaign (*center*) and the component model (*right*)

schema offers support for its integration in the building overall performance framework and in the consequent regulating logics. BIM is, at present, the full object-based model backed by semantic information, which can support the implementation of the functional component model. BIM applications typically store this building model data using the IFC Building Product Model, which is an open data model. The IFC model specification is registered by the International Organization for Standardization (ISO) and is an official International Standard (ISO 16739:2013).

4 Validated Building Operation

A large modern building contain extensive internal systems to manufacture the indoor environment. Heating and cooling, ventilation, lighting, fire alarms, security and networking are among the most commonly systems deployed in buildings. Those systems exist as closed-loop systems, are frequently provided by different vendors and have little interoperability or extensibility beyond the scope of the original system design. The current trend is to harmonize existing systems with the built environment in order to implement a comprehensive Building Management System (BMS). The BMS optimal performance is achieved by introducing an innovative building operating system functional layer (BOS) capable of supporting a coordinated building resource management [3]. The achievement of key performances can be monitored and addressed, and the overall operation of the building can be validated by means of the BOS. Doing so is, at present, a significant challenge, since such a system must bring together legacy systems with their own quirks, provide a path forward for new, native devices, and provide improved and simplified interfaces at multiple levels of abstraction. The BOS will provide the necessary levels of abstractions between the application layer (e.g. high-level AAL caregivers support) and the systems, making application portable among different buildings, by avoiding semantic mismatches between the application logics and the surrounding built environment. Figure 3 shows the building operating system conceptual and computational components:

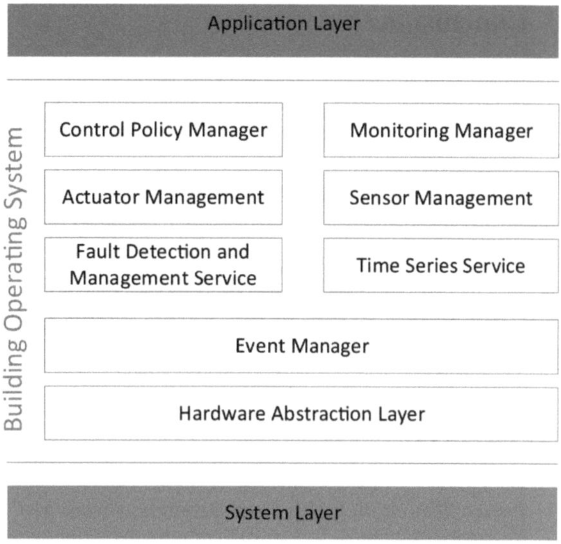

Fig. 3 The building operating system component architecture

- *Hardware abstraction layer*—provides access abstraction by means of shared namespace, shared interface design and system semantics, allowing for abstract qualification of the system performance and of the relationships between the underlying sensors, actuators, and equipment.
- *Real-time time series services*—provide processing and archiving resources to manage time alignment of the different events occurring within the building; this service is fundamental to build a coherent time aligned status of the building, since the processes that co-occur in the built environment have different time scales and are consequently monitored with different sampling times.
- *Fault Detection and Management Service*—provides system misbehavior detection and management.
- *Sensor management*—provides sensor management and virtualization services in order to transform sensor measurement and/or to implement for sensor fusion processes.
- *Actuator management*—provides actuator virtualization service through actuator fusion procedures that are aimed to apply complex control procedures encompassing different systems.
- *Control Policy and Monitoring management*—provide high-level interface to the implementation of control and monitoring policies.

5 Conclusions

This paper discussed the issues concerning the integration of Ambient Assisted Living and, more in general, of automation technologies in the standard building performance frame. It has been argued that this could be achieved through two substantial technological progresses, namely: Performance Oriented Design and Validated Building Operation. The object-oriented approach to building modelling and the definition of an integrated building automation architecture, the Building Operating System, have been identified as the enabling technologies to achieve flexible and effective automation of the built environment.

References

1. Preiser, W., Ostroff, E.: Universal Design Handbook. McGraw Hill, New York (2001)
2. National BIM Standard—United States. National Building Information Model Standard Project Committee, http://www.buildingsmartalliance.org/
3. Dawson-Haggerty S., Krioukov A., et al.: BOSS: building operating system services. In: Proceedings of the 10th USENIX Symposium on Networked Systems Design and Implementation (NSDI), Usenix (2013)

Architectures for Alternative Mobility and Relational Fields: Innovative People-Oriented Approaches

Fausto Pugnaloni and Cecilia Carlorosi

Abstract Infrastructures, effect of development strategies, have always represented an opportunity to rewrite society. The subject has been highly controversial since the 60s'. Nowadays, it is at the core of the cultural debate centered on sustainable development, around which research is connecting the concepts of city livability, resource-efficient mobility, sustainable urban society. The problem of urban congestion amplifies the issue of suburban disintegration, with the needs, uses and purposes of the city and of its inhabitants being more and more neglected. With reference to a sustainable territory transformation, the challenge of the research is double: on one hand, it aims at elaborating hypothesis regarding the effectiveness of the possible scenarios of urban alternative mobility—ought to become more efficient; on the other hand, it seeks to integrate mobility systems with the project of territorial interchanges, grouping the population together into poles, linked by environmentally-friendly collective transports. The scenarios offered within the open field of urban phenomenology intend to create ideal social and environmental conditions, to optimize the way urban spaces are used and to create a mobility system able to guarantee an accessible and safe urban environment.

1 Introduction

Mobility represents a fertile multidisciplinary field and a transversal sector of architectural/urban subjects and of political/social matters.

The trend towards urban sprawl and, consequently, the segregation of some specific social categories have the effect of setting a series of limits: these can be

F. Pugnaloni (✉) · C. Carlorosi
DICEA—Department of Civil and Building Engineering, and Architecture, Università Politecnica delle Marche, via Brecce Bianche, Ancona, Italy
e-mail: f.pugnaloni@univpm.it

C. Carlorosi
e-mail: c.carlorosi@univpm.it

S. Longhi et al. (eds.), *Ambient Assisted Living*, DOI: 10.1007/978-3-319-01119-6_39,

overcome by the current trends and services of shared mobility, influencing the characteristics of the built environment and the relationships among people.

The attempt to try and make architecture an ethic discipline is based on the main goal of *sustainable development*, to which the themes of city livability, resource-efficient mobility, sustainable urban society also refer to.

Within this area of research, urban livability takes on a particular meaning if intended as a social problem. The way disabled people perceive the accessibility to our urban heritage is used as a critical mass, in order to identify the development of an urban society based on the principles of solidarity and social coherence.

To this end, the structural funds allocated by the European Union as tools of the regional policies aim at fostering the economic and social cohesion for the next programming period 2014/2020, with specific reference to the strengthening of infrastructures and related services [5].

2 The Street Belongs to the Automobile

Throughout history, the car has had different meanings in terms of cultural evolution. In the 50s, it totally revolutionized the transport sector and, consequently, the life and habits of Italian people. It eventually became a primary asset and the symbol of economic progress, able to connect people and cut distances, therefore contributing to the creation of "the global village", where cultures, tourist flows and professional dynamism overlap (Fig. 1).

"Today's world is globalized and mobility—virtual or physical—has become a priority. Over the last decades the concept of "flow" has developed, and so has the concept of "flow of people". Here, the man comes into play: he starts to build, creating a series of limits around him, paths leading to something" (Fuksas). On the other hand, nowadays cars seem to be responsible for many pathologies, as the modern sociology of mobility demonstrates. Chronic tiredness, backache, respiratory problems, allergies, as well as anxiety attacks and panic disorders. Each year, in Italy, 20 million people get sick in their car. "Traffic makes our cities more and more congested, polluted and unlivable and it is one of the main causes of diseases and problems. Over the past year, 47 % of Italians had problems caused by an excessive use of the car. [...] Not to mention the constantly increasing aggressive behaviours, acknowledged by 49 % of car users. This is what was revealed by the survey called *Driving in the traffic*: *problems and consequences on human health*, carried out by Doxa- Osservatorio TomTom on 738 Italians aged between 18 and 65".[1]

[1] Demoscopic survey carried out by the Doxa Institute. The data were collected in February 2011 using the C.A.T.I. Survey system—Computer Assisted Telephone Interview.
(source: www.gotogo.it).

Improving the traffic flow means fostering the possibility to enjoy what the city centre has to offer, acting against social disintegration: in the era of the transformations of the virtual world and of social networks, cities live their own historical and social role.

If, on one hand, the personal use of virtual networks and devices allow to enjoy cities more (PCs, mobile/smart phones, navigators); on the other hand, the gaps in the culture of mobility highlight the insufficiency of the existing means of transport, totally inadequate in our cities as well as in the entire territory.

As Gino Finizio explains, the consumer world is personal; conversely, "the concept of *personal* is not used in mobility: in an urban area, each car is used by an average of 1,3 passengers; the average weight of the vehicles driving in urban areas is 1,120 kg, against the average 100 kg transported; 80 % of the power needed for the drive is only used for moving the vehicle alone" [6].

Among the solutions offered to the problem of urban mobility: new future vehicle architectures, single-seaters with reduced weight—max 400 kg—and limited power consumption, which will probably be available on the market soon.

Irrespective of the experimentation on alternative vehicles, however, there is a strong need to identify the image of the *car city*: future cities will have to be functional to the needs of contemporary living and to the morphological social changes of the 20th century. On the other hand, the demand for urban comfort must also be given an answer, through the creation of easily accessible urban spaces, both for citizens and for tertiary activities.

3 Organic Urban Restoration

The problem of urban congestion amplifies the issue of suburban disintegration, with the needs, uses and purposes of the city and of its inhabitants being more and more neglected.

In urban centres the social aspect is often left behind [7]. People have now accepted a situation aimed at improving the quality of life, although it often changes the rules of those who use the services: parking difficulties and growing traffic tend to increasingly discourage motorists from entering city centres.

Today's cities are "ill": cars invade squares and streets [4]; the main problem is represented by private cars literally congesting our modern cities, "built around" cars and therefore now become unlivable, both in terms of resource consumption and quality of life.

In Italy, a considerable amount of CO_2 emissions present in the atmosphere is produced by transport. According to *The Energy and the Environment Report 2007* by ENEA, the transport sector is responsible for 1/3 of the total amount of the non-renewable resources consumed (1/3 is used by industry and the rest by the tertiary and residential sectors).

Nowadays, however, some innovations in terms of traffic management can be applied to integrate the traditional tools in the planning of a new model of city, based

Fig. 1 The main arteries for city traffic (*source Magic Motorways,* Bel Geddes [1])

on a civil progress of humanity and on an organic development both of big and small-sized cities.

With reference to territory development, the possible strategies are chiefly based on mild approaches. Taking into consideration the increasing demand for mobility in urban areas that risk congestion and whose direct cost is estimated by the International Association of Public Transport at around 2 % of the GNP, the alternative given by the sustainable practices appears to be an adequate tool for the principles of sustainable development without establishing the critical situations that heavy infrastructures generate in the territory.

Hence, basing the planning of a *car city* on the identification of the spaces and the purposes of a territory particularly suitable for car use, the analysis of the interaction between the infrastructures and their effects on the final user highlights important aspects for the rearrangement of the territory.

4 New People-Oriented Approaches

Italy, more than any other country, boasts an enormous heritage in terms of urban spaces and a centuries-old tradition in the use of streets and squares.

However, the adaptation of its historic centres to the new needs of our fast, contemporary life have gradually led to the thickening of the infrastructures which fragment the urban areas—followed by gutting and demolition interventions—to a totally passive reaction on the side of the citizens and to an ever stronger awareness that the vital nature of the city must be preserved. This goes hand in hand with a sufficient level of socio-economic services and with the harmonic development of the systems of the built environment and of mobility.

The building of infrastructures and the current mobility networks are to be modified in order to connect the different parts of a territory without neglecting the real needs of the different users (commuters, tourists, older people).

With reference to a sustainable territory transformation, the challenge of the research is double: on one hand, it aims at elaborating hypothesis regarding the effectiveness of the possible scenarios of urban alternative mobility—ought to become more efficient; on the other hand, it seeks to integrate mobility systems with the project of territorial interchanges, grouping the population together into poles, linked by environmentally-friendly collective transports.

4.1 Easylife and Alternative Mobility

The new programming policies can start to use the results of the study on mobility based on selected groups of users and on efficient public transport systems other than the car, which can foster the exchange of ideas and services.

New widespread forms of shared mobility are developing, hand in hand with the public systems (Fig. 2).

- Interactive digital maps
  Nowadays, *Infomobility services* and digital maps in metropolitan areas allow to spot all sorts of architectural barriers.
  Different organizations are currently committed in social progress and city livability;
  MapAbility.org, for instance, gathers data on the accessibility of some selected

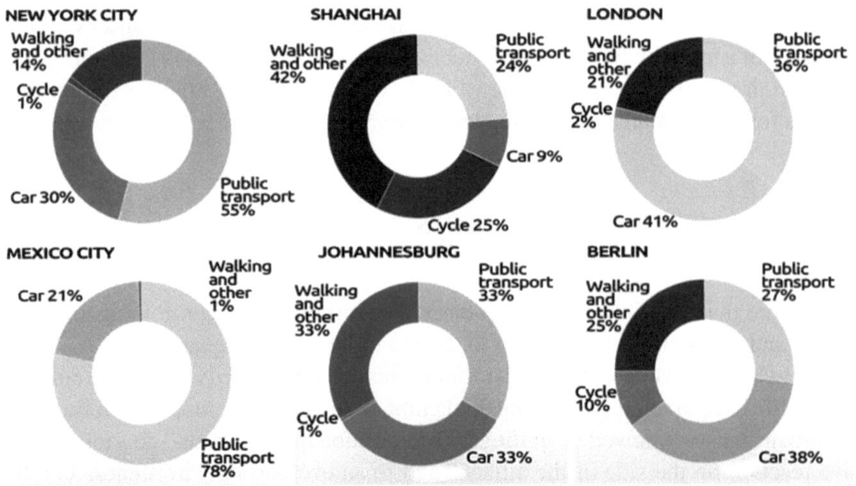

Fig. 2 *How people move*. The diagrams illustrate how people get to work in the six Urban Age city. (*source The endless city,* Burdett and Sudjic [2])

cities in order to create digital maps available for on-line consultation,[2] thus allowing a real-time planning of the accessible itineraries, possible barriers and useful services, promoting accessible tourism and urban mobility.
The gathering of data is carried out interactively through the GPS systems included in any modern smartphone. The data collected by means of photographic surveys undergo a direct on-line classification, followed by the creation of a series of itineraries through specific computer tools. The result is a detailed digital map of all the accessible streets of a city, that can be consulted on line: from any mobile phone, PC or iPad, anyone will be able to detect all sorts of architectural barriers.

- *City parks*
  The *High Line* in New York outlines an extremely encouraging perspective regarding the transformation of an existing object in a sustainable way.
  Built in the 1930s as a sky train some 9 m off the ground, the existing rail line extended for 21 km and was reconverted in 2006 into a public park of 2.33 km, becoming an icon of Manhattan.
  The new linear park, which kept the tracks and has a considerable number of points that intercept the road, has affected its surroundings, giving rise to numerous commercial activities, creating economic satellite activities for the pre-existing setting, and inducing also the residents to adorn communal areas with greenery.
- *Walkways*
  One of the most interesting cases regarding urban liveability is represented by the *Walkways* system created in the centre of Hong Kong in the 1960s, where the edge of the road has been shifted to a higher level, disconnecting the pedestrian

[2] www.MapAbility.org.

walkways from the heavily congested roads, enriching the scenographic look that distinguishes this city and supporting its vertically stratified planning on various levels.

- *Smart mobility*
  Many projects in Italy are aimed at reducing the volume of traffic inside the historical town center.
  The *Minimetro project* was implemented in Perugia (Umbria) between 2002 and 2008 and is represented by the urban infrastructure which connect the outskirts of the west of the city with the old town centre.
  The "people mover" system uses carriages with a capacity of maximum 20 people, which are pulled by cable between the end stations, making 5 intermediate stops and moving on two parallel tracks. The category of *people mover systems* defines short range public transportation solutions, generally automated, like for example moving walkways, escalators and lifts.

4.2 New Relational Fields

The approaches identified, although useful in terms of solutions to specific urban mobility problems, can be interpreted as part of a wider re-assessment of the mobility system.

Such proposal operates on a territorial scale and is based on the project of a series of intermodal infrastructural poles dedicated to collective and shared mobility (Fig. 3).

The decongestion of car traffic can become possible only when activating a co-modality logic and favouring the exchange between private and public means of transport.

Here are the goals of this strategy:

- Create new possibilities to access the territory, through autonomous poles collaborating for the collective functioning of the system;
- Create innovative addresses based on resource-efficient mobility and multiple means of transport.
- Optimize the duration of the itineraries;
- Make suburban villages more accessible;
- Recover "light infrastructures": green trails and abandoned paths;
- Recover the value of "minor" dimensions, improving their accessibility;
- Plan livable cities through adequate structures, meeting the needs of modern living and the demand for urban comfort;
- Solve the problem of the isolation of historical centres, through a stronger dialogue among the existing poles.

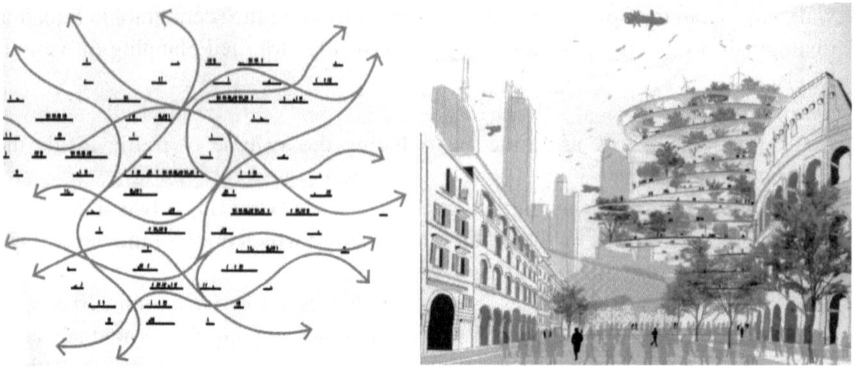

Fig. 3 *Energy Forest*. Sou Fujimoto Architects, 2013. Energy Forest is an energy station project, a place where green solutions meet both people and cars. (*source* [3, 8, 9])

The principle of inter-scalarity regulates the overall system, conjugating the comprehension of large-scale phenomena with small-scale solutions. The common theme all along the scale is the setting up of fast and integrated means of transport in order to alleviate private traffic congestion through adequate structures for the social growth.

A key-element is represented by **multimodal centers,** which operate on a territorial scale: they gather the flows coming from the main street axis and guarantee the intersections with the surrounding urban poles by means of different functions and services.

These poles include numerous functions in order to meet the needs of all sorts of users: they analyze all the journeys and suggest alternative energy-saving ways to reach the next destination, thus also creating meeting points, places for social activities and opportunities for a better flow distribution. This proposal allow people to re-appropriate the spaces that render the public life a pleasant and shared experience.

On the urban scale, the flows are connected by **urban hubs** (lifts, escalators, passenger interchange hubs): within the system, they aim at creating new urban connections and integrated systems of transport encouraging pedestrianization, in order to bring the individual back to the centre of the urban space, constituted by the city's streets and squares.

The project of new urban layouts, in light of the above-mentioned analyzed approaches (see Sect. 4.1), allows to adapt the needs to preserve and safeguard urban areas to the new use of the space, easing the use of urban areas and their numerous functions.

The urban hubs will allow a significant step forward in terms of viability as well as of noise and air pollution reduction, thus improving the quality of life of the citizens and the accessibility of the meeting points.

5 Conclusions

The experiment described is based on the firm belief that urban planning and architecture produce the best results when they meet people's daily needs—both collective and individual — and shape the city.

Among the greatest advantages offered, there is the wide range of choices, which, however, cannot be made the most of without the possibility to move easily.

The final result, within the open field of urban phenomenology, is the creation of ideal social and environmental conditions, the optimization of the way urban spaces are used and the creation of a mobility system able to guarantee an accessible and safe urban environment.

References

1. Bel Geddes, N.: Magic Motorways. Random house, New York (1940)
2. Burdett, R., Sudjic, D.: The Endless City. Phaidon Press, London (2010)
3. Ciorra, P. (edited by): Energy—Oil and post-oil architecture and grids. Catalogue of the exhibition by Pippo Ciorra (Rome, Maxxi, 2013). Electa, Milano (2013)
4. Di Gioia, D.: La città vivibile. Cacucci Editore, Bari (1992)
5. European Union: Cohesion Policy 2014–2020—Investing in growth and jobs. Publications Office of the European Union, Luxembourg (2011)
6. Finizio, G.: Minimo & sostenibile—la città ha raggiunto la montagna, umanesimo disegnativo. Skira, Milan (2012)
7. Gehl, J.: Vita in città, spazio urbano e relazioni sociali. Maggioli Editore, Milano (2012)
8. www.gotogo.it
9. www.MapAbility.org

MIX
Papier aus verantwortungsvollen Quellen
Paper from responsible sources
FSC® C105338

If you have any concerns about our products,
you can contact us on
ProductSafety@springernature.com

In case Publisher is established outside the EU,
the EU authorized representative is:
Springer Nature Customer Service Center GmbH
Europaplatz 3, 69115 Heidelberg, Germany

Printed by Libri Plureos GmbH
in Hamburg, Germany